AF751066

BIOETHANOL

Edited by **Marco Aurelio Pinheiro Lima**
and **Alexandra Pardo Policastro Natalense**

Bioethanol
Edited by Marco Aurelio Pinheiro Lima and Alexandra Pardo Policastro Natalense

Published by InTech

Janeza Trdine 9, 51000 Rijeka, Croatia

Copyright © 2012 InTech

All chapters are distributed under the Creative Commons Attribution 3.0 license, which allows users to download, copy and build upon published articles even for commercial purposes, as long as the author and publisher are properly credited, which ensures maximum dissemination and a wider impact of our publications. After this work has been published by InTech, authors have the right to republish it, in whole or part, in any publication of which they are the author, and to make other personal use of the work. Any republication, referencing or personal use of the work must explicitly identify the original source.

As for readers, this license allows users to download, copy and build upon published chapters even for commercial purposes, as long as the author and publisher are properly credited, which ensures maximum dissemination and a wider impact of our publications.

Notice

Statements and opinions expressed in the chapters are these of the individual contributors and not necessarily those of the editors or publisher. No responsibility is accepted for the accuracy of information contained in the published chapters. The publisher assumes no responsibility for any damage or injury to persons or property arising out of the use of any materials, instructions, methods or ideas contained in the book.

Publishing Process Manager Dragana Manestar

Technical Editor Teodora Smiljanic

Cover Designer InTech Design Team

Additional hard copies can be obtained from orders@intechopen.com

Bioethanol,
Edited by Marco Aurelio Pinheiro Lima and Alexandra Pardo Policastro Natalense
p. cm.
ISBN 978-953-51-0008-9

We are IntechOpen, the world's leading publisher of Open Access books

Built by scientists, for scientists

3,550+
Open access books available

112,000+
International authors and editors

115M+
Downloads

151
Countries delivered to

Our authors are among the
Top 1%
most cited scientists

12.2%
Contributors from top 500 universities

CLARIVATE ANALYTICS BOOK CITATION INDEX INDEXED

WEB OF SCIENCE™

Selection of our books indexed in the Book Citation Index
in Web of Science™ Core Collection (BKCI)

Interested in publishing with us?
Contact book.department@intechopen.com

Numbers displayed above are based on latest data collected.
For more information visit www.intechopen.com

For EU product safety concerns:
IN TECH d.o.o., Prolaz Marije Krucifikse Kozulić 3, 51000 Rijeka, Croatia,
info@intechopen.com or visit our website at intechopen.com.

Meet the editors

Dr Marco Aurelio Pinheiro Lima achieved his Undergraduate and Masters degrees in Physics at Universidade de São Paulo (USP), São Paulo. He obtained PhD in Chemistry at California Institute of Technology (Caltech), Pasadena. He is a Full-time Professor at the Physics Institute of UNICAMP, Campinas, since 1998. His area of research includes low energy electron and positron scattering from molecules. He has delivered many invited talks in international conferences (USA, England, Italy, Austria, Dennmark, Japan, India, Venezuela, Argentina, and Germany) and has published more than 130 scientific papers in refereed journals. Since 2008. he is the Director of the Brazilian Bioethanol Science and Technology Laboratory (Laboratório Nacional de Ciência e Tecnologia do Bioetanol – CTBE/CNPEM), Campinas.

Dr Alexandra Pardo Policastro Natalense is theoretical physicist. She has obtained her PhD at Unicamp in electron scattering by molecules and pos-doctorate at Texas A&M University in molecular photoionization. Currently she is the assessor of Bioethanol Science and Technology National Laboratory (CTBE) director and conducts research on the breaking of hemicelulose by electron impact. She is a head hunter for the construction of the human resource team of CTBE since 2010.

Contents

Preface

Nine billion. This is the estimated number of people who will inhabit our planet in 2050. In a few decades, we will have nearly a quarter more than humans circling the globe in search of food, shelter, clothing and other manufactured products. Among the new individuals, the United Nations (UN) estimates that 98% will live in developing countries, with the highest level of economic growth, which in turn will result in a considerable expansion in per capita consumption worldwide.

On the one hand we have a significant increase in energy demand and consumption, resulting from population and income growth, and on the other, there is a considerable uncertainty about the world's available supply of natural resources to support this development. Intergovernmental Panel on Climate Change (IPCC) recent reports have shown strong evidence of the impact of human activity on the climate of the planet. Estimates of the entity warn about a potential increase in global average temperature by up to 5 or 6°C by the end of this century. The raise in temperature itself would cause drastic changes in many ecosystems, but the reports also mention the intensification of extreme weather events such as hurricanes.

This apparently catastrophic scenario for the maintenance of the human species on Earth, opens up several possibilities for what is now called "green" or low carbon economy. We are talking about creating new businesses and industries geared to develop products and services with low consumption of natural resources and reduced emission of greenhouse gases. Within this category of business, biofuels is a highlight and the central theme of this book.

Biofuels are now the main alternative to automotive fossil fuels due to the fact that they are produced from renewable sources such as sugar cane, corn, cassava, oil seeds, agricultural waste, algae, etc.. Ethanol from sugar cane produced in large scale in Brazil, for example, illustrates the benefits of these products. Its production costs are low, which makes it competitive with oil derivatives. Each unit of fossil energy used in ethanol production is reversed in eight to nine units stored in the fuel. Finally, one of the most important qualities, each cubic meter of sugarcane bioethanol used as fuel reduces from 1.7 to 1.8 tonnes of CO_2 (equivalent) emitted into the atmosphere. Due to the flex-fuel engine technology, more than 90% of light vehicles produced in Brazil are now able to run on 100% fuel ethanol.

The successful history of Brazilian ethanol is undoubtedly the first successful case of production and use of a biofuel in large scale, but is far from being the last. From the beginning of the century, two main factors have made the world turn its attention to research on biofuels. The first, already mentioned, is the increasing debate on climate issues. The second is the raise in the price of the oil barrel. In 1970, before the first shock in the price of fossil fuel, a barrel costs about \$ 3. In 2008 the price was above \$ 120. These facts stimulated scientists around the world to focus their research on themes that could result in the diversification of the energy matrix in many countries.

The globalization of the research on biofuels may bring a number of advancements to the industry and has already awakened a wish the market: to also convert cellulose into ethanol. Materials not used in the production of biofuels, such as sugarcane bagasse, corn stover and forest residues can be a significant source of additional ethanol, provided that appropriate industrial technologies are developed. In the case of sugarcane ethanol for example, data from the Brazilian Bioethanol Science and Technology Laboratory (CTBE) indicate that the conversion of bagasse and straw would increase the current production of bioethanol in Brazil in about 50%.

However, the challenges to make this technological potential an industrial reality are numerous and need investment in research and development (R&D). There are technological barriers with respect to the initial treatment of the raw material, production of microorganisms that break down cellulose into fermentable sugars, the fermentation of five-carbon sugars (pentoses), among others.

The new global market of bioenergy that has been structured in recent years has yet another relevant route for exploration: the biorefinery. Similar to the oil industry, it uses different types of processes to transform the same raw material in different products used by many industrial sectors, such as food, pharmaceutical, chemical, etc.. Companies and research institutes have studied and developed processes that convert biomass into raw materials for their production chain, potentially replacing substances that were produced from petroleum. Thus, in most cases, the environmental benefits and the reduction of dependence on fossil fuels are evident. Some studies even indicate that the use of biomass within the biorefinery concept may improve the profitability of cellulosic ethanol technology (second generation) and favor the integration of this new technology with the current first generation process.

The first section of this book presents some results for first generation ethanol production, i. e., from starch and sugar raw materials, which include cassava, sorghum, and sugarcane. In the second section, the chapters present results on some of the efforts being made around the world in order to develop an efficient technology for producing second-generation ethanol from different types of lignocellulosic materials. While efficient ethanol production technologies are being developed, one can also start thinking about different uses for it. In addition to the more straightforward use as fuel, it is worth to study other applications. The chapter in the third section points to the use of hydrogen in fuel cells, where this hydrogen could be produced from ethanol.

Acknowledgements

The editors would like to acknowledge Luiz Paulo Juttel for contributing with the information in the preface and the board of referees that made the technical revision of the chapters:
Antonio Bonomi
Arnaldo César da Silva Walter
Carlos Eduardo Vaz Rossell
George Jackson de Moraes Rocha
Manoel Regis Lima Verde Leal
Marcos Silveira Buckeridge
Oscar Antonio Braunbeck

Marco Aurelio Pinheiro Lima and **Alexandra Pardo Policastro Natalense**
Brazilian Bioethanol Science and Technology Laboratory (CTBE),
Brazil

Part 1

First Generation Bioethanol Production (Starch and Sugar Raw-Materials)

1

Cassava Bioethanol

Klanarong Sriroth[1], Sittichoke Wanlapatit[2]
and Kuakoon Piyachomkwan[2]
[1]*Dept. of Biotechnology, Faculty of Agro-Industry,*
Kasetsart University, Bangkok
[2]*Cassava and Starch Technology Research Unit,*
National Center for Genetic Engineering and Biotechnology (BIOTEC)
Thailand

1. Introduction

1.1 Cassava

Cassava (*Manihot esculenta* Crantz) is a shrubby perennial crop in the Family of Euphorbiaceae. It is also named others, depending upon geographic regions such as yucca in Central America, mandioca or manioca in Brazil, tapioca in India and Malaysia and cassada or cassava in Africa and Southeast Asia. Cassava is mostly cultivated in tropics of Africa, Latin America and Asia, located in the equatorial belt, between 30° north and 30° south. The crop produces edible starch-reserving roots which have long been employed as an important staple food for millions of mankind as well as animal feed. Due to the fact of ease of plantation and low input requirement, cassava is mostly cultivated in marginal land by poor farmers and is sometimes named as the crop of the poor. In these planting areas, cassava plays an essential role not only as food security, but also income generation. In addition to a primary use for direct consumption and animal feed, starch-rich roots are good raw materials for industrial production of commercial tapioca starch, having excellent characteristics of high whiteness, odorless and tasteless and when cooked, yielding high paste viscosity, clarity and stability. The distinct attributes of extracted cassava starch, either as native or modified form, are very attractive for a broad range of food and non-food application including paper, textile, pharmaceutical, building materials and adhesives. Furthermore, cassava starch is extensively utilized for a production of sweeteners and derivatives including glucose syrup, fructose syrup, sugar alcohols (e.g. sorbitol, mannitol), and organic acids (e.g. lactic acid, citric acid). The application of cassava as renewable feedstock is now expanded to biorefinery, i.e. a facility that integrates processes and equipment to produce fuels, power, chemicals and materials from biomass (Fernando et al., 2006). With this regard, cassava is signified as a very important commercial crop that can have the value chain from low-valued farm produces to high-valued, commercialized products.

1.2 Cassava agronomy and plantation

Cassava is well recognized for its excellent tolerance to drought and capability to grow in impoverished soils. The plant can grow in all soil types even in infertile soil or acid soil (pH

4.2-4.5), but not in alkaline soil (pH > 8). Despite of that, cassava prefers loosen-structured soil such as light sandy loams and loamy sands for its root formation. As the drought - tolerant crop, cassava can be planted in the lands having the rainfall less than 1,000 mm or unpredictable rainfall. Rather than seeding, the plants are propagated vegetatively from stem cuttings or stakes, having 20-cm in length and at least 4 nodes. To ensure good propagation, good-quality stakes obtained from mature plants with 9-12 months old should be used. The appropriate time of planting is usually at an early period of rainy seasons when the soil has adequate moisture for stake germination. When planted, the stakes are pushed into the soil horizontally, vertically or slanted; depending on soil structure. For loosen and friable soil, the stakes are planted by pushing vertically ("standing"), or slanted approximately 10 cm in depth below the soil surface with the buds facing upward. This planting method gives higher root yields, better plant survival rates and is easy for plant cultivation and root harvest (Howeler, 2007). The horizontal planting is suited for heavy clay soils. Planting with 100 x 100 cm spacing (or 10,000 plants/hectare) is typical, however, less spacing (100 x 80 cm or 80 x 80 cm) and larger spacing (100 x 120 cm or 120 x 120 cm) are recommended for infertile sandy soil and fertile soil, respectively. At maturity stage with 8-18 months after planting, the plants with two big branches (i.e. dichotomous branching) or three branches (i.e. trichotomous branching) are 1-5 m in height with the starch-accumulating roots extending radially 1 m into the soil. Mature roots are different in shapes (as conical, conical-cylindrical, cylindrical and fusiforms), in sizes (ranging from 3 to 15 cm in diameter, as influenced by variety, age and growth conditions) and in peel colors (including white, dark brown and light brown). Although the roots can be harvested at any time between 6-18 months, it is typically to be harvested on average at 10-12 months after planting. Early or late harvesting may lower root yields and root starch contents. Still, the actual practice of farmers is depending on economic factors, i.e. market demand and root prices. Root harvesting can be accomplished manually by cutting the stem at a height of 40 - 60 cm above the ground and roots are then pulled out by using the iron or woody stalk with a fulcrum point in between the branches of the plant. Plant tops are cut into pieces for replanting, leaves are used for making animal fodder and roots are delivered to the market for direct consumption or to processing areas for subsequent conversion to primary products as flour, chips and starch.

1.3 Cassava production

Since 2004, the world production of cassava roots has been greater than 200 million tons and reaches 240 million tons in 2009 (Food and Agriculture Organization [FAO], 2011; Table 1). The major cassava producers are located in three continental regions which are Nigeria, Brazil and Thailand, accounting approximately for 20, 11 and 12% of total world production, respectively. In the last two decades, the world production of cassava continuously increases (Table 1), as primarily driven by the market demand, in particular an expansion of global starch market. The growth rate of root production in the last decade (2000-2009) is even greater than the previous one (1990-1999) due to markedly rising demand of cassava for bioethanol production in Asia especially in China and Thailand. Interestingly, the root productivity of cassava has been dramatically increased in some countries including Vietnam, India, Indonesia and Thailand by 8.46, 7.46, 6.22 and 5.85 tons/hectare in the past 10 years. The root productivity of India is the greatest (34.37 tons/hectare), followed by Thailand (22.68 tons/hectare) and Vietnam (16.82 tons/hectare) while the world average is

Region	Production (x 1,000 tons)[1]							
	1990-94 average	1995-99 average	20000-04 average	2005	2006	2007	2008	2009
World	160,769	163,220	187,540	207,332	222,879	226,312	232,463	240,989
Africa								
Angola	1,868	2,743	6,366	8,606 (4.15)	8,810 (3.95)	9,730 (4.30)	10,057 (4.33)	12,828 (5.32)
Congo	648	770	869	935 (0.45)	1,000 (0.45)	950 (0.45)	1,000 (0.43)	n.a.
Ghana	5,216	7,148	9,356	9,567 (4.61)	9,638 (4.32)	10,218 (4.32)	11,351 (4.88)	12,231 (5.08)
Nigeria	27,073	32,053	34,669	41,565 (20.05)	45,721 (20.51)	43,410 (20.51)	44,582 (19.18)	n.a.
Tanzania	7,281	5,666	4,651	5,539 (2.67)	6,158 (2.76)	6,600 (2.76)	6,600 (2.84)	n.a.
Latin America and Caribbean								
Brazil	23,420	20,686	22,973	25,872 (12.48)	26,639 (11.95)	26,541 (11.73)	26,703 (11.49)	26,031 (10.80)
Asia								
India	5,530	5,895	6,135	7,463 (3.60)	7,855 (3.52)	8,232 (3.64)	9,056 (3.90)	9,623 (3.99)
Indonesia	16,263	15,742	17,601	19,321 (9.32)	19,987 (8.97)	19,988 (8.83)	21,593 (9.29	22,039 (9.15)
Thailand	20,011	16,757	19,097	16,938 (8.17)	22,584 (10.13)	26,916 (11.89)	25,156 (10.82)	30,088 (12.49)
Viet Nam	2,421	2,051	4,213	6,716 (3.24)	7,783 (3.49)	8,193 (3.62)	9,396 (4.04)	8,557 (3.55)
Others	51,038	53,709	61,609	64,809	66,705	65,534	66,969	119,593

[1] The numbers in parenthesis represent the percentage of total world production.
n.a. = not available
Source: Food and Agriculture Organization of the United Nations [FAO], 2011

Table 1. Annual production of cassava roots by major producers.

12.64 tons/hectare (Table 2). The world leading cassava producers, i.e. Nigeria and Brazil, however, do not have much improvement in root productivities in the past 10 years; only by 2.10 tons/hectare (from 9.70 to 11.80 tons/hectare since 2000 to 2008) and by 0.35 tons/hectare (from 13.55 to 13.90 tons/hectare since 2000 to 2009), respectively.

The production of cassava can be simply increased by expanding planting areas. Nevertheless, in most regions, no new marginal land is accessible as well as forestry areas are not allowed for area expansion. Moreover, in some countries, there is a competition for land uses among other economic crops such as sugar cane and maize in Thailand. The sustainable and effective means of increasing root production should be achieved by an increase in root productivity. Yields or root productivities of cassava roots vary significantly with varieties, growing conditions such as soil, climate, rainfall as well as agronomic practices. Better root yields can be obtained by well-managed farm practices including time of planting (early of a wet season), land preparation (plowing by hand or mechanically and ridging), preparation of planting materials (ages of mother plants, storage of stems, length & angle of cuttings, chemical treatment), planting method (position, depth of planting and spacing), fertilization (type of fertilizers - chemical vs. organic, dose , time and method of fertilizer application), erosion control, weed control, irrigation and intercropping (Howeler, 2001; 2007). The agronomic practices implemented by farmers vary markedly from regions to regions, depending greatly on farm size, availability of labor, soil and climatic conditions as well as socio-economic circumstances of each region (Table 3). It is very interesting to note that the highest root productivity was reported in India (i.e. 40 tons/hectare) which was irrigated cassava rather than rainfed one, with a highest amount of fertilizer application. In some planting areas such as in Thailand, irrigation is now introduced instead of relying only on rainfall. Yet, the investment cost is high and farmer's decision is upto market demand, price of cassava roots as well as other competitive crops. By effective farm management, it is expected that the root productivity can be increased twice, from 25 to 50 tons/hectare. By combining that with varietal improvement, the root productivity can be potentially improved upto 80 tons/hectare (Tanticharoen, 2009).

The production cost of cassava is classified into fixed costs and variable costs. The fixed costs include land rent, machinery, depreciation cost and taxes. The variable costs are consisted of labor costs (for land preparation, planting material preparation, planting, fertilizer & chemical application, weeding, harvesting and irrigation) and others including planting materials, chemicals (herbicides, sacks), fuels and tools. Except China, all countries demonstrate that the labor cost is greater than 40% of total production cost. In particular, the labor cost as well as the fixed costs of cassava plantation in India is quite high comparatively to other countries, making their production cost quite high. A semi-mechanized practice for cassava plantation is therefore developed in some countries such as Brazil and Thailand in order to minimize the labor cost, and hence total production cost.

1.4 Cassava attributes

Cassava plants photosynthesize and store solar energy in a form of carbohydrate, mainly as starch in edible, underground roots. The roots are very moist having the water content around 59-79% w/w (Table 4). On dry solid basis, starch is a major component of cassava roots, accounting upto 77-94% w/w, the rests are protein (1.7-3.8% w/w), lipid (0.2-1.4% w/w), fiber (1.5-3.7% w/w as crude fiber, i.e. cellulose and lignin) and ash (1.8-2.5% w/w) (Table 4). Some sugars, i.e. sucrose, glucose and fructose are also found in storage roots at 4-8% w/w (dry basis). In addition to cellulosic fiber, the roots also contain non-starch

Regions	Root productivity (tons/ hectare)									
	2000	2001	2002	2003	2004	2005	2006	2007	2008	2009
Nigeria	9.70	9.60	9.90	10.40	11.00	10.99	12.00	11.20	11.80	n.a.
Brazil	13.55	13.54	13.77	13.44	13.63	13.61	14.05	14.01	14.14	13.90
Thailand	16.86	17.54	17.07	19.30	20.28	17.18	21.09	22.92	21.25	22.68
Indonesia	12.53	12.94	13.25	14.88	15.47	15.92	16.28	16.64	18.09	18.75
Ghana	12.28	12.34	12.25	12.69	12.42	12.76	12.20	12.76	13.51	13.81
India	26.91	26.70	27.27	26.21	27.04	30.50	32.11	32.22	33.54	34.37
Viet Nam	8.36	12.01	13.17	14.28	14.98	15.78	16.38	16.53	16.91	16.82
World	10.38	10.68	10.71	10.93	11.29	11.33	12.12	12.11	12.45	12.64

n.a. = not available

Source: Food and Agriculture Organization of the United Nations [FAO], 2011

Table 2. World average root productivity (tons/ hectare) and those of major producers.

	Thailand	Indonesia[1]	India[2]	Vietnam[3]	China
Cassava area (hectare/ farmer)	2-3	0.3-1.0	0.5-1.0	0.2-0.9	0.2-0.4
Intercrops	None (95%), Maize (5%)	Maize+rice- soybean/peanut	None/vegetables	None/maize	None/peanut
Land preparation	Tractor (3 + 7 disc)	Manual/animal/tracto r	Tractor	Animal/tractor	Manual/animal
Fertilizer use -Organic (ton/hectare) -Inorganic (kg $N+P_2O+K_2O$/hectare)	 Little 30-120	 Low, 3-10 Medium, N-only	 10-20 High	 0-5 0-60	 3-5 NPK
Planting time	March-May (70%) Sept-Nov	Oct-Dec (90%)	Jan-Mar (90%) Sept-Oct	Feb-May (80%) Oct-Nov	Feb-Apr (90%)
Harvest time	Dec-May Aug-Dec	Jul-Sept	Oct-Jan	Feb-Mar Sept-Oct	Nov-Jan
Planting space (m)	0.8x1.2, 0.8x0.8	1.0x0.8, 2.0x0.5	1.0x1.0	1.2x0.8, 0.8x0.8	1.0x1.0, 0.8x0.8
Planting method	Vertical	Vertical	Vertical	Horizontal	Horizontal
Weed control	Hoe 2-3x small tractor/ Paraquat	Hoe 1-2x	Hoe 4-5x	Hoe 2-3x	Hoe 2-3x
Harvest method	Hand/tractor	Hand	Hand	Hand	Hand
Main varieties	KU 50, Rayong 90, Rayong 60, Rayong 5	Adira 4, local varieties	H-226, H-165, local varieties	KM 94, KM 60, 34, HL 23	SC 205, SC 201, SC 124
Labor use (m-days/hectare)	50-60	150-300	200-350	100-200	90-180
Yield (tons/ hectare)	23.40	20	40[4]	25	20
Production cost					
-Variable costs (USD/hectare) (Labor cost) (Other costs: Fertilizers, chemicals, cuttings, transportation)	365.91 (167.18) (198.73)	265.92 (185.37) (80.55)	663.85 (421.70) (242.15)	384.67 (213.60) (171.07)	427.62 (167.40) (260.22)
- Fixed costs (USD/hectare)	48.89	46.67	236.50	60.00	94.94
- Total production costs					
USD/hectare	414.80	312.59	900.35	444.67	520.56
USD/ton fresh roots	17.73 27.33[5]	15.63	22.51	17.79	26.03

[1] Information of Java and Sumatra [2] Information of Tamil Nadu [3] Information of South Vietnam
[4] Irrigated cassava [5] Source: Office of Agricultural Futures Trading Commission [AFTC], 2007
n.a. = not available Source: Howeler, 2001

Table 3. Agronomic practices and production cost of cassava plantation in some Asian countries.

Composition[1]	Grains						Tubers	Roots	Cassava chips
	Maize	Wheat	Barley	Sorghum	Rye	Rice[2]	Potato[3]	Cassava[4]	
Moisture	12-15	11-14	11-14	11-14	11-14	14	78	59-70	14
Starch	65-72	62-70	52-64	72-75	52-65	68[5]	77[5]	77-94[5]	77
Sugar	2.2	n.a.	n.a.	n.a.	n.a.				n.a.
Protein	9-12	12-14	10-11	11.2	10-15	6.6	10	1.7-3.8	3.1
Lipid	4.5	3	2.5-3	3.6	2-3	1.9	0.4	0.2-1.4	1.1
Fiber/ Cell wall materials	9.6	11.4	14	n.a.	n.a.	16.1	1.8	1.5-3.7[6]	3.1
Ash	1.5	2	2.3	1.7	2	4.0	4.5	1.8-2.5	1.4

[1] %w/w (dry basis) except moisture content reported as %w/w (wet basis)
[2] As paddy rice (Juliano, 1993)
[3] Source: Treadway, 1967
[4] Source: Breuninger et al., 2009
[5] As starch and sugar content
[6] As crude fiber content.
n.a. = not available
Source: Monceaux, 2009

Table 4. Chemical composition of starch-accumulating edible parts of various starch crops.

polysaccharides, i.e. hemicellulose and pectic substances as evidenced by a presence of monosaccharide including rhamnose, fucose, arabinose, xylose, mannose, galactose, glucose in hydrolyzed cell wall materials (Kajiwara & Maeda, 1983; Menoli & Beleia, 2006; Charles et al., 2008). Some minerals such as sodium, calcium, potassium, magnesium, iron, copper, zinc, manganese and phosphorus are detected in fresh roots as well (Balagopalan et al., 1988; Rojanaridpiched, 1989; Charles et al., 2005).

Unlike grains of cereals having low moisture content (11-15%), cassava roots contain very high moisture contents and are very perishable. This is a constraint for cassava utilization as roots are subjected to deterioration and spoilage by microorganism attacks during storage. Fresh roots can be stored only a few days and should be transformed to products as soon as they are harvested. To prolong their shelf-life, the roots can be simply chopped and sun-dried; the final product is named as cassava chip with the moisture content approximately 14% (Table 4). Cassava roots also contain much lower protein contents than cereals.

The starch content of mature roots can range significantly, depending on genetic traits and environmental factors during plant development, as well as harvest time or ages after planting. Roots collected from crops being planted with the drought during initial state of growth have much lower starch contents and root yields than those from crops without the drought (Pardales and Esquibel, 1996; Santisopasri et al., 2001; Sriroth et al., 2001). Immature or young roots (less than 8 months) provide low starch yields due to low starch contents and root yields. The genetic and environmental growth condition can also influence starch qualities in term of starch composition (amylose and amylopectin content), ease of cooking as indicated by gelatinization or pasting temperature and cooked paste viscosity (Moorthy and Ramanujam, 1986; Asaoka et al., 1991;1992; Defloor et al., 1998; Sriroth et al., 1999; Santisopasri et al., 2001).

2. Use of cassava for bioethanol production

2.1 Bioethanol production

Instead of chemical synthesis, the bioprocess, i.e. fermentation of simple sugars by microorganism is nowadays used extensively to produce ethanol from renewable sugar-containing biomass. Important ones are sugar crops, starch crops, and lignocellulosic materials derived from agricultural residues. The two former ones are recognized as the first generation feedstock for bioethanol production while the last one is the second generation feedstock. When ethanol is produced by yeast fermentation of sugar feedstock such as sugar cane, molasses, sugar beet and sweet sorghum, yeast can directly consume simple sugars and convert them to ethanol. However, starch and cellulose feedstock are a polymer of glucose and cannot directly be utilized by yeast. They have to be converted or depolymerized to glucose prior to yeast fermentation. Depolymerization or hydrolysis of starch is much simpler and more cost effective than that of cellulosic materials and can be achieved by acid or enzyme or a combination of both.

Starch is a polysaccharide comprising solely of glucose monomers which are linked together by glycosidic bonds. It is composed of two types of glucan namely amylose, a linear glucose polymer having only α-1,4 glycosidic linkage and amylopectin, a branched glucose polymer containing mainly α-1,4 glycosidic linkage in a linear part and a few α-1,6 at a branch structure. Most starches contain approximately 20-30% amylose and the rest are amylopectin. Some starches contain no amylose such as waxy corn starch, waxy rice starch, amylose-free potato, amylose-free cassava and some have very high amylose contents upto 50-70% as in high amylose maize starches. These two polymers organize themselves into semi-crystalline structure and form into minute granules, which are water insoluble. Starch granules are less susceptible to enzyme hydrolysis. Upon cooking in excess water, the granular structure of starch is disrupted, making glucose polymers become solubilized and more susceptible to enzyme attacks. At the same time, the starch slurry becomes more viscous. This process is known as gelatinization and the temperature at which starch properties are changed is named as gelatinization temperatures. Different starches have different gelatinization temperatures, implying different ease of cooking. Cassava starch has a lower cooking temperature, relatively to cereal starches; the pasting temperatures for cassava, corn, wheat and rice are 60-65, 75-80, 80-85 and 73-75°C (Swinkels, 1998; Thirathumthavorn & Charoenrein, 2005).

The starch hydrolysis by enzymes is a two-stage process involving liquefaction and saccharification. Liquefaction is a step that starch is degraded by an endo-acting enzyme namely α-amylase, which hydrolyzes only α-1,4 and causes dramatically drop in viscosity of cooked starch. Typically, liquefying enzymes can have an activity at a high temperature (> 85°C) so that the enzyme can help reduce paste viscosity of starch during cooking. The dextrins, i.e. products obtained after liquefaction, is further hydrolzyed ultimately to glucose by glucoamylase enzyme which can hydrolyze both α-1,4 and α-1,6 glycosidic linkage. Glucose is then subsequentially converted to ethanol by yeast. By the end of fermentation, the obtained beer with approximately 10%v/v ethanol, depending on solid loading during fermentation, is subjected to distillation and dehydration to remove water and other impurities, yielding anhydrous ethanol (Figure 1).

2.2 Cassava feedstock

Cassava roots can be used as the feedstock for bioethanol production. During the harvest season, roots are plenty and cheap. However, roots contain very high moisture contents and

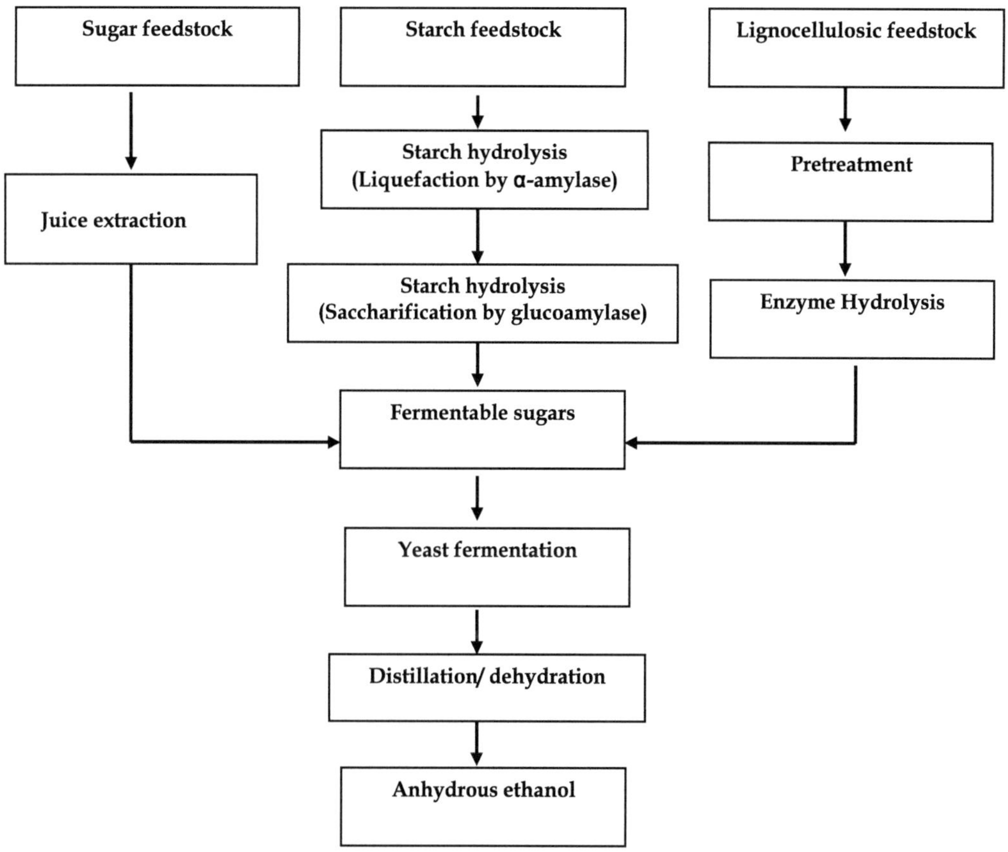

Fig. 1. Schematic diagram of bioethanol production by fermentation process of sugar, starch and lignocellulosic feedstock.

are prone to spoilage over the storage time. Mostly, roots are transformed readily to a dried form called cassava chips nearby the plantation areas. To produce chips, harvested roots are cut into pieces manually or by small machine and then sun-dried. The dried chips contain low moisture contents (< 14%), are less bulky, less costly for transportation and can be stored for a year in the warehouse. In addition, dried cassava chips have comparable characteristics as corn grains and can be processed by adopting conversion technology of corn grains. Cares must be taken when storing dried chips as heat can be generated and accumulated inside the heap. Therefore, the warehouse should have a good air ventilation system to prevent overheating and burning of chips. When used, the chips have to be transferred, using the rule of first-in and first-out, to the process line. Dusts are produced, resulting in starch loss as well as severe air pollution. The major concern of using chips is soil and sand contamination, being introduced from roots and during drying on the floor. Sand and soil can cause machine corrosion and result in shorter machine shelf life. They have to be removed in the production process. In Thailand where chips are used for many applications including animal feed and bioethanol production, farmers are encouraged to produce a premium quality of chips that meets with the standard regulation announced by Ministry of Commerce (Table 5).

Parameter	Value
Starch content	
- by Polarimetric method	Not less than 70%by weight
- by Nitrogen Free Extract, NFE	Not less than 75% by weight
Fiber	Not greater than 4% by weight
Moisture	Not greater than 13% by weight
Sand and soil	Not greater than 2% by weight
Unusual color and odor	Not detected
Spoilage and molds	Not detected
Living insects	Not detected

Table 5. Standard quality of premium grade of cassava chips, announced by Ministry of Commerce, Thailand.

When cassava is used for bioethanol production, different forms including fresh roots, chips and starch can be used. Table 6 summarizes advantages and disadvantages of using different forms of cassava feedstock. The factory has to make and manage an effective feedstock plan as the feedstock cost can account upto 70%of total ethanol production cost. Types of feedstock used for bioethanol plants depend on many concerns including plant production capacity, plant location, nearby cassava growing areas, amount of feedstock available and processing technology. Ethanol plants that are not close to cassava farms prefer to use dried chips to reduce costs of transportation and storage, while those locating next to cassava fields can use chips and roots. When using both feedstocks, the plants have to somewhat adjust the process in particular feedstock preparation.

2.3 Cassava feedstock preparation

2.3.1 Cassava chip

Similar to bioethanol production of corn grains, there are two processes for preparing cassava chips which are "Dry Milling" and "Wet Milling". In Dry Milling process (Figure 2), chips are transferred to the hopper and a metal and stone detector. The chips are then milled and sieved to obtain fine powders. Coarse powders are remilled. The fine powder having all components including fibers is slurried with water and proceeds to cooking and enzyme hydrolysis. The heat to cook slurry for liquefaction process is usually from direct steam instead of a jet cooker due to the difficulties of handling particles and contaminants in slurry. Owing to contamination of sand, conveyor system and grinding system require special treatment. Furthermore, after passing through syrup making process, an extra separation unit or hydrocyclone is required to remove sand and other impurities. The dry milling process is suitable for batch fermentation. Most of existing factories in China and some factories being installed in Thailand apply this dry milling process as it uses less equipment and investment (Sriroth & Piyachomkwan, 2010b).

As corn grains are composed of many valuable components including protein, lipid and starch, wet milling process has been developed as a separation technique in order to fractionate starch and other high-valued products including corn gluten meal with high protein content, corn gluten feed and corn germ for oil extraction. The grains are initially

Form	Advantages	Disadvantages
Fresh roots	- Low cost during harvest - Less costly to remove soil & sand - Contain fruit water having some nutrients and minerals that are advantageous to yeast fermentation	- Not available for whole year, seasonally harvested - Bulky, costly to transport - Cannot be stocked / short shelf-life due to high moisture content/ high perishability - Difficult to adjust total dry solid content in a fermentor - Limit total dry solid content for high solid loading or very high gravity (VHG) fermentation
Dried Chips	- Extended shelf-life - Can be stored - Less costly for transportation - Can be processed by applying conversion technology of corn grains	- Higher cost than fresh roots - Must be dried before stored - High soil & sand contamination - Limit total dry solid content for high solid loading or very high gravity (VHG) fermentation
Starch	- Less costly to stock - Less costly to transport - Easy to adjust total solid content in a reactor and prepare high solid loading slurry	- High feedstock cost - High production cost - Loss of nutrients during starch extraction process - High demand in other production of valued products

Table 6. Advantages and disadvantages of different forms of cassava feedstock.

cleaned and soaked in steeping water containing some chemicals such as sulfur dioxide, the most typically used one, and lactic acid to soften the grains. The soften kernels are milled to be suitable for degermination process and separated germ is used for oil extraction. Degermed ground kernels are then passed through fine mills so that the fiber can be readily separated. The protein is further fractionated from the defibered starch slurry by centrifugal separators. After fractionation of each component, starch slurry is further processed to cooking and enzyme hydrolysis for ethanol production. In wet milling process of cassava chips (Figure 3), the starch slurry is prepared from dried chips by modifying typical cassava starch production process. Unlike wet milling of corn grains with water, the chips are milled to fine powder before slurried with water. The process is sometimes named as "Starch milk" process of which the starch is then extracted from chips by a series of extractors. After depulping, the starch slurry is then concentrated by a separator and subjected to a jet cooker for liquefaction. Currently, only a few plants are using this process, because this process requires a high investment. Factories have modified the process by reducing the extractor and stipping tank unit. Wet milling generates high starch losses in the solid waste. However, the process is more controllable and can be practically applied to high solid loading and continuous fermentation process (Sriroth & Piyachomkwan, 2010b).
In contrast to wet milling process, dry-milling process does not fractionate each component, yielding a by-product of mixed components. Although more valuable products are co-produced by wet milling process, this process is capital and energy intensive and results in a lower yield of ethanol as compared to dry-milling process; one ton of corn yields 373 and 388 liters of ethanol when processed by wet- and dry -milling, respectively (FO Licht, 2006).

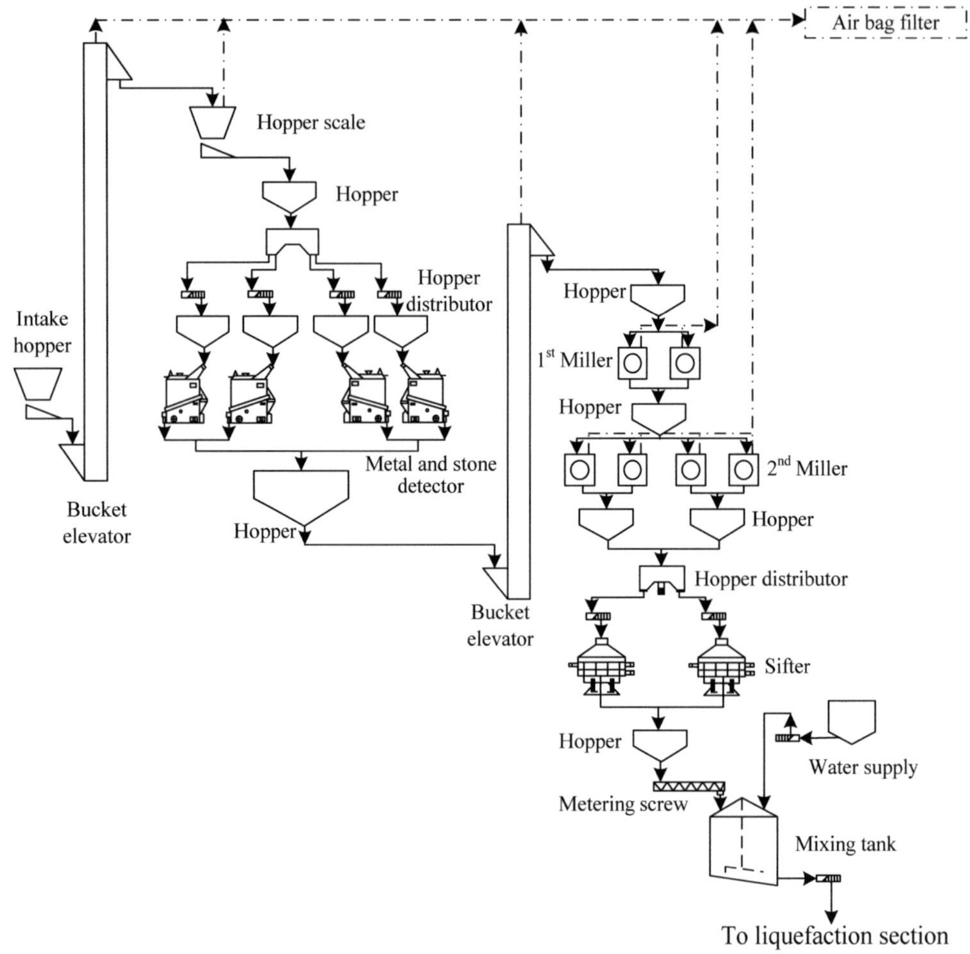

Fig. 2. "Dry milling" process for bioethanol production from dried cassava chips.

A modified dry-milling process has been developed recently by quickly removing germ or both germ and fiber prior to fermentation (Singh and Eckhoff, 1997; Wahjudi et al., 2000; Huang et al., 2008). This combined process improves cost reduction as compared to wet-milling process while increases value addition to dry-milling process. Although, cassava chips are corn analog and can be processed either by wet-milling or dry-milling process, the chips do not contain other valuable components. Dry-milling process is therefore generally applied for bioethanol production.

2.3.2 Cassava roots

During cassava harvest season, fresh roots are plenty available and the price is low. Therefore, it is common to use them to make slurry by grinding and then mix with cassava chip. Alternatively, cassava roots are used as a main raw material and then cassava chips are used to adjust the solid concentration. Similar to dried chips, there are two processes for preparing cassava fresh roots for bioethanol production, namely "With fiber" and "De-fiber" process.

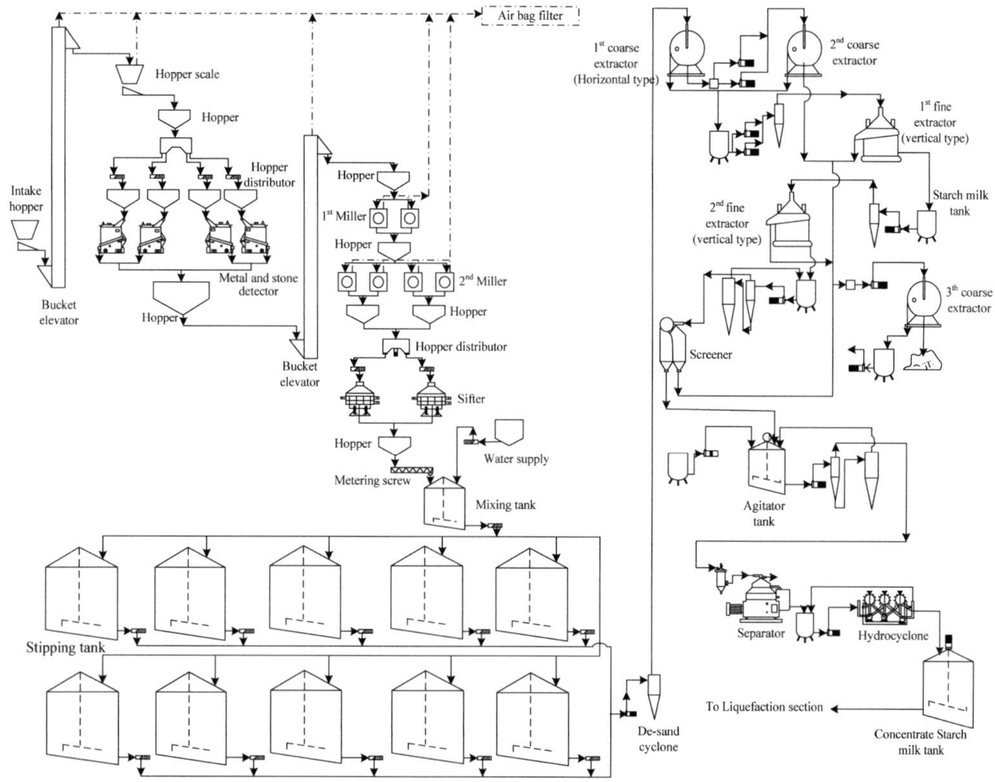

Fig. 3. "Wet milling" or "Starch Milk" process for bioethanol production from dried cassava chips.

In "With Fiber" process (Figure 4), the roots are transferred to the root hopper, in which soil and sand are effectively removed by root peelers. The roots are then washed and subjected to the chopper and rasper. The puree of milled roots is then slurried without fiber removal and used for liquefaction. This process requires less equipment and investment cost and is recommended for batch-type fermentation (Sriroth & Piyachomkwan, 2010b). However, with the presence of cell wall materials, ground fresh roots has developed semi-solid like characteristic and should be slurried with water to reduce viscous behavior. This causes dilution of solid loading in a fermentor, yielding a low ethanol concentration in final beer. A pretreatment of ground fresh roots with appropriate cell wall degrading enzymes has been introduced to handle that inferior flowability (Martinez-Gutierrez et al., 2006; Piyachomkwan et al., 2008), allowing potential use of fresh roots with Very High Gravity (VHG), i.e. high solid loading (> 30%) process and resulting in a higher ethanol concentration (upto 14.6% w/w or 18%v/v) in beer (Thomas et al., 1996). By almost doubling the ethanol concentration in the final beer, the VHG process can not only minimize the energy consumed during the downstream distillation process, but also improve the plant capacity. This concept can be applied to improve fermentation of other feedstock as well.

Similar to wet milling process of cassava chips, in "De-fiber" process (Figure 5), the starch slurry is prepared from fresh roots by modifying a typical cassava starch production

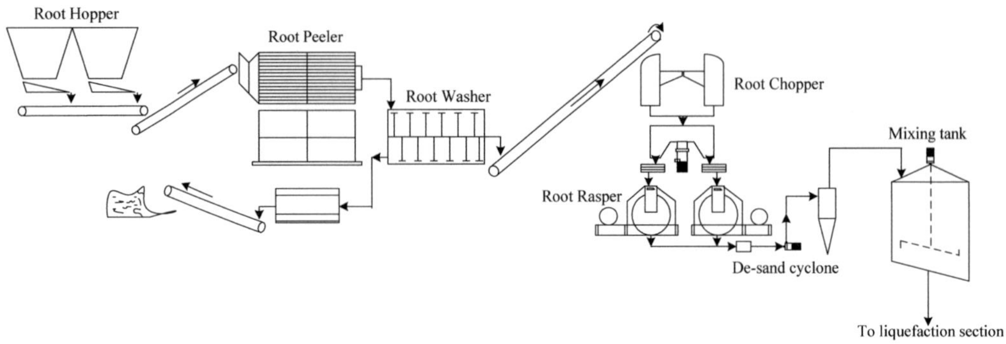

Fig. 4. "With fiber" process - starch slurry preparation from fresh cassava roots for bioethanol production.

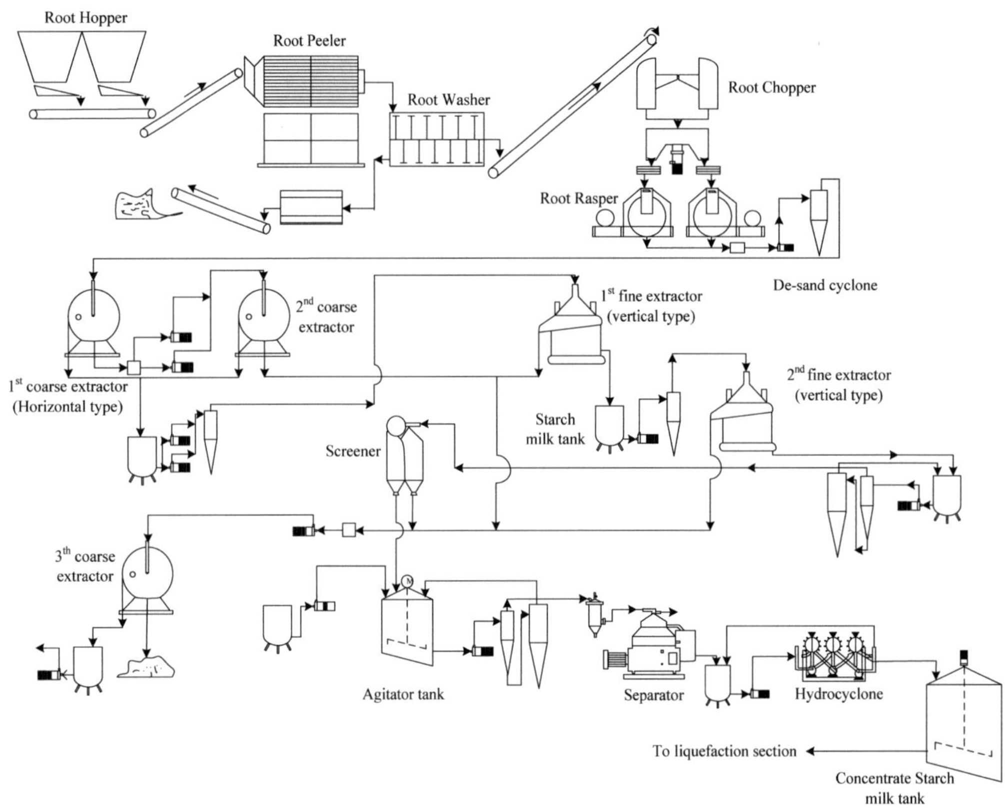

Fig. 5. "De-fiber" process - starch slurry preparation from fresh cassava roots.

process. After desanding and washing, roots are subjected to the chopper and rasper. The pulp is removed and starch is extracted by a series of extractors. After depulping, the starch slurry is then concentrated by a separator and subjected to a jet cooker for liquefaction. This process requires a higher investment cost and also generates high starch losses in the pulp. However, defiber process is more controllable and can be readily applied to current well-established technology of ethanol production from other materials. It is also practical for applying in high solid loading and continuous fermentation process (Sriroth & Piyachomkwan, 2010b).

2.4 Cassava bioethanol production

As described previously, the ethanol production from cassava feedstock involves 5 main steps (Figure 6a) which are

- Feedstock preparation: the main purpose of this step is to make cassava feedstock be physically suitable for downstream processing, i.e. cooking, starch hydrolysis, fermentation and distillation & dehydration. Details are different regarding to types of feedstock and milling process. In general, the preparation includes impurity removal (washing and peel removal of fresh roots, metal detector, sand and soil removal of slurry by hydrocyclone), size reduction by milling or rasping and fiber separation.
- Cooking: The starch is cooked to rupture the granular structure and hence improve its susceptibility to enzyme hydrolysis. Cooking is achieved at a temperature greater than a gelatinization temperature. During cooking, the high viscosity of the slurry is developed due to starch gelatinization and swelling of some particles. Cooking is, therefore, commonly performed in a presence of liquefying enzymes, i.e. α-amylase to liquefy cooked slurry.
- Starch hydrolysis: Starch is enzymatically hydrolyzed to glucose by α-amylase and subsequently by glucoamylase. The liquefaction by α-amylase is usually conducted at high temperatures at which the starch become gelatinized. After liquefaction, the liquefied slurry is cooled down to an optimum temperature for glucoamylase hydrolysis which is about 50-55°C, depending on enzyme types.
- Yeast fermentation: Glucose is then fermented by yeast. By the end of fermentation, the obtained beer contains approximately 10%v/v ethanol, depending on solid loading during fermentation.
- Distillation and dehydration: The beer is subjected to distillation to concentrate the ethanol to 95% and then dehydration to remove water, yielding anhydrous ethanol (99.5%).

Nowadays, the production process of bioethanol from starch feedstock is developed to significantly reduce processing time and energy consumption by conducting saccharification and fermentation in a same step; this process is called "Simultaneous Saccharification Fermentation", or SSF process (Figure 6b). In this SSF process, the liquefied slurry is cooled down to 32°, afterward glucoamylase and yeast are added together. While glucoamylase produces glucose, yeast can use glucose to produce ethanol immediately. No glucose is accumulated throughout the fermentation period (Figure 7) (Rojanaridpiched et al, 2003).

The material balance of ethanol process with a production capacity of 150,000 liters/day, a recommended size of ethanol plants for optimum production costs, feasible feedstock management and effective waste water treatment, from dried cassava chips by Simultaneous Saccharification and Fermentation (SSF) process is estimated from production data collected during pilot trials (at 2,500 L working volume) and factory survey (Figure 8) (Sriroth et al., 2006). The conversion ratio of feedstock (kg) to ethanol (liter, L) is about 2.5:1 for dried chips or 6:1 for fresh roots, this conversion factors are starch-quantity dependent. Based on the pilot production data, the estimated production cost, excluding the feedstock cost, of ethanol from cassava chips by SSF process is about 0.259 USD/L (Rojanaridpiched et al., 2003; Sriroth et al., 2006) which is close to a value reported by FO Licht to be 0.24 USD/L (FO Licht, 2006). The estimated production cost of cassava chips are detailed in Table 7 (Sriroth et al., 2010a).

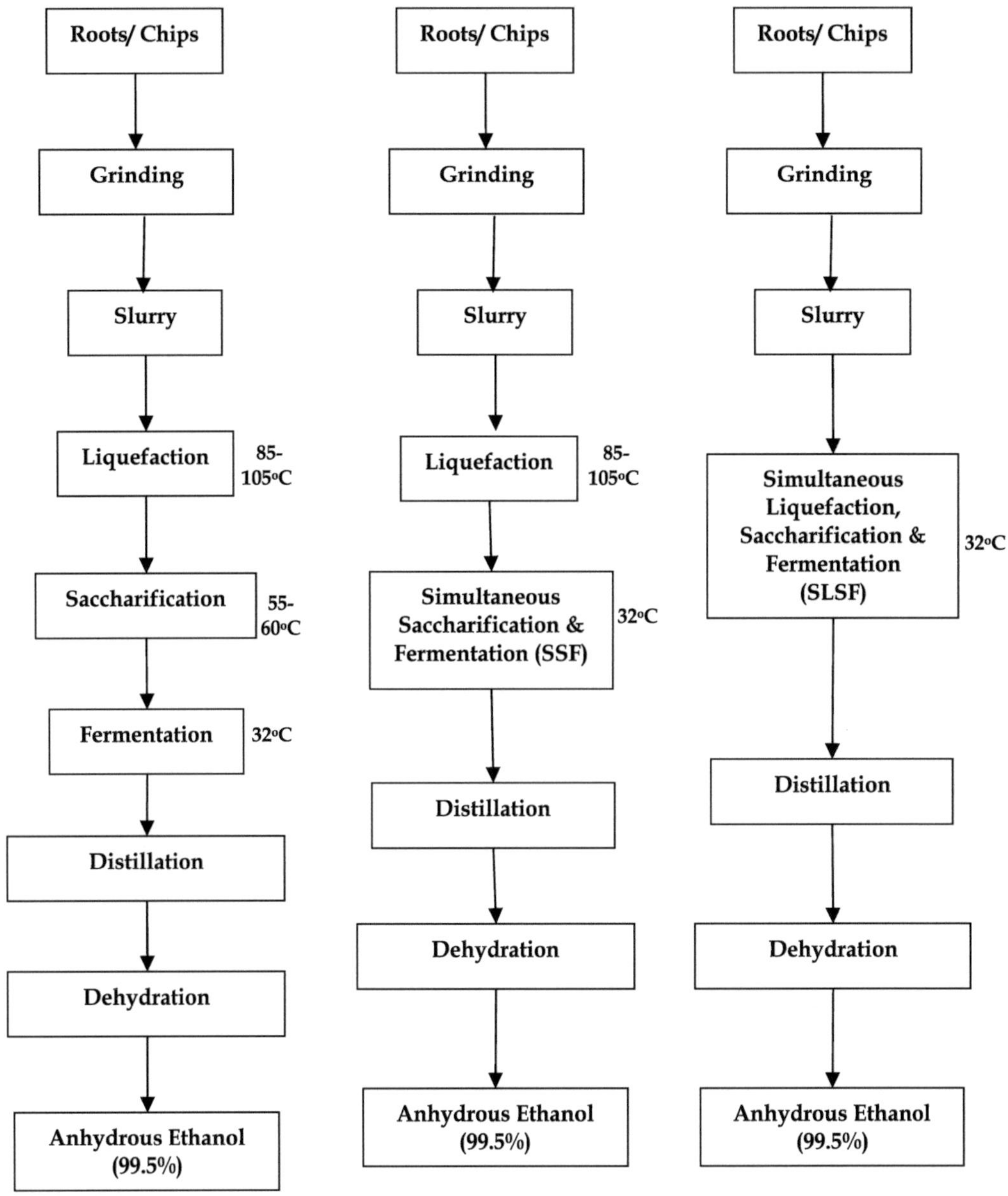

Note: The temperatures are enzyme- and yeast-type dependent.

Fig. 6. (a) Conventional, (b) SSF, Simultaneous Saccharification and Fermentation and (c) SLSF, Simultaneous Liquefaction, Saccharification and Fermentation process of ethanol production from cassava feedstock.

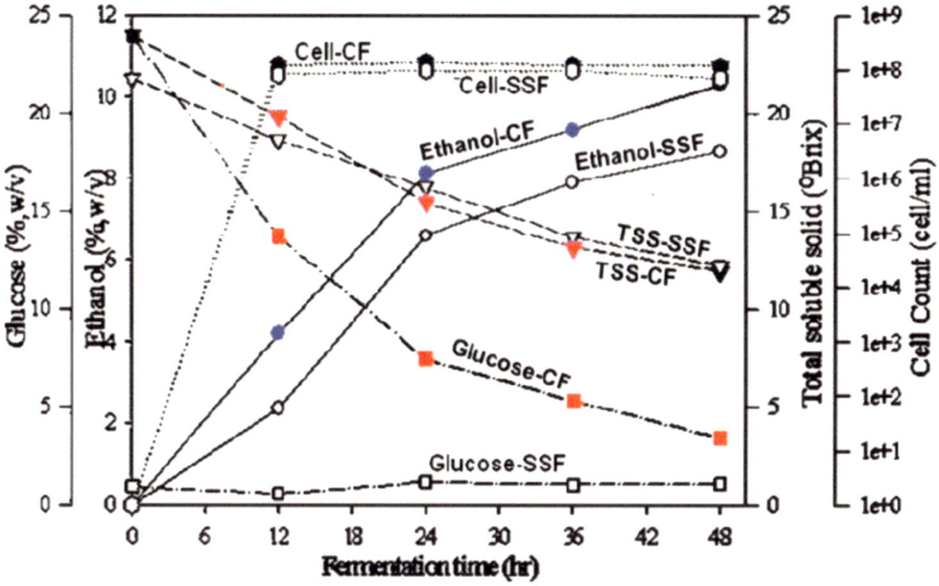

Fig. 7. Changes of total soluble solid (TSS, °Brix by a refractometer), cell counts, glucose and ethanol contents (by High Performance Liquid Chromatography using Bio-Rad Aminex HPX-87H column), during ethanol fermentation from cassava chips by conventional fermentation (CF) and Simultaneous Saccharification and Fermentation (SSF) process. (Experimental condition for CF; a slurry of 25% dry solid was initially liquefied by α-amylase at 95°C, saccharified by glucoamylase at 60°C and then fermented by yeast (*Saccharomyces cerevisiae*) at 32°C) and for SSF; a slurry of 25% dry solid was liquefied in a similar manner and then subjected to SSF by adding a mixture of glucoamylase enzyme and yeast at 32°C).

In SSF process, the starch in cassava material has to be initially cooked and liquefied prior to SSF process. Recently, a granular starch hydrolyzing enzyme has been developed to produce fermentable sugars from native or uncooked corn starch and is then applied to cassava chips (Piyachomkwan et al., 2007; Sriroth et al., 2008). This enzyme can attack directly to uncooked starch granules (Figure 9), allowing liquefaction, saccharification and, in the presence of yeast, fermentation to occur simultaneously in one step at the ambient temperature without cooking; this process is Simultaneous Liquefaction, Saccharification Fermentation or SLSF (Figure 6c). Figure 10 demonstrates the ethanol production from cassava chips using conventional, SSF and SLSF process. It is interesting to note that by SLSF process, the total soluble solid and glucose content do not change over fermentation as starch is used in a native, granular insoluble form. The fermentation efficiency of SLSF process is reported to be comparable with cooked process (Table 8). SLSF process is energy-saving, easy to operate and can be applied economically to produce sustainable energy, at a small scale, for community.

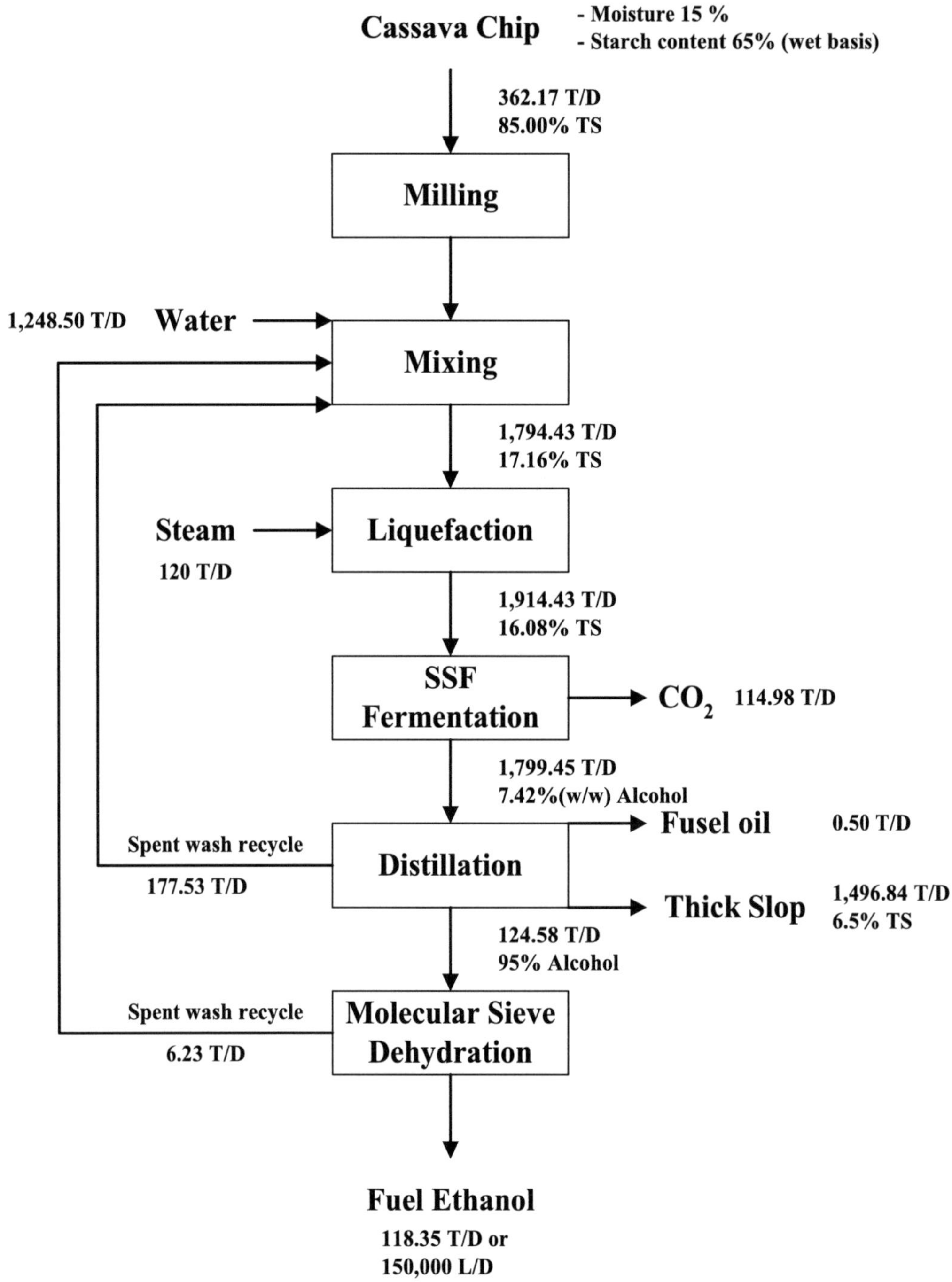

Fig. 8. Mass balance of ethanol production from cassava chip by SSF (Simultaneous Saccharification and Fermentation) process; T/D = Tons/Day, TS = Total Solid, L/D =Liter/day (Fermentation efficiency 90%, Distillation efficiency 98.5%) Source: Sriroth et al., 2006

Descriptive	Estimated production costs (USD/L)	
	Thailand[1]	China[2]
Materials and chemicals	0.032[b]	0.201[c]
Energy	0.109	0.013
Wage and addition	0.014	0.005
Depreciation	0.036	0.016
Maintenance	0.002	0.011
Miscellaneous	0.007	0.003
Fiscal charges	0.029	0.013
Land rent expense	-	0.0003
Selling expense	0.014	0.003
Waste treatment	0.014	-
Insurance	0.002	-
Water	-	-
Total cost	0.259	0.267

[1] the conversion ratio = 400 L ethanol/ton of cassava chips and the material and chemical cost excludes the raw material cost.
[2] the conversion ratio = 460 L ethanol/ton of cassava chips and the material and chemical cost includes raw material cost.
Source: Sriroth et al., 2010a

Table 7. Estimated production costs of ethanol from cassava chips by Simultaneous Saccharification and Fermentation (SSF) process.

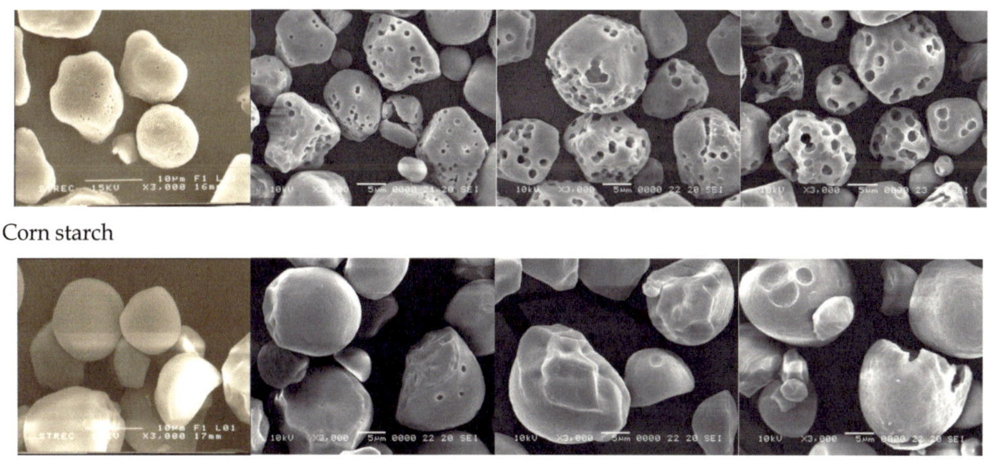

Corn starch

Cassava starch

O hr 12 hr 24 hr 48hr

Fig. 9. Scanning electron microscopic (SEM, at 3,000x magnification) pictures of corn and cassava starches treated with granular starch hydrolyzing (using 30% db starch in 0.05M acetate buffer, pH 4.5 and incubating with 0.5% granular starch hydrolyzing enzyme (Stargen™, Danisco-Genencor, USA, at 32°C).
Source: Sriroth et al., 2007

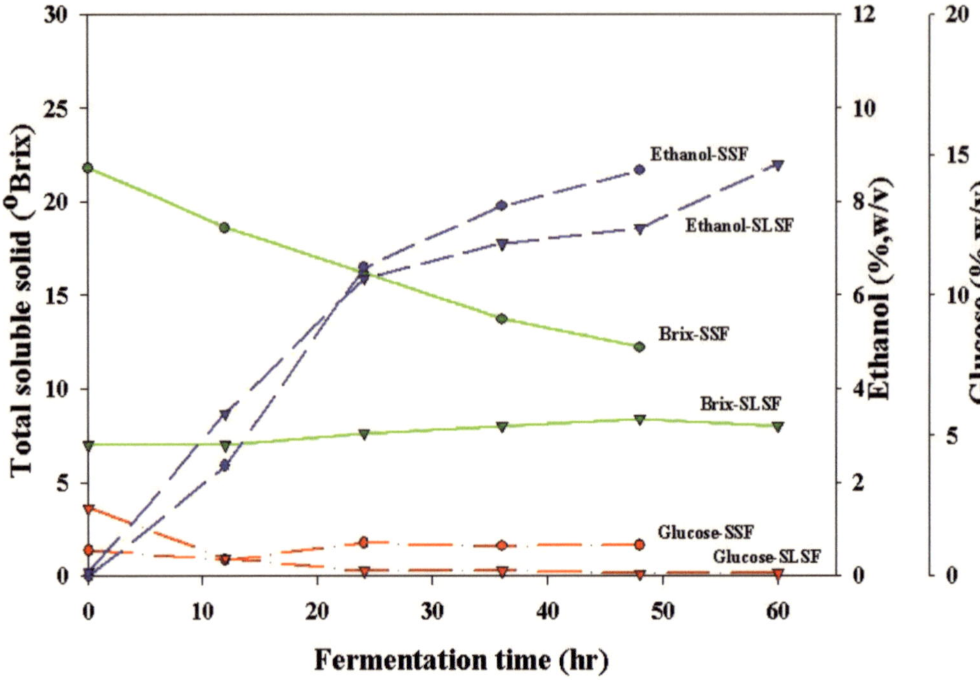

Fig. 10. Changes in total soluble solids (°Brix), glucose and ethanol content (%w/v) during ethanol production from cassava chips by Simultaneous Saccharification and Fermentation (SSF; the slurry of 25% dry solid was liquefied by 0.1% Termamyl 120L (Novozymes) at 95-100°C, 2 hr followed by simultaneous saccharification and fermentation with 0.1% AMG (Novozymes) and *Saccharomyces cirivisiae* yeast at 32°C for 48 hrs) and Simultaneous Liquefaction, Saccharification and Fermentation (SLSF; the slurry of 25% dry solid was liquefied, saccharified and fermented with 0.25% granular starch hydrolyzing enzyme (Stargen™, Danisco-Genencor, USA) and *Saccharomyces cirivisiae* yeast at 32°C, 60 hr. Source: Rojanaridpiched et al., 2003 ; Sriroth et al., 2007

2.5 Cassava bioethanol wastes and their utilization

During cassava bioethanol production, wastes are generated; the quantity and quality are depending significantly on feedstock quality and processing types. Since dry milling process is more widely used for bioethanol production from cassava feedstock, the information provided here is based on dry milling process of cassava chips. Similar to dry milling process for bioethanol production of corn grains, both of solid and liquid wastes are obtained at the end of distillation. The waste can be generated as a whole stillage containing both solid and liquid waste if the whole beer is subjected to the mash column without fiber separation. This process is applied in order to minimize ethanol loss in the solid pulp if fiber separation is accomplished prior to distillation. Recently, the process is adjusted by separating the fiber first and the fiber is washed to collect ethanol in pulp. At the production capacity of 150,000 liters of anhydrous ethanol/day, the total whole stillage is produced approximately 1,400-1,600 m^3/day, being wet cake 100-200 ton/day and the stillages 1,200-1,400 m^3/day (Sriroth et al., 2006).

Parameters	Values	
	SSF[1]	SLSF[2]
Slurry		
Volume (L)	2,053	2,200
% Total solid (w/v)	24.18	24.24
% Starch content of chips	80.4%	74.49%
pH	4.68	4.45
Beer after fermentation		
Fermentation time (hrs)	48	60
Volume (L)	2,166	2,258
Total soluble solid (°Brix)	12.2	7.4
Glucose content (%w/v)	1.09	1.24
Ethanol content (%w/v)	8.66	8.18
Cell counts (x 10^7 cell/ml)	6.82	1.15
Yield		
g ethanol/g dried chips	0.378	0.344
g ethanol/g starch	0.470	0.462
%Fermentation efficiency[3]	82.88	82.11

[1] Using 25% dry solid of chips, liquefied by 0.1% Termamyl 120L (Novozymes) at 95-100°C, 2 hr followed by simultaneous saccharification and fermentation with 0.1% Rhizozyme (Alltech) or AMG (Novozymes) and *Saccharomyces cirivisiae* at 32°C for 48 hrs.
[2] Using 25% dry solid of chips, liquefied, saccharified and fermented with 0.25% granular starch hydrolyzing enzyme (Stargen™, Danisco-Genencor, USA) and *Saccharomyces cirivisiae* at 32°C, 60 hr.
[3] as a percentage of theoretical yield
Source: Rojanaridpiched et al., 2003 ; Sriroth et al., 2007

Table 8. Parameters and results of ethanol production from cassava chips by SSF and SLSF process.

2.5.1 Solid waste

The wet cake has the total solid around 20-30% and contains a mixed component of cassava feedstock since no fractionation of cassava components is employed in dry milling process. The wet cake can be used to produce Dry Distillers Grains With Solubles or DDGS as developed in corn ethanol industry. However, cassava roots do not contain a high protein content as corn grains, cassava DDGS contains less protein contents (around 11-14% and 30% dry basis for cassava and corn DDGS; Sriroth et al., 2006). Though the solid waste from cassava chips is not as valuable as corn DDGS, it can be utilized in many ways:

- To produce biogas: this waste treatment has been practiced in China. The solid waste in the thick slop is sent to Biomethylation process. The results are successfully reported by Dai et al. (2006).
- To feed the burner: Another alternative design for Thai factories is that the solid waste from the thick slop is separated by a decanter so that the moisture content is around 50% of total solids (50% H_2O). This semi-dry solid is then used as the feedstock for fuel production by direct burning.
- To supplement in animal feed: The solid waste contains some fibers, proteins and ash and can be used as animal feed fillers.

2.5.2 Liquid waste (stillage)

Whilst the slop from molasses is very dark in color, cassava liquid waste has a light yellowish color with a lower COD (40,000-60,000) and BOD (15,000-30,000) values. The characteristics of waste water from the ethanol factories using cassava and molasses as feedstock are shown in Table 9. In consideration of this, the waste water from cassava-based process is much easier to handle than the waste obtained from molasses. This implies less investment and operational costs. In China, cogeneration of biogas obtained from waste water treatment in ethanol factory operating with cassava is reported to be able to cover all electricity needs in ethanol production process and still have some excess to supply to the grid (Dai et al., 2006). The practice for using thin stillage in Thailand is also for biogas production.

Characteristic	Factory using cassava chips	Factory using molasses
1. Chemical Oxygen Demand (COD, mg/L)	40,000-60,000	100,000-150,000
2. Biological Oxygen Demand (BOD, mg/L)	15,000-30,000	40,000-70,000
3. Total Kjedahl Nitrogen (TKN, mg/L)	350-400	1,500-2,000
4. Total Solids (mg/L)	60,000-65,000	100,000-120,000
5. Total Suspended Solid (TSS, mg/L)	3,000-20,000	14,000-18,000
6. Total Volatile Solids (mg/L)	20,000-40,000	n.a.
7. Total Dissolved Solids (mg/L)	50,000	105,000-300,000
8. pH	3.5-4.3	4.1-4.6

Source: Sriroth et al., 2006; n.a. = not applicable

Table 9. Characteristics of stillage obtained from ethanol factories in Thailand.

3. Lesson learned from Thai cassava bioethanol industry

The ethanol industry in Thailand has been active since 1961 as one of the Royal Project of His Majesty the King. Later, as an oil-importing country, Thailand has lost economic growth opportunity and energy security due to limited oil supply and price fluctuation. The seek for alternative energy for liquid fuel uses for transportation sector has been developed as a part of National Energy Policy and ethanol was then upgraded as national policy in 1995, initially in order to replace a toxic Octane Booster, i.e. Methyl tert-butyl ether (MTBE) in gasoline. By that time, the consumption rate of gasoline was 20 million liters per day which required 10% Octane Booster or 2 million liter per day; this formula is equivalent to Gasohol E10 (for octane 91 and 95), a blend of unleaded gasoline with 10% v/v anhydrous ethanol. With a rising concern of Global Warming and Clean Development Mechanism (CDM), gasohol containing higher ethanol components has been currently developed; E20 & E85. Presently, there are 47 factories legally licensed to produce biofuel ethanol with a total capacity of 12.295 million liters/day or 3,688.5 million liters per annum (at 300 working days). Two feedstocks, namely sugar cane molasses and cassava are their primary raw materials. A total of 40 factories use only a single feedstock; 14 factories using molasses with a total production capacity of 2.485 million liters/day, 25 factories using cassava with a total production capacity of 8.590 million liters/ day and 1 factory using sugar cane with a total production capacity of 0.2 million liters/ day. A multi-feedstock process using both molasses and cassava is, however, preferred in some factories (7 factories with a total production capacity of 1.020 million liters/day) to avoid feedstock shortage (Department of Alternative Energy Development and Efficiency, DEDE, 2009). A complication of Thai bioethanol industry is generated due to the fact that there are two feedstock types being used in other industries and also other alternative energy for transportation, i.e. LPG (Liquefied petroleum gas) and CNG (Compressed natural gas), being promoted by the government.

3.1 Feedstock supply

In Thailand, cassava is considered as one of the most important economic crops with the annual production around 25-30 million tons. The role of cassava in Thailand is not only as a subsistent cash crop of farmers, but it also serves as an industrial crop for the production of chips and starch, being supplied for food, feed and other products. This can be indicated by a continuous increase in root production since 2000 and be greater than 20 million tons since 2006. With the national policy on bioethanol use as liquid fuel, it significantly drives a rise in root demand. Various scenarios have been proposed to balance root supply and demand, in order to reduce the conflict on food vs. fuel security. Under the normal circumstance, root surplus should be used for bioethanol production, which initiates another industrial demand of roots and helps stabilize root price for farmers. Figure 11 is an example of projecting plan for root consumption by various industries, which corresponds to the targeted root production, proposed by Ministry of Agriculture and current root demand for chip and starch production. Another scenario is to reduce the amount of exporting chips and allocate those locally to existing industries. Meanwhile, the campaign for increasing root productivity (ton per unit area) by transferring good farming and agricultural practices has been distributed throughout the countrywide. In spite of that root shortage occurs in the last few years, caused by unexpected climatic change and widespread disease, i.e. mealy bugs. This, in fact, critically affects starch industries at a much greater extent than ethanol industry. Nevertheless, the starch industry is more competitive for higher root prices than ethanol industry. This situation of an unusual reduction of root supply emphasizes the need of increasing root production. A short-term policy on increasing root productivity from 25 tons/hectare by good farm management and cultivation practice has continuously pursued and expected to be 50 tons/hectare. Furthermore, long-term plan on R&D for varietal improvement is also greatly significant in order to develop varieties with higher root productivity (potentially be upto 80 tons/hectare), good disease resistance and good adaptation to climatic change such as higher growing temperatures or very dry condition.

3.2 Ethanol demand in biofuel use

Presently, there are 17 factories operating with the total production capacity of 2.575 million liters/day but most of them have operated under their full production capacity due to oversupply of ethanol. The influencing factor for decreasing ethanol demand is other alternative energy for transportation, i.e. Liquid Petroleum Gas (LPG) and Compressed Natural Gas (CNG), being promoted by the government. LPG is a primary fuel for household use such as for cooking that is why it is important to control the price of LPG (being low at 18 Baht per kilogram or 0.59 USD per kilogram). This promotes an increase usage of LPG in automobile sector, as indicated by increasing automobile engine change from gasoline to LPG. For CNG, there are several policies being released to promote the use

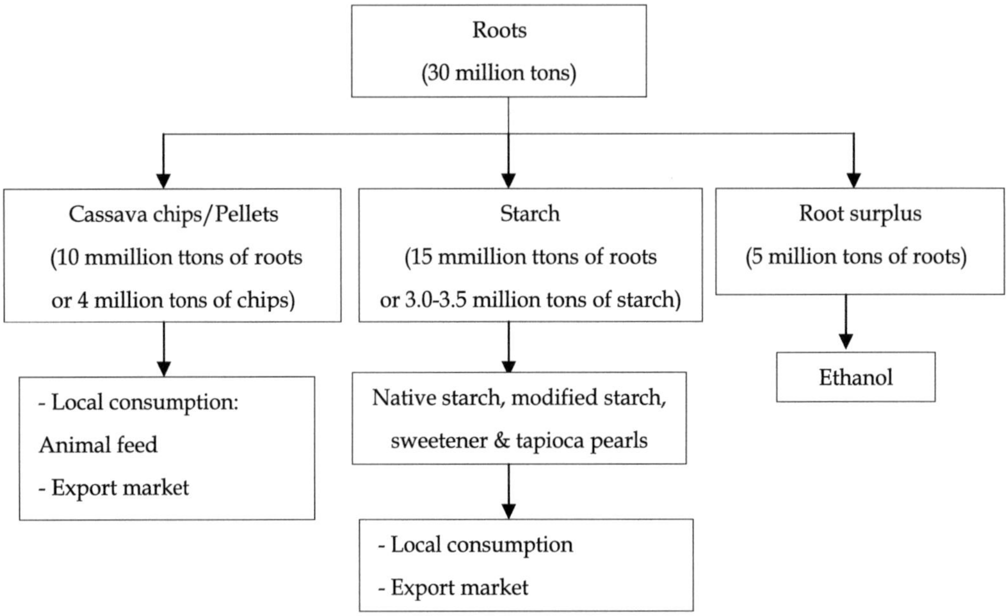

Fig. 11. Schematic diagram of cassava root consumption projection plan in Thailand.

of CNG in automobile sector in order to reduce the amount of gasoline consumption. Firstly, there is an exemption of tax for CNG fuel tank. Secondly, government agrees to cover the cost of changing car engine to CNG-using engine for taxi countrywide and tax reduction for CNG fuel cost. With the cost of production and fuel itself, the actual price is at 14.75 Baht per kilogram but the selling price is only at 8.50 Baht per kilogram. This difference requires a significant amount of subsidized oil fund to compensate the gap. At present, the government led by Ministry of Energy, has considered different mechanisms to intensify ethanol demands in transportation sector by promoting use of gasohol with a higher ethanol component (E85 and E100) for Flexible Fuel Vehicle (FFV), use of ethanol in motorbikes and use of ethanol as diesohol for trucks. These applications need technical support to acquire consumer's confidence. In addition, supporting policy and effective mechanism for exporting bioethanol can help expand market demand.

3.3 Regulation and pricing

To establish the local market for bioethanol demand in transportation sector, Thailand has released the regulation of denatured ethanol for gasohol uses (Table 10) to ensure high quality fuel for automobile use. No regulation for biofuel uses is announced by the government. In stead, the utilization of bioethanol as liquid fuel has been promoted by price incentive system. The retail price of gasohol E10 (for octane 95) is cheaper than gasoline around 0.33 USD/liter by exemption and reduction of excise & municipal tax and Oil Fund charge.

At the initial phase of trading ethanol locally, the price of ethanol for domestic market is referred to the price of imported ethanol from Brazil (FOB price of Brazilian Commodity Exchange Sao Paulo) with the additional cost of freight, insurance, loss, survey and currency exchange rate. Thai cassava ethanol industry has used two feedstock, i.e. molasses and cassava; the former one being utilized at a higher production capacity. This leads to shortage

No.	Description/Details	Value	Analytical method
1	Ethanol plus higher saturated alcohols, %vol.	> 99.0	EN 2870 Appendix 2 Method B
2	Higher saturated (C3-C5) mono alcohols, %vol.	< 2.0	EN 2870 Method III
3	Methanol, %vol.	< 0.5	EN 2870 Method III
4	Solvent washed gum, mg/100 mL	< 5.0	ASTM D 381
5	Water, %wt.	< 0.3	ASTM E 203
6	Inorganic chloride, mg/L	< 20	ASTM D 512
7	Copper, mg/kg	< 0.07	ASTM D 1688
8	Acidity as acetic acid, mg/L	< 30	ASTM D 1613
9	pH	> 6.5 and < 9.0	ASTM D 6423
10	Electrical conductivity, μS/m	< 500	ASTM D 1125
11	Appearance	Clear liquid, not cloudy, homogenous, and no colloidal particles	
12	Additive (if contains)	Agree with consideration of Department of Energy Business	

Source: Modified from Department of Energy Business, Ministry of Energy, Thailand (2005).

Table 10. The Thai standard of denatured ethanol for gasohol use as announced by the Department of Energy Business.

of molasses and price increase. As a result, the reference price based on Sao Paulo does not reflect the real ethanol situation of the country, both in term of production and uses. Subsequently, the pricing formula of biofuel ethanol has been revised. The reference price of bioethanol for fuel uses, as approved by The National Energy Policy Committee, Ministry of Energy, has taken into account for the cost of raw materials and produced quantities of fuel ethanol from both feedstocks, i.e. molasses and cassava, using the conversion ratios of 4.17 kg molasses (at 50°Brix) and 2.63 kg cassava chips (with starch content > 65%) for 1L of anhydrous ethanol. In addition, the structure of ethanol reference prices includes the production costs of each feedstock, which are 6.125 and 7.107 Baht/L for molasses and cassava, respectively. This monthly-announced ethanol reference price reflects the actual cost of local ethanol producers.

$$P_{Eth} = \frac{(P_{Mol} \times Q_{Mol}) + (P_{Cas} \times Q_{Cas})}{Q_{Total}}$$

Where

P_{Eth} = Monthly reference price of ethanol (Baht/L)

P_{Mol} = Price of molasses-based ethanol (Baht/L)

P_{Cas} = Price of cassava-based ethanol (Baht/L)

Q_{Mol} = Quantity of molasses-based ethanol (million L/day)

Q_{Cas} = Quantity of cassava-based ethanol (million L/day)
Q_{Total} = Total ethanol quantity (million L/day)
(For Q_{Mol}, Q_{Cas} and Q_{Total} using the value of one month previously, e.g. for the 5th month reference price, use the value of 3th month)

$$P_{Mol} = R_{Mol} + C_{Mol}$$

Where
P_{Mol} = Price of molasses-based ethanol (Baht/L)
R_{Mol} = Raw material cost of molasses, using a previous 3-month average export price announced by Thai Customs Department and the conversion ratios of 4.17 kg molasses (at 50°Brix) for 1L of anhydrous ethanol, e.g. using the average export price of 1st, 2nd, and 3rd month to calculate the price of 5th month
C_{Mol} = Production cost of molasses-based ethanol (6.125 Baht/L)

$$P_{Cas} = R_{Cas} + C_{Cas}$$

Where
P_{Cas} = Price of cassava-based ethanol (Baht/L)
R_{Cas} = Raw material cost of cassava, using the root price of one month previously, the conversion of 2.38 kg (25% starch) fresh roots for 1kg of chips with the production cost of 300 Baht/ ton chips, and the conversion ratios of 2.63 kg cassava chips (with starch content > 65%) for 1L of anhydrous ethanol
C_{Cas} = Production cost of cassava-based ethanol (7.107 Baht/L)

(Note: 1 USD = 30 Baht)

4. Conclusion

Cassava is not only a traditional subsistence food crop in many developing countries, it is also considered as an industrial crop, serving as a significant raw material base for a plenitude of processed products. Important ones are starches, modified starches and sweeteners for application in food, feed, paper, textile, adhesive, cosmetics, pharmaceutical, building and biomaterial. Consequently, the demand of cassava has been rising continuously and thereby contributes to agricultural transformation and economic growth in developing countries. Recently, in some countries such as Thailand, China and Vietnam, cassava is also used as the energy crop for producing bioethanol, an environmentally friendly, renewable alternative fuel for automobile uses. The promise of using cassava for bioethanol use is supported by many reasons including distinct plant agronomic traits for high tolerance to drought and soil infertility, low input requirement relatively to other commercial crop, and potential improvement of root yields. In addition, roots are rich in starch and contain low impurities. Although, fresh roots contain high moisture contents and are perishable, simple conversion to dried chip can be achieved by farmers at a low cost. Chips as corn analog are less costly to transport, store and process. High energy input for ethanol production from starch materials become less concerned as low energy consumption processes are developed including SSF, SLSF for uncooked process and VHG for a higher ethanol concentration. Improved waste treatment and utilization is also significant in order to minimize overall production cost. With those development, the use of cassava as an energy crop raises more concerns for food and fuel security. Both are critical to agricultural countries that mainly import fossil oil fuel and have lost their economic growth. To overcome that concern, the development of sufficient feedstock supply is considered as the first priority. A short-term and long term plans for root yield and productivity improvement by good cultivation practice and varietal improvement have been presently implemented in some regions. By that with a combination of zero-waste process concept, effective policies and market mechanism, the use of cassava as a food crop, industrial crop and energy crop become sustainable and beneficial to mankind.

5. References

Asaoka, M., Blanshard, J. M. V. & Rickard, J. E. (1991). Seasonal Effects on the Physico-chemical Properties of Starch from Four Cultivars of Cassava. *Starch/Starke*, Vol. 43, pp. 455-459.

Asaoka, M., Blanshard, J. M. V., Rickard & J. E. (1992). Effects of Cultivar and Growth Season on the Gelatinisation Properties of Cassava (*Manihot esculenta*) starch. *J Sci Food Agri.*, Vol. 59, pp. 53-58.

Balagopalan, C., Padmaja, G., Nanda, S. K. & Moorthy, S. N. (1988). Cassava in Food, Feed, and Industry. CRC Press, Inc., Florida. 205 p.

Breuninger, W. F, Piyachomkwan, K. & Sriroth, K. (2009). Tapioca/cassava Starch: Production and use. In: *Starch chemistry and technology*, BeMiller, J. & Whistler, R. 3rd ed. Academic press, New York. p 544.

Charles, A. L., Sriroth, K. & Huang, T. (2005). Proximate Composition, Mineral Contents, Hydrogen Cyanide and Phytic Acid of 5 Cassava Genotypes. *Food Chemistry*, Vol. 92, No. 4, pp. 615-620.

Charles, A. L., Huang, T. C. & Chang, Y. H. (2008). Structural Analysis and Characterization of a Mucopolysaccharide Isolated from Roots of Cassava (*Manihot esculenta* Crantz L.). Food Hydrocolloids, Vol. 22, No. 1, pp. 184-191.

Dai, D., Hu, Z., Pu, G., Li, H. & Wang, C. (2006). Energy Efficiency and Potential of Cassava Fuel Ethanol in Guangxi Region of China. *Energy Conversion & Management*, Vol. 47, pp. 1686-1699.

Defloor, I., Swennen, R., Bokanga, M., & Delcour, J.A. (1998). Moisture Stress During Growth Affects the Breadmaking and Gelatinisation Properties of Cassava (*Manihot esculenta* Crantz) Flour. *J Sci Food Agri.*, Vol. 76, pp. 233-238.

Department of Alternative Energy Development and Efficiency [DEDE]. (2009). Alternative energy: Gasohol. Ministry of Energy. Available source: http://www.dede.go.th.

FO Lichts. (2006). *An F.O. Licht Special Study: World Ethanol Market: The Outlook to 2015.* F.O.Licht Calvery Road, Tunbridge Wells, Kent, UK. p. 197.

Food and Agriculture Organization (FAO). 2011. Available from http://www.faostat.fao.org/site/339.

Fernando, S; Adhikari, S; Chandrapal, C; & Murali, N. (2006). Biorefineries: Current Status, Challenges, and Future Direction. *Energy & Fuels*, Vol. 20, pp. 1727-1737.

Howeler, R. H. (2001). Cassava Agronomy Research in Asia: Has it benefited Cassava Farmers?, *Proceedings of 6th Regional Workshop: Present Situation and Future Research and Development Needs*, pp. 345-382, Ho Chi Minh city, Vietnam, Feb 21-25, 2000.

Howeler, R. H. (2007). Agronomic Practices for Sustainable Cassava Production in Asia, *Proceedings of 7th Regional Workshop: Cassava Research and Development in Asia: Exploring New Opportunities for an Ancient Crop*, pp. 288-314, Bangkok, Thailand, Oct 28-Nov 1, 2002.

Huang, H., Ramaswamy, S., Tschirner, U. W., & Ramarao, B.V. (2008). A Review of Separation Technologies in Current and Future Biorefineries. *Separation and Purification Technology*, Vol. 62, pp. 1-21.

Juliano, B.O. (1993). Rice in Human Nutrition. FAO Food and Nutrition Series No. 26. The International Rice Research Institute (IRRI) and food and agriculture organization of United Nations [FAO].

Kajiwara, S. & Maeda, H. (1983). The Monosaccharide Composition of Cell Wall Material in Cassava Tuber (*Manihot utilissima*). *Agric. Biol. Chem.*, Vol. 47, No. 10, pp. 235-2340.

Martinez-Gutierrez, R., Destexhe, A., Olsen, H. S. & Mischler, M. (2006). Mash Viscosity Reduction. US Patent 20060275882.

Menoli, A. V. & Beleia, A. (2006). Starch and Pectin Solubilization and Texture Modification During Pre-cooking and Cooking of Cassava Root (*Manihot esculenta* Crantz). *LW-Food Science and Technology*, Vol. 40, No. 4, pp. 744-747.

Monceaux, D. A. (2009). Alternative feedstocks for fuel ethanol production. In: *The Alcohol Textbook: A Reference for the beverage, fuel and industrial alcohol industries*, Ingledew, W. M., Kelsall, D. R., Austin, G. D. & Kluhspies, C. Nottingham University Press, Nottingham. ISBN 978-1-904761-65-5. pp. 47-71.

Moorthy, S. N. & Ramanujam, T. (1986). Variation in Properties of Starch in Cassava Varieties in Relation to Age of the Crop. *Starch/Starke*, Vol. 38, pp. 58-61.

Office of Agricultural Futures Trading Commission, AFTC. (2007). Cassava. http://www.aftc.or.th.

Pardales, J. R., & Esquibel, C. B. (1996). Effect of Drought During the Establishment Period on the Root System Development of Cassava. *Jpn J Crop Sci.*, Vol. 65, No. 1, pp. 93-97.

Piyachomkwan, K., Wansuksri, R., Wanlapatit, S., Chatakanonda, P. & Sriroth, K. (2007). Application of Granular Starch Hydrolyzing Enzymes for Ethanol Production. In: *Starch: Progress in Basic and Applied Science*, Tomasik, P., Yuryev, V. P. & Bertoft, E., Polish Society of Food Technologists, Poland, pp. 183-190.

Piyachomkwan, K., Wansuksri, R., Wanlapatit, S. and Sriroth, K. (2008). Improvement of Utilizing Fresh Cassava Roots as Feedstock for Fermentation Process by Cocktail Enzymes. National Center for Genetic Engineering and Biotechnology.

Rojanaridpiched, C. 1989. Cassava: Cultivation, Industrial Processing and Uses. Kasetsart University. Bangkok. 439p.

Rojanaridpiched, C., Kosintarasaenee, S., Sriroth, K., Piyachomkwan, K., Tia, S., Kaewsompong, S. & Nitivararat, M. (2003). Development of Ethanol Production

Technology from Cassava Chip at a Pilot Plant Scale. National Research Council of Thailand.
Santisopasri, V., Kurotjanawong, K., Chotineeranat, S., Piyachomkwan, K., Sriroth, K. & Oates, C.G. (2001). Impact of Water Stress on Yield and Quality of Cassava Starch. *Industrial Crops and Products*, Vol. 13, No. 2, pp. 115-129.
Singh, V. & Eckhoff, S. R. (1997). Economics of Germ Preparation for Dry-Grind Ethanol Facilities. *Cereal Chemistry*, Vol.. 74, No. 4, pp. 462-466.
Sriroth, K., Santisopasri, V., Petchalanuwat, C., Piyachomkwan, K., Kurotjanawong, K. & Oates, C. G. (1999). Cassava
Starch Granule Structure-function Properties : Influence of Time and Conditions at Harvest on Four Varieties of Cassava Starch. *Carbohydrate polymers*, Vol. 38, No. 2, pp. 161-170.
Sriroth, K., Piyachomkwan, K., Santisopasri, V. & Oates, C. G. (2001). Environmental Conditions During Root Development: Indicator of Cassava Starch Quality. *Euphytica*, Vol. 20, No. 1, pp. 95-101.
Sriroth, K., Piyachomkwan, K., Wanlapatit, S., Thitipraphunkul, K. and Laddee, M. (2006). Study on Utilization of By-Products from Ethanol Process for Value Addition. The Department of Alternative Energy Development and Efficiency, Ministry of Energy.
Sriroth, K., Piyachomkwan, K., Keawsompong, S., Chatakanonda, P., Wanlapatit, S & Wansuksri, R. (2007). Improvement of Ethanol Production from Cassava by Application of Granular Starch Hydrolyzing Enzymes. National Research Council of Thailand, Bangkok.
Sriroth, K., Piyachomkwan, K., Keawsompong, S. & Wanlapatit, S. (2008). Production of Bioethanol from Cassava by Single Step Uncooked Process. Thai Patent, Priority Information Application No. 0801003407.
Sriroth, K., Piyachomkwan, K., Wanlapatit, S. & Nivitchayong, S. (2010a). The Promise of a Technology Revolution in Cassava Bioethanol : From Thai Practice to the World Practice. *Fuel.* Vol. 89, pp. 1333-1338.
Sriroth, K. & Piyachomkwan, K. (2010b). Processing of Cassava into Bioethanol. *Proceedings of 8th Regional Workshop: A New Future for Cassava in Asia: Its Use as Food, Feed and Fuel to benefit the poor*, pp. 740-750, Vientiane, Lao PDR, Oct 20-24, 2008.
Swinkels, J. J. M. (1998). Industrial starch chemistry, AVEBE Brochure. The Netherlands
Tanticharoen, M. (2009). A Study on Potential Improvement of Crop Yields of Sugarcane, Cassava and Palm Oil for Biofuel Production: Application of technology and planting area expansion. Thailand Research Fund. 182p. Bangkok.
Thirathumthavorn, D. & Charoenrein, S. (2005). Thermal and Pasting Properties of Acid-Treated Rice Starch. *Starch/ Starke*, Vol. 57, pp. 217-222.
Thomas, K.C., Hynes, S. H. & Igledew, W. M. (1996). Practical and Theoretical Considerations in the Production of High Concentrations of Alcohol by Fermentation. *Process Biochemistry*, Vol. 31, No. 4, pp. 321-331.
Treadway, R.H. (1967). Manufacture of Potato Starch. In: *Starch: Chemistry and Technology*, Whistler, R.L. and Paschall EF, New York. p 90.

Wahjudi, J., Xu, L., Wang, P., Buriak, P., Singh, V., Tumbleson, M. E., Rausch, K. D. & Eckhoff, S. R. (2000). The "Quick Fiber" Process: Effect of Temperature, Specific Gravity and Percentage of Residual Germ. *Cereal Chemistry*, Vol. 77, No. 5, pp. 640-644.

2

Single-Step Bioconversion of Unhydrolyzed Cassava Starch in the Production of Bioethanol and Its Value-Added Products

Azlin Suhaida Azmi[1,2], Gek Cheng Ngoh[1],
Maizirwan Mel[2] and Masitah Hasan[1]
[1]*Department of Chemical Engineering, University of Malaya, Kuala Lumpur*
[2]*Biotechnology Engineering, Kulliyah of Engineering, International Islamic University Malaysia, Kuala Lumpur*
Malaysia

1. Introduction

The global economic recession that began in 2008 and continued into 2009 had a profound impact on world income (as measured by GDP) and energy use. Since then the price of the energy supply by conventional crude oil and natural gas production has been fluctuating for years which has resulted in the need to explore for other alternative energy sources. One of the fastest-growing alternative energy sources is bioethanol, a renewable energy which can reduce imported oil and refined gasoline, thus creates energy security and varies energy portfolio. Global biofuel demand is projected to grow 133% by 2020 (Kosmala, 2010). However, the biofuel supply is estimated deficit by more than 32 billion liters over the same period and the deficit is worse for ethanol than biodiesel. Ethanol may serve socially desirable goals but its production cost is still remained as an issue. Extensive research has been carried out to obtain low cost raw material, efficient fermentative enzyme and organism, and optimum operating conditions for fermentation process. In addition to that, researchers have been trying to capitalize certain features of the plant equipment and facilities to increase the throughput of ethanol and other high value by products as well as to apply suitable biorefinery for the product recovery. At the same time, effort has been made to reduce utilities costs in water usage, cooling or heating, and also consumables usage via minimizing the effluent production.

Aimed to provide an alternative means for ethanol production, this book chapter introduces a single-step or direct bioconversion production in a single reactor using starch fermenting or co-culture microbes. This process not only eliminates the use of enzymes to reduce the production cost but also yield added value by-products via co-culture of strains. Before further elaboration on this single-step fermentation, we will visit the conventional process, the substrate preparation and microbe used. By this way a clear picture of the differences between conventional process and the proposed single-step fermentation with the advantages and disadvantages of both processes will be discussed.

1.1 Conventional process of starch fermentation

Traditionally, production of ethanol from starch comprises of three general separate processes namely; liquefaction using *α*-amylase enzyme, which reduces the viscosity of the starch and fragments the starch into regularly sized chains, followed by saccharification whereby the starch is converted into sugar using glucoamylase enzyme. Each of the process operated at different temperature and pH optima with respect to the maximum enzyme reaction rate. The final process involved the fermentation of sugar into ethanol using yeast. The simplified flow of the process can be summarized as in Figure 1.

Fig. 1. Conventional Starch Fermentation.

1.2 Substrate and the preparation

In this chapter, starch as carbon source will be primarily discussed in the application for single-step or direct bioconversion. Starch is a polysaccharide and the most abundant class of organic material found in nature. Sources of starch that are normally used in the production of ethanol are derived from seeds or cereals such as corn, wheat, sorghum, barley, soy and oat. Other sources of starch can be from tuber or roots such as potato, yam or cassava. By using starch as substrate for bioethanol production has distinct advantages in terms of its economical pretreatment and transportation compared to other types of biomass. For example cassava or tapioca tuber that has received an enormous attention in the production of biofuel in particular bioethanol in East Asia region such as China, Thailand, Malaysia and Indonesia (Dai et al., 2005; Hu et al., 2004; Nguyen et al., 2006). Cassava is a perennial woody shrub, ranks second to sugarcane and is better than both maize and sorghum as an efficient carbohydrate producer under optimal growing conditions. It is also the most efficient producer under suboptimal conditions of uncertain rainfall, infertile soil and limited input encountered in the tropic (Fregene and Puonti-Kaerlas, 2002).

Before undergo conventional or traditional fermentation, starch regardless of its sources required to be hydrolyzed. Two types of hydrolysis usually applied are mineral acid hydrolysis and enzymatic hydrolysis. The mineral acid or acid-base involved in the hydrolysis can be of diluted or concentrated form. The dilute acid process at 1-5% concentration is conducted under high temperature and pressure and has fast reaction time in the range of seconds or minutes. The concentrated acid process on the other hand uses relatively mild temperatures and the reaction times are typically much longer as compared to those in the dilute acid hydrolysis. The biggest advantage of dilute acid processes is their fast reaction rate, which facilitates continuous processing for hydrolysis of both starch and cellulosic materials. Their prime disadvantage is the low sugar yield and this has opened up a new challenge to increase glucose yields higher than 70% (especially in cellulosic material) in an economically viable industrial process while maintaining high hydrolysis rate and minimizing glucose decomposition (Xiang et al., 2004; McConnell, 2008). The concentrated acid hydrolysis offers high sugar recovery efficiency, up to 90% of both hemicelluloses and cellulose sugars. Its drawback such as highly corrosive and volatility can be compensated by low temperatures and pressures employed allowed the use of relatively low cost materials such as fiberglass tanks and piping. Without acid recovery, large quantities of lime must be used to neutralize the acid in the sugar solution. This neutralization forms large quantities of calcium sulfate, which requires disposal and creates additional expense. In addition to that, this type of hydrolysis has resulted in the production of unnatural compounds that have adverse effect on yeast fermentation (Tamalampudi et al., 2009).

Enzymatic hydrolysis of starch required at least two types of enzymes. This is due to that the starch or amylum comprises of two major components, namely amylose, a mainly linear polysaccharide consisting of α-1,4-linked D-glucopyranose units and the highly branched amylopectin fraction that consists of α-1,4 and α-1,6-linked D-glucopyranose units (Knox et al., 2004). Depending on type of plants, starch typically contains 20 to 25% amylose (van der Maarel et al., 2002) and 75 to 80% amylopectin (Knox et al., 2004). These two type linkage, α-1,4 and α-1,6-linked required an efficient starch hydrolysis agent or enzyme that can fraction α-1,4 and promote α-1,6 debranching activity. Since starch contains amylose and amylopectin, single or mono-culture cells are usually added during fermentation stage where starch has already been hydrolyzed to reducing sugar by hydrolysis agent such as acid-base or microbial enzymes in pretreatment and saccharification steps. The microbial enzyme of α -amylase cleaves α-1,4 bonds in amylose and amylopectin which leads to a reduction in viscosity of gelatinized starch in the liquefaction process. The process is the hydration of starch by heating the starch in aqueous suspension to give α-amylase an access to hydrolyze the starch. Dextrin and small amount of glucose and maltose are the end products. Exoamylases such as glucoamylase is then added during saccharafication which hydrolyses 1,4 and 1,6-alpha linkages in liquefied starch (van der Maarel et al., 2002). Enzyme has an advantage over acid-based hydrolysis. Amylolytic enzymes hydrolysis work at milder condition with the temperature lower 110°C (Cardona et al., 2010). However, enzyme is expensive especially cellulosic enzyme where it was reported the most expensive route accounted for approximately 22%-40% of total total cost (Wooley et al., 1999; Yang and Wyman, 200; Rakshit, 2006). Furthermore, fermentation of high concentration of starch to obtain high yield of ethanol is unfeasible due to reducing sugar inhibition on enzyme. This was shown in the work of Kolusheva and Marinova (2007) where the high reducing sugar produced from hydrolysis of high concentration not only inhibited the enzyme activity but also the fermenting yeast.

1.3 Microbes

Many investigators offer direct fermentation of starch using amylolytic microorganisms as an alternative to the conventional multistage that employs commercial amylases for liquefaction and saccharification, and followed by yeast fermentation. By using the amylolytic microorganism, ethanol production cost can be reduced via recycling some of the microorganism back to fermentors, thereby maintaining a high cell density, which facilitates rapid conversion of sugar into ethanol. However, there are very few types of amylolytic yeasts that are capable of efficiently hydrolyzing starch. Recombinant microbes and mix of amylolytic microorganism with glucose fermenting yeast in co-culture fermentation can resolve this setback. To further minimize contaminations and process handling cost, a single step or direct bioconversion of cassava or tapioca starch to bioethanol in one reactor (i.e. simultaneously saccharification and fermentation) in place of separate multistage processes will be focused upon in this chapter.

2. Single-step bioconversion

The idea of single-step bioconversion is to integrate all processes such as liquefaction, saccharafication and fermentation in one step and in one bioreactor. This alternative process will reduce contamination and the operation cost resulted from multistage processes of ethanol production. This also will reduce energy consumption of the overall process. The one-step bioconversion can be done by using recombinant clone or by co-culture or consortium of microorganisms that able to degrade or digest starch into intermediate product such as oligosaccharides and reducing sugar by starch fermenting microorganism(s). Then, the fermentation followed by fermenting the intermediate products into ethanol by microbe in the mixture. This process not only eliminates the use of enzymes to reduce the production cost but also yield added value by-products via co-culture of microbes. Besides, it also has a distinctive advantage as far as biorefinery is concerned. Unlike enzymes which normally required purification before recycled and added into the process, microbial growth can replace cells that have been removed. Even if cell separation and recycle are required, the processes are simpler compared to the more complex and sophisticated enzyme separation and purification process such as enzyme membrane reactor (Iorio et al., 1993) using ultrafiltration, extraction in aqueous two-phase system (ATPS) of water-soluble polymers and salts and/or two different water soluble polymers (Minami and Kilikian, 1998; Bezerra *et al.*, 2006) and selective precipitation (Rao et al., 2007).

2.1 Carbon source

The cost of raw material is important and cannot be overlooked since it governs the total cost which represents more than 60% of total ethanol production cost (Ogbonna et al., 2001). Using cassava (*Manihot esculenta*) or tapioca starch as substrate in bioethanol production will reduce the production cost since cassava plants are abundant, cheap and can easily be planted. It is a good alternative at low production cost. It is a preferred substrate for bioethanol production especially in situation where water availability is limited. It tolerates drought and yields on relatively low fertility soil where the cultivation of other crops would be uneconomical especially on idle lands. Furthermore, the starch has a lower gelatinization temperature and offers a higher solubility for amylases in comparison to corn starch (Sanchez and Cardona, 2008).

Cassava is one of the richest fermentable substances and most popular choice of substrates for bioethanol production in the Asian region. The fresh roots of cassava contain 30% starch and 5% sugars while the dried roots contain about 80% fermentable substances. Its roots can yield up to an average 30-36 t/ha. Several other varieties of its non edible tubers maybe selected based on the cyanide content which can be categorized as sweet, bitter, non-bitter and very bitter cassava contains 40-130 ppm, 30-180 ppm, 80-412 ppm and 280-490 ppm, respectively (Food Safety Network, 2005). Since fresh cassava tubers cannot be kept long, it needs to be processed immediately or produced ethanol from dried root. Alternatively, its roots can be milled and dried to form pallet or flour. This will prolong its storage time and save storage spaces. Cassava tuber also can be kept in soil after maturating for several months unharvest without deteriorating. Besides the tuber, cassava waste also can be utilized for ethanol production due to its high content of cellulose, hemicelluloses and starch respectively at 24.99%, 6.67% and 30-50% (w/w) (Ferreira-Leitão et al., 2010).

One of the advantages of using starch such as cassava is that most of the plants can be intercropped with other plants such as cover crops (legume plant) or tree crops (such as cocoa plant and palm oil plant) which can simultaneously grow together (Aweto and Obe, 1993; Polthanee et al., 2007). Polthanee et al. (2007) discussed four possible ways of intercropping practice. They are i) mixed inter-cropping, simultaneous growing of two or more crop species; ii) row-intercropping where simultaneous growing of two or more crop species in a well-defined row arrangement in an irregular arrangement; iii) strip inter-cropping, simultaneous growing of two or more crop species in strips wide enough to allow independent cultivation but, at the same time, sufficiently narrow to induce crop interactions and iv) Relay inter-cropping, planting one or more crops within an established crop in a way that the final stage of the first crop coincides with the initial development of the other crops. This will improve the land productivity and better land usage without the need to explore new land which might lead to deforestation. Figures 2 and 3 show the row-

Fig. 2. Soyabean in four-year-old oil palm (Ismail et al., 2009)

Fig. 3. Cassava intercrop with oil palm (Ismail et al., 2009)

intercropping of immature oil palm plantation intercrop with soyabean and cassava, respectively by Malaysian Palm Oil Berhad (MPOB), Malaysia.

2.2 Preparation method

The first step in the pre-separation process of starchy root or cassava tuber is to remove the adherent soil from roots by washing in order to prevent any problem later caused by the soil and sand. The process is followed by disintegration of cell structure to break down the size mechanically (i.e. milling) or thermally (i.e. boiling or steaming) or by combination of both processes. Slurry will be produced from the disintegration process which contains a mixture of pulp (cell walls), fruit juice and starch. This slurry can be cooked directly to gelatinized starch. When it is required, it can also be separated to produce flour by exploiting the difference in density using hydrocyclone and/or centrifuge separators as presented in Table 1.

Component	Density g/ml
Starch	1.55
Cell walls (fibers)	1.05
Water	1.00
Soil, sand	above 2

Table 1. Density of root components, water and soil (International Starch Institute, 2010).

For direct fermentation from starch to ethanol, there are two techniques normally employed in preparing starch medium which are non-cooking and low-temperature cooking fermentation. In non-cooking technique no heating is required however an aseptic chemical or method may be required to avoid contamination. Since it is uncooked, some aeration or agitation may also be required to avoid sedimentation of the starch particle. In low-cooking temperature fermentation, the medium is either semi or completely gelatinized first prior to inoculation of fermenting microorganism. Gelatinized starch forms a very viscous and complex fermentation media. It contains nutrients that required by microorganisms to grow and to produce different fermentation products. During fermentation, various physical, biochemical and physical reactions take place in the media. The nature and composition of the fermentation media will also affect the efficiency of the fermentation process. Many difficulties in designing and managing biological processes are due to the rheologically complicated behavior of fermentation media. Due to that, a pseudoplastic of a non-Newtonian behavior of starch solution is essential for cooked or gelatinized starch. This pseudoplastic property of gelatinized starch is important because it has suspending properties at low shear rates and its viscosity becomes sufficiently low when it is processed at higher rates of shear. Any fermentation medium which does not apply any viscosity reduction agent such as enzyme, its viscous nature combined with non-Newtonian flow will affect the mass heat transfer, dissolved oxygen homogeneity, mixing intensity, cell growth rate and eventually, the product accumulation state. Thus, it is imperative to minimize the viscosity to eliminate these problems. Starch slurry or flour concentration, temperature, agitation speed and cooking/gelatinization time are the major factors affecting media preparation. Optimization study of these conditions is useful prior to single-step fermentation of consortium or co-culture microorganisms. Table 2 gives the gelatinization temperature for different sources of starch. This information is helpful to prepare cooked or gelatinized starch for direct bioconversion at low temperature cooking.

Starch	Gelatinization Temperature Range (°C)
Potato	59-68[a,b,c]
Cassava/ tapioca	58.5-70[a,c]
Corn	62-80[a,b]
Paddy, rice and brown rice	58-79[a]
Sorghum	71-80[a]
Waxy corn	63-72[b]
Wheat	52-85[a,d]

[a]Turhan and Sağol (2004), [b] Whistler and Daniel (2006), [c]Tulyathan et al. (2006), [d] Sağol et al. (2006)

Table 2. Starch gelatinization temperature range

2.3 Direct starch fermentation without enzyme

In the industry whereby ethanol is produced from starch, temperature around 140°C-180°C is applied to cook the starch during hydrolysis using α-amylase prior to liquefaction. This high-temperature completely sterilizes harmful microorganisms and increases the efficiency of saccharification for high ethanol yield (Shigechi et al, 2004a, b). Consequently, this resulted in high energy consumption and added cost to amylolitic enzymes used in the process which ultimately increased the overall production cost. Several methods have been developed to reduce the energy consumption by applying milder liquefaction and/or saccharafication temperatures (Kolusheva and Marinova, 2007; Majovic et al., 2006; Montesinos and Navarro, 2000; Paolucci-Jeanjean et al, 2000) and also by exercising non-cooking fermentation (Shigechi et al., 2004b; Zhang et al, 2010). However, these types of fermentation usually required longer process time and sometimes may demand for additional volume of enzyme to maintain same productivity. The cost of enzyme will upset the total process cost.

To overcome this shortcoming, an alternative method of direct fermentation from starch may be employed to reduce the cost of enzyme. However, there are relatively few fermentation microorganisms that are capable of converting starch directly to ethanol since they do not produce starch-decomposing enzymes. One of the attempts to resolve this problem is by constructing recombinant microbes to coproduce α-amylase and glucoamylase with incorporating low temperature cooking of starch prior to fermentation by many research teams as shown in Table 3.

Several investigators reported that direct fermentation of starch using amylolytic microorganism offers a better alternative to the conventional multistage using commercial amylases for liquefaction and saccharification followed by fermentation with yeast (Abouzied and Reddy, 1986; Verma et al., 2000; Knox et al., 2004). By using this amylolytic microorganism in direct fermentation, the ethanol production cost can be reduced via recycling some of the microorganism back to fermentors, thereby maintaining a high cell density, which facilitates rapid conversion of substrate into ethanol. Furthermore by using cell exhibiting amylolytic activities, unlike using liquid enzyme that needs to be replenished or recycled unless if it is in immobilized system, the cell can multiply and reproduce with the enzymes. Fermentation using recombinant microbes, the starch medium can be prepared at low temperature cooking or uncook as a raw starch.

Another attempt of the direct fermentation without utilising any enzyme is by using co-culture microbes in the process. Instead of having enzyme separated and purified in different processes and subsequently to be used for hydrolysis in another separated process

Author	Transformed/ recombinant strain	Source of α-amylase	Source of glucoamylase	Type of starch	Starch concentration (g/L)	Max. ethanol concentration (g/L)
Altıntaş et al (2002)	*Saccharomyces cerevisiae*	*Bacillus subtilis*	*Aspergillus awamori*	Pure starch in 2.5 L fedbatch	40	29.7
Ülgen et al. (2002)	*Saccharomyces cerevisiae*	*Bacillus subtilis*	*Aspergillus awamori*	Starch	5- 80	47.5 (fed-batch culture) 15.6 (batch culture)
Knox et al. (2004)	*Saccharomyces cerevisiae*	*Lipomyces kononenkoae*	*Saccharomycopsis fibuligera*	Pure starch (Merck)	55	21
Shigechi et al. (2004a)	*Saccharomyces cerevisiae*	*Bacillus stearothermophilus*	*Rhizopus oryzae*	Corn starch cook at 80°C	50 90	18 30
Shigechi et al. (2004b)	*Saccharomyces cerevisiae*	*Streptococcus bovis*	*Rhizopus oryzae*	Raw corn starch in shake flask	200 g/L total sugar	61.8
Öner et al. (2005)	Respiration-Deficient Recombinant *S. cerevisiae*	*Bacillus subtilis*	*Aspergillus awamori*	Starch	5% starch + 0.4% (wt/vol) glucose	6.61
Khaw et al. (2007)	*S. cerevisiae* (non- and flocculent)	Not stated	Not stated	Raw corn starch	100	8
Kotaka et al. (2008)	*S. cerevisiae* (Sake yeast strain)	Not required	*Aspergillus oryzae Rhizopus oryzae*	Corn starch	50	18.5
He et al. (2009a)	*Zymomonas mobilis*	Not required	*Aspergillus awamori*	Raw Sweet potato	20.00 50.00	10.53 13.96

Table 3. Recombinant microbes for direct fermentation at low cooking temperature.

which contribute to higher expense, co-culture fermentation is worth to be considered as it might reduce the cost by omitting the unnecessary steps. While recombinant microorganism is constructed to provide the amylase activities, co-culture is simply selecting the microorganisms that naturally possess these amylase activities and combine them to work together to produce ethanol from starch.

Not many research works dedicated and related to co-culture fermentation for direct bioconversion of starch to ethanol. From just a few, same conclusions were drawn on the fermentation yield of the co-culture was better compared to mono-culture with improvement in the ethanol fermentation process. For instance study done by Verma et al. (2000), the co-culture fermentation of liquefied starch to ethanol can be carried out effectively with fermentation efficiency up to 93% compared to 78% and 85% when two-step bioconversion process using α-amylase and glucoamylase were applied to hydrolyze starch. Abuzied and Reddy (1986) observed that higher cell mass was produced in monoculture than in co-cultures which suggesting that substantially more carbon is used for cell production in monoculture, whereas in the co-culture most of the substrate carbon is utilized for ethanol production. Studies on co-culture microorganisms and systems are summarized in Table 4. The co-culture fermentation can either be simultaneous or subsequent mode for direct fermentation of low-temperature-cooking starch.

Strains for co-culture fermentation can also be obtained inexpensively from dry starter such as Ragi Tapai or Ragi Tape. This is similar to other oriental starter such as Ragi in Malaysia and Indonesia, Bubod in Philipine, Loog-pang in Thailand, Nurok in Korea, Koji in Japan, Banh Men in Vietnam, Chinese yeast in Taiwan and Hamei and Marcha in India. It is a dry-starter culture prepared from a mixture of rice flour and water or sugar cane juice/extract (Merican and Yeoh, 2004, Tamang et al., 2007). Clean rice flour is mixed with water or sugar cane juice to form thick paste. Sometime spices such as chilies, pepper, ginger and garlic which are assumed to carry desirable microorganism or may inhibit the development of undesirable microorganism are added to the paste (Basuki *et al.*, 1996; Merican and Yeoh, 2004). Then the thick paste is shaped into hemispherical balls. Ragi from previous batch is inoculated either on thick paste before or after it is shaped into hemispherical balls. Hesseltine *et al.* (1988) reported that at least one yeast and one *Mucoraceous* mold (*Mucor, Rhizopus,* and *Amylomyces*) were present with one or two of cocci bacteria in every sample of the dry starter. Apart from the *Rhizopus sp.* which capable of producing lactic acid besides fermentable sugar and ethanol (Soccol *et al.*, 1994), lactic acid bacteria are among the integral of the dry starter such as *Pediococcus pentosaceus, Lactobacillus curvatus, Lactobacillus plantarum* and *Lactobacillus brevis* (Sujaya et al., 2002; Tamang et al., 2007).

The traditional fermented food of tapai or tape' usually contains ethanol at concentration of 1.58% with high sugar content at concentration of 32.06%. Microaerophilic condition is required for the fermentation condition since fungi are unable to grow under anaerobic conditions and will result in unhydrolyzed starch. At lower temperature of 25°C, higher alcohol content will be produced after 144 h whereas at temperature of 37°C the tapai produces higher sugar content and becomes sweeter. (Merican and Yeoh, 2004). Tapai may contain up to 5% (v/v) of ethanol concentration (Basuki et al., 1996).

The benefit of using strains from dry starter such as ragi is that its application to produce fermented food such as tapai, is proven edible. Moreover, with addition of *S. cerevisiae* into the medium, the residue from ethanol recovery will contain yeast extract which can be processed as animal feed since it is edible and contain valuable nutrient that suitable for animal consumption as compared to fermentation using microbe such as *Escherichia coli.*

Direct fermentation has several advantages. First, to have multistage processes carried out in one reactor in which the glucose is produced during saccharification and simultaneously is fermented to ethanol can reduce contaminations and process handling cost. Second, direct fermentation reduces energy consumption. The starch medium can be prepared either at low-cooking temperature or by using the raw starch (uncooked starch). Even though some aseptic chemical or method may be required especially in raw starch fermentation, the cost incurred is still lower than the cost of energy consumption used in conventional fermentation.

Third, by applying direct fermentation, it is able to reduce inhibition of reducing sugar on fermenting yeast. In conventional fermentation, when starch is hydrolyzed using enzyme or mineral acid, certain amount of reducing sugar will be produced depending on the starch concentration. High level of reducing sugar in the fermentation medium (above 25% (w/v)) exerts osmotic pressure to the cells and limits their fermenting activity. This value may vary with different fermenting yeasts. However in direct fermentation, the osmotic pressure can be reduced by simultaneous converting starch to sugar and sugar to ethanol. This is particularly true in the recombinant clone which can co-express both the degrading enzymes. In the case of co-culture fermentation, the suitable inoculation time for the second microorganism needs to be determined. This is to avoid high yield of reducing sugar in

Author	1st microorganism	2nd microorganism	Co-culture System/ Fermentation procedure	Type of starch and concentration	Maximum ethanol concentration
Hyun and Zeikus (1985)	*Clostridium thermohydrosulfuricum*	*Clostridium thermosulfurogenes*	14 L microfermentor	5 % Starch with TYE medium (contains vitamin solution, ammonium chloride, magnesium chloride and trace mineral)	>120 mM
Abouzied and Reddy (1986)	*Aspergillus niger*	*Saccharomyces cerevisiae*	Simultaneous co-culture (500 mL shake flask)	Potato starch recovered from waste water of a potato chip manufacturing plant. (5% (w/v) starch)	5%(w/v)
Abouzied and Reddy (1987)	*Saccharomycopsis fibuligera*	*Saccharomyces cerevisiae*	Co-Culture fermentation (500 ml shake flask)	Similar to Abouzied and Ready (1986)	5%(w/v)
Reddy and Basappa (1996)	*Endomycopsis fibuligera* NRRL 76	*Zymomonas mobilis* ZM4	Shake flask	22.5% (w/v) cassava starch	10.5% (v/v)
Jeon et al. (2007)	*Aspergillus niger*	*Saccharomyces cerevisiae*	Separate fermentation in serial bioreactors (1.5 - 3.0 L).	Potato starch 55 g/L/day	22 g/L/day
He et al. (2009b)	*Paenibacillus sp.*	*Zymomonas Mobilis*	Simultaneously vs. subsequently co-cultured at 48 h of fermentation time. (100 mL shake flask)	50.0 g/L raw sweet potato starch (5% w/v starch)	6.6 g/L (120 h fermentation, pH 6.0) From subsequent co-culture
Yuwa-Amornpitak (2010)	*Rhizopus sp.*	*Saccharomyces cerevisiae*	Subsequently co-culture at 24, 48 and 72 h.	6%	14.36 g/L at 24 h subsequent co-culture

Table 4. The co-culture microorganisms in direct fermentations without enzyme addition.

medium before the second inoculation. When reducing sugar inhibition is avoided, fermentation of high starch concentration can be achieved for high ethanol yield and thus it reduces the water use. Subsequently this will reduce energy consumption in ethanol-water separation.

Direct fermentation is not limited to starch as it had been reported that different sugars from lignocellulosic hydrolysates such as mixture of glucose and pentose sugar for instance; xylose (Murray and Asther, 1984; Kordowska and Targonski, 2001; Qian et al., 2006) were fermented by glucose and pentose-fermenting microorganisms.

2.4 Ethanol by-products

During the fermentation process, several by-products are produced together with ethanol. In co-culture fermentation which involves different strains, different side-products besides ethanol are produced. The list of by-products and their applications in industry are listed in Table 5.

By-product name	Application
L-Lactic Acid (LA)	**Food and baverage** (acidulent, pH regulator, emulsifier, flavor enhancer & preservative), **cosmetics** (skin rejuvenating agent, moisturizer & exfoliant), **industrial** (degreasing agent, solvent & complexant), **pharmaceuticals** (sanitizer, drug delivery & administration, intermediate for optical active drug), **animal feed** (feed additive for farmed animals to reduce intestinal infection) (Hyflux ltd., 2008)
Polylactic acid (PLA)	**Food packaging** (disposable service ware, food containers & cartons), **medical** (suture threads, bone fixation & drug delivery), **non-woven** (diapers, specialty wipes & geotextiles), **fiberfill** (mattresses, pillows & comforters), **woven fibers** (apparel, socks, decorative fabrics), **specialist applications** (automotive heads & door liners) (Hyflux ltd., 2008)
Acetic acid (ethanoic acid)	Vinegar, chemical reagent, industrial chemical, food industry (food additive code E260 as acidity regulator)
Acetoin	Food flavoring and fragrance
Carbon dioxide	Carbonated water, dry ice, fire extinguisher, photosynthesis
Glycerol	Cosmetic and toiletries, paint and varnishes, automotive, food and beverages, tobacco, pharmaceutical, paper and printing, leather and textile industries or as a feedstock for the production of various chemicals (Pagliaro and Rossi, 2010; Wang *et al.*, 2001).

Table 5. Ethanol by-products and their applications.

The production of by-products somehow reduces the ethanol yield due to the competition from other metabolic conversions. The inhibition of lactic and acetic acids on yeast for ethanol production in corn mash was examined when both the acids synergistically reduced the rates of ethanol synthesis and the final quantities of ethanol produced by the yeast (Graves et al., 2007). The inhibitory effects of the acids were more apparent at elevated temperatures. So, a reduction in the formation of by-products is needed to achieve higher ethanol yield.

Alternatively, a fermentation process should not be only aimed for higher conversions of raw materials and ethanol productivity, but should rather take the advantage of the byproducts released during the transformation of feed stocks and convert them into valuable co-products. To reduce the inhibition effect, in-situ separation can be applied to separate the valuable co-product from the process. In this way, economical and environmental criteria can be met. However, depending on the objective and the economic analysis of the particular ethanol plant, the by-products may either generate extra revenue for the plant or just an inhibition the conversion process.

Among the ethanol byproducts, glycerol and lactic acid are used extensively by industries and can increase the production profit. These fermentative products have attracted interest owing to their prospect environmental friendliness and of using renewable resources instead of petrochemical. These byproducts have broad applications which can generate lucrative profit for the processes i.e. lactic acid. The global market for lactic acid is predicted to reach 328.9 thousand tonnes by 2015 (Plastics Today, 2011). The world consumption of lactic acid is stimulated by its applications in key industries such as cosmetics, biodegradable plastics and food additives. Lactic acid is used as a pH balancer in shampoos and soaps, and other alpha hydroxy acid applications, was expected to elevate the consumption in the market. Polymer lactic acid (PLA) for biodegradable plastics has properties similar to petroleum derived plastic and was expected to increase the demand for environmental friendly packaging. Food additives will continue to be the largest application area for lactic acid globally, but biodegradable plastics represent the fastest growing end-use application.

Glycerol or glycerin is a simple alcohol produced by *S. cerevisiae* during glucose fermentation to ethanol to maintain the redox balance. The global market for glycerin is forecasted to reach 4.4 billion pounds by 2015 (PRWeb, 2010). The increased demand for glycerin was reported to originate from various end-use area such as oral care, personal care, pharmaceutical and food and beverage. In fact, there are over 1,500 end-uses for the chemical. In most products, glycerin is used in very small portions with exception in a few end-uses which require a significant amount of glycerin in their formulation. Glycerin is also used in several novel applications such as propylene glycol, syngas and epichlorohydrin and it is expected to improve the glycerin demand.

Glycerol also can be potentially used as fuel additives for diesel and biodiesel formulation that assist to a decreasing in particles, hydrocarbons, carbon monoxide and unregulated aldehydes emissions. It can also act as cold flow improvers and viscosity reducer for use in biodiesel and antiknock additives for gasoline (Rahmat et al., 2010). Since glycerol is also produced in the fermentation broth, it is attractive as an entrainer to reduce the use of fresh entrainer in extractive distillation of azeotrope mixture of ethanol-water system.

3. Conclusion

Cassava is an attractive alternative as the carbon substrate for ethanol production especially where water availability is limited as it can tolerate drought and yields on relatively low fertility soil. The conventional method for the ethanol production involves liquefaction, saccharification and fermentation steps which are time consuming and cost ineffective, in view of the use of enzymes. Therefore, direct fermentation with integrated steps that incorporating recombinant or co-culture strains in a single reactor offers a more convenient method for the production of ethanol and its high value by products. By co-culture fermentation, high starch concentration can be used to reduce water usage in fermentation and subsequently in ethanol-water separation system. Furthermore, the fermentation medium can be prepared at lower temperature or raw starch can be used for direct fermentation to reduce the energy consumption. From the safety, economic and production process aspects, single-step bioconversion using co-culture microorganisms is a better alternative as far as production of ethanol and its by products from starch is concerned. The ethanol by-products such as lactic acid and glycerol can be value added co-products to generate extra revenue.

4. References

Abouzied, M.M., Reddy, C.A., (1986). *Direct Fermentation of Potato Starch to Ethanol by Cocultures of Aspergillus niger and Saccharomyces cerevisiae.* Applied and Environment Microbiology, pp. 1055-1059.

Abouzied, M.M., Reddy, C.A., (1987). *Fermentation of starch to ethanol by a complementary mixture of an amylolytic yeast and Saccharomy cescerevisiae.* Biotechnology Letters 9 (1), pp. 59-62.

Aljundi, I.H., Belovich,J.M., Talu, O. (2005). *Adsorption of lactic acid from fermentation broth and aqueous solutions on Zeolite molecular sieves.* Chemical Engineering Science 60, pp. 5004 - 5009.

Altintaş, M.M., Ülgen, K.Ö., Kirdar, B., Önsan, Z.I., Oliver, S.G., (2002). *Improvement of ethanol production from starch by recombinant yeast through manipulation of environmental factors.* Enzyme and Microbial Technology 31, pp. 640–647.

Aweto, A.O., Obe, O.A., (1993). *Comparative effects of a tree crop (cocoa) and shifting cultivation on a forest soil in Nigeria.* The Environmentalist 13 (3), pp. 183-187.

Basuki T., Dahiya D.S., Gacutan Q., Jackson H., Ko S.D., Park K.I., Steinkraus K.H., Uyenco F.R., Wong P.W., Yoshizawa K. (1996). Indegenous fermented foods in which ethanol is a major product, In: *Handbook of indigenous fermented food.* Steinkraus K.H., Dekker, New York, pp. 363-508.

Bezerra, R.P., Lins Borba, F.K.S., Moreira, K.A., Lima-Filho J.L., Porto, A.L.F., Chaves, A.C. (2006) *Extraction of Amylase from Fermentation Broth in Poly(Ethylene Glycol) Salt Aqueous Two-Phase System.* Brazilian Archives of Biology and Technology, Vol. 49, n. 4, pp. 547- 555.

Cardona, C.A., Sánchez, Ó.J., Guttierrez, L.F. (2010). Chapter 5: Hydrolysis of Carbohydrate Polymers, In: *Process synthesis for fuel ethanol production,* CRC Press. ISBN: 978-1-4398-1597-7. US Energy Information Admistration (2010). *International Energy Outlook 2010: World Energy Demand and Economic Outlook,* 15th March 2011, Available from:
http://www.eia.doe.gov/oiaf/ieo/world.html

Dai D, Hu Z, Pu G, Li H, Wang C. (2005). *Energy efficiency and potential of cassava fuel ethanol in Guangxi region of China.* Energy Conv. Manage. 47(13-14), pp. 1689-1699.

Fregene, M., puonti-Kaerlas, J. (2002). Chapter 10: Cassava Biotechnology, In: *Cassava: Biology, Production and Utilization,* Hillocks, R.J., Thresh, J.M., Belloti, A.C., CAB International, pp. 179-207.

Food Safety Network (2005). *Cassava fact sheet.* 20th April 2011, Available at
http://www.foodsafety.ksu.edu/articles/533/cassava_factsheet.pdf

Gao, M.T., Shimamura, T., Ishida, N., Takahashi, H. (2011). *pH-uncontrolled lactic acid fermentation with activated carbon as an adsorbent.* Enzyme and Microbial Technology, doi:10.1016/j.enzmictec.2010.07.015.

Graves, T., Narendranath, N.V., Dawson, K., Power, R., (2007). *Interaction effects of lactic acid and acetic acid at different temperatures on ethanol production by Saccharomyces cerevisiae in corn mash.* Applied Microbiology and Biotechnology 73, 1190–1196.

He, M.X., Feng, H., Bai, F., Li, Y., Liu, X., Zhang, Y.Z. (2009a). *Direct production of ethanol from raw sweet potato starch using genetically engineered Zymomonas mobilis.* African Journal of Microbiology Research Vol. 3(11), pp. 721-726.

He, M.X., Li, Y., Liu, X., Bai, F., Feng, H., Zhang, Y.-Z. (2009b). *Ethanol production by mixed-culture of Paenibacillus sp. and Zomomonas mobilis using raw starchy material from sweet potato.* Annals of Microbiology 59(4), pp. 749-754.

Hesseltine, C.W., Rogers, R., Winarno, F.G., (1988). *Microbiological studies on amylolytic oriental fermentation starters.* Mycopathologia 101, 141-155.

Hu, Z., Pu, G., Fang, F., W., C., 2004. *Economics, environment, and energy life cycle assessment of automobiles fuel by bio-ethanol blend in China.* Renewable Energy 29, pp. 2183-2192.

Huang, H.-J., Ramaswamy, S., Tschirner, U.W., Ramarao, B.V. (2008). *A review of separation technologies in current and future biorefineries.* Separation and Purification Tecnology 62, pp. 1-21.

Hyflux Ltd. (2008). 21st May 2009, Available via http://www.hyflux.com/biz_renewables.html

International Starch Institute (2010). *Technical Memorandum on Cassava Starch.* 20th April 2011, Available at http://www.starch.dk/isi/starch/cassavastarch.asp

Iorio, G., Calabro, V., Todisco, S. (1993) *Enzyme Membrane Reactors,* In: *Membrane processes in separation and purification,* Crespo, J.G., Boddeker, K.W., pp. 149-167, Kluwer Academic Publishers, ISBN 0-7923-2929-5, Netherland.

Ismail, S., Khasim, N., Raja Omar, R.Z. (2009). *Double-row avenue system for crop integration with oil palm.* MPOB Information Series, ISSN 1511-7871.

Jeon, B.Y., Kim, S.J., Kim, D.H., Na, B.K., Park, D.Y., Tran, H., T., Zhang, R., Ahn, D.H. (2007) *Development of a serial bioreactor system for direct ethanol production from starch using Aspergillus niger and Saccharomyces cerevisiae.* Biotechnology and Bioprocess Engineering 12, pp. 566-573.

Khaw, T.S., Katakura, Y., Ninomiya, K., Maoukamnerd, C., Kondo, A., Ueda, M., Shioya, S. (2007). *Enhancement of Ethanol Production by Promoting Surface Contact between Starch Granules and Arming Yeast in Direct Ethanol Fermentation.* Journal of Bioscience and Bioengineering 103 (1), pp. 95-97.

Knox, A.M., du Preez, J.C., Kilian, S. (2004). *Starch fermentation characteristics of Saccharomyces cerevisiae strains transformed with amylase genes from Lipomyces kononenkoae and Sacchamycopsis fibuligera.* Enzyme and Microbial Technology 34, pp. 453-460.

Kolusheva, T., Marinova, A., (2007). *A study of the optimal conditions for starch hydrolysis through thermostable a – amylase.* Journal of the University of Chemical Technology and Metallurgy, 42, 1, pp. 93-96.

Kosmala, F. (2010). *Global biofuel demand projected to grow 133% by 2020, production must increase by 32 billion liters: New report (September 24, 2010).* 13th April 2011, Available from
http://www.soyatech.com/news_story.php?id=20284

Kotaka, A., Sahara, H., Hata, Y., Abe, Y., Kondo, A., Kato-Murai, M., Kuroda, K., Ueda, M. (2008). *Note: Efficient and direct fermentation of starch to ethanol by sake yeast strains displaying fungal glucoamylases* Biosci. Biotechnol. Biochem., 72 (5), pp. 1376–1379.

Majovic L., Nikolic, S.,Rakin, M., Vukasinović, M., (2006). *Production of bioethanol from corn meal hydrolyzates.*Fuel 85, pp. 1750-1755.

McConnell, C. (2008). *Acid Hydrolysis.* Qittle. 1st April 2011, Available at
http://doyouqittle.com/2008/03/08/acid-hydrolysis/

Merican, Z., Yeoh Q.L., (2004). *Tapai processing in Malaysia: A technology in transition,* in: Steinkraus, K.H. (Eds), *Industrialization of indigenous fermented foods,* Marcel Dekker Inc., New York, pp. 247-270.

Minami, N.M., Kilikian, B.V. (1998) *Separation and purification of glucoamylase in aqueous two-phase systems by a two-step extraction.* Journal of Chromatography B, 711, pp. 309–312.

Montesinos, T., Navarro, J.M. (2000*). Production of alcohol from raw wheat flour by Amyloglucosidase and Saccharomyces cerevisiae.* Enzyme and Microbial Technology 27, pp. 362–370.

Murray W.D., Asther M. (1984). *Ethanol fermentation of hexose and pentose wood sugars produced by hydrogen-fluoride solvolysis of aspen chip.* Biotechnology Letter 6 (5), pp. 323-326.

Nguyen, T.L.T., Gheewala, S.H., Garivait, S., (2006). *Life cycle Cost Analysis of Fuel Ethanol Produce from Cassava in Thailand.* The 2nd Joint International Conference on "Sustainable Energy And Environment (SEE 2006). 29th March 2011, Available from
http://www.jgsee.kmutt.ac.th/see1/cd/file/C-024.pdf

Ferreira-Leitão, V., Gottschalk, L.M.F., Ferrara, M.A., Nepomuceno, A.L., Molinari, H.B.C., Bon, E.P.S. (2010). *Biomass residue in Brazil: Availability and potential.* Waste Biomass Valor 1, pp. 65-76.

Ogbonna, J.C., Mashima, H., Tanaka, H. (2001). *Scale up of fuel ethanol production from sugar beet juice using loofa sponge immobilized bioreactor.* Bioresource Technology 76, pp. 1-8.

Öner, E.T., Oliver, S.G., Kirdar, B. (2005*). Production of Ethanol from Starch by Respiration-Deficient Recombinant Saccharomyces cerevisiae.* Applied and Environmental Microbiology, pp. 6443–6445.

Pagliaro M., Rossi M. (2010). *The future of glycerol.* 2nd Ed. RSC Publishing, Cambridge, UK.

Paolucci-Jeanjean,D., Belleville, M.P., Zakhia, N., Rios, G.M. (2000). *Kinetics of Cassava Starch Hydrolysis with Termamyl® Enzyme.* Biotechnology and Bioengineering 68 (1).

Plastics Today (2011). *Bioplastic demand spurs global growth in lactic acid production (10th January 2011).* 13th April 2011, Available from
http://www.plasticstoday.com/articles/bioplastic-demand-spurs-global-growth-lactic-acid-production

Polthanee, A., Wanapat, S., Wanapat, M., Wachirapokorn, C. (2007). *Cassava-Legumes inter-cropping: A potential food-feed system for dairy farmers.* International Workshop

Current Research and Development on Use of Cassava as Animal Feed, Thailand, pp. 1-9.

PRWeb (2010). *Global glycerin market to reach 4.4 bilion pounds by 2015, according to a new report global industry analysyts, Inc.* 29th October 2010, Available from http://www.prweb.com/releases/glycerin_natural/oleo_chemicals/prweb4714434.htm

Qian M., Tian S., Li X., Zhang J., Pan Y., Yang X. (2006). *Ethanol production from dilute-acid softwood hydrolysate by co-culture.* Applied Biochemistry and Biotechnology 134(3), 273-283.

Rahmat, N., Abdullah, A.Z., Mohamed, A.R. (2010). *Recent progress on innovative and potential technologies for glycerol transformation into fuel additives: A critical review.* Renewable and Sustainable Energy Reviews 14, pp. 987–1000.

Rakshit, S.K. (2006). Thermozymes, in: Pandey, A., Webb, C., Soccol, C.R., Larroche, C. (editors), Enzyme Technology. Springer, pp. 603-612.

Rao D., Swamy, A. V. N. , SivaRamaKrishna, G. (2007) *Bioprocess technology Strategies, Production and Purification of Amylases: An overview* . The Internet Journal of Genomics and Proteomics, Vol. 2, n. 2, http://www.ispub.com/journal/the_internet_journal_of_genomics_and_proteomics/volume_2_number_2_22/article/bioprocess_technology_strategies_production_and_purification_of_amylases_an_overview.html

Reddy, O.V.S., Basappa, S.C. (1996). *Direct fermentation of cassava starch to ethanol by mixed Cultures of endomycopsis fibuligera and zymomonas mobius :Synergism and limitations.* Biotechnology Letters 18 (11), pp. 1315-1318.

Sağol, S., Turhan, M., Sayar, S. (2006). *A potential method for determining in situ gelatinization temperature of starch using initial water transfer rate in whole cereals.* Journal of Food Engineering 76, pp. 427–432.

Sanchez, O.J., Cardona, C.A. (2008). *Trend in biotechnological production of fuel ethanol from different feedstocks.* Bioreseource Technology 99, pp. 5270-5295.

Shigechi, H., Fujita, Y., Koh, J., Ueda, M, Fukuda, H., Kondo, A. (2004a). *Energy-saving direct ethanol production from low-temperature-cooked corn starch using a cell-surface engineered yeast strain co-displaying glucoamylase and a-amylase.* Biochemical Engineering Journal 18, pp. 149–153.

Shigechi, H., Koh, J., Fujita, Y., Matsumoto, T., Bito, Y., Ueda, M., Satoh, E., Fukuda, H., Kondo, A. (2004b). *Direct production of ethanol from raw corn starch via fermentation by use of a novel surface-engineered yeast strain codisplaying glucoamylase and a-amylase.* Applied and Environmental Microbiology, pp. 5037–5040.

Soccol, C.R., Stonoga, V.I., Raimbault,M., (1994). *Production of L-lactic acid by Rhizopus species.* World Journal of Microbiology and Biotechnology 10, pp. 433-435.

Sujaya, I.N., Amachi, S., Saito, K., Yokota, A., Asano, K., Tomita, F. (2002). *Specific enumeration of lactic acid bacteria in ragi tape by colony hybridization with specific oligonucleotide probes.* World Journal of Microbiology & Biotechnology 18, pp. 263–270,

Tamalampudi, S., Fukuda, H., Kondo., A., (2009). Chapter 8: Bioethanol from starchy biomass. Part II Hydrolysis and fermentation, In: *Handbook of Plant-Based Biofuels*, Pandey A., pp. 105 - 119. ISBN: 978-1-56022-175-3.

Tamang, J.P., Dewan, S., Tamang, B., Rai, A., Schilinger, U., Holzapfel, W.H. (2007). *Lactic acid bacteria in Hamei and Marcha of North East India.* Indian Journal of Microbiology 47, pp.119–125

Tulyathan, V., Chimchom, K., Ratanathammapan, K., Pewlong, C., Navankasattusas, S. (2006). *Determination of starch gelatinization temperatures by means of polarized light intensity detection.* J. Sci. Res. Chula. Univ. 31 (1). 4th April 2011, Available from http://www.thaiscience.info/Art0069cle%20for%20ThaiScience/Article/2/Ts-2%20determination%20of%20starch%20gelatinization%20temperatures%20by%20means%20of%20polarized%20light%20intensity%20detection.pdf

Turhan, M., Sağol, S. (2004). *Abrupt changes in the rates of processes occurring duringhydrothermal treatment of whole starchy foods around the gelatinization temperature–a review of the literature.* Journal of Food Engineering 62, pp. 365–371.

Ülgen, K.Ö., Saygili, B., Önsan, Z.İ., Kirdar, B. (2002). *Bioconversion of starch into ethanol by a recombinant Saccharomyces cerevisiae strain YPG-AB.* Process Biochemistry 37, pp. 1157–1168.

van der Maarel, M.J.E.C., van der Veen, B., Uitdehaag, J.C.M., Leemhuis, H., Dijkhuizen, L. (2002). *Properties and applications of starch-converting enzymes of the a-amylase family.* Journal of Biotechnology 94, pp. 137–155.

Verma, G., Nigam, P., Singh, D., Chaudry, K., (2000). *Bioconversion of starch to ethanol in a single-step process by coculture of amylolytic yeasts and Saccharomyces cerevisiae 21.* Bioresource Technology 72, pp. 261-266.

Wang, Z.-X., Zhuge, J., Fang, H., Prior, B.A. (2001). *Glycerol production by microbial fermentation: A review.* Biotechnology Advances 19, pp. 201-223.

Wasewar, K.L., Heesink, A.B.M., Versteeg, G.F., Pangarkar, V.G. (2002). *Equilibria and kinetics for reactive extraction of lactic acid using Alamine 336 in decanol.* Journal of Chemical and Biotechnology 77, pp. 1068-1075.

Whistler, R.L., Daniel, J.R. (2006). *Starch*, in *Kirk Othmer Encyclopedia of Chemical Technology Volume 22, Fifth Edition* (Ed. A. Seidel), John Wiley & Sons, Inc., Hoboken, USA. ISBN: 978-0-471-48501-8.

Wooley, R., Ruth, M., Glassner, D., Sheehan, J. (1999). *Process Design and Costing of Bioethanol Technology: A Tool for Determining the Status and Direction of Research and Development.* Biotechnol. Prog. 15, pp. 794-803.

Xiang, Q., Lee, Y.Y., Torget, R. T., (2004). *Kinetics of Glucose Decomposition during Dilute-Acid Hydrolysis of Lignocellulosic Biomass.* Applied Biochemistry and Biotechnology, vol. 113–116, 1127- 1139.

Yang, B., Wyman, C.E. (2008). *Pretreatment: the key to unlocking low-cost cellulosic ethanol.* Biofuels, Bioproduct and Biorefining 2, pp. 26–40.

Yuwa-Amornpitak, T. (2010). *Ethanol production from cassava starch by selected fungi from Tan-Koji and Saccharomyces cerevisiae.* Biotechnology, 9, pp. 84-88.

Zhang, L., Chen , Q., Jin, Y., Xue, H., Guan,J., Wang, Z., Zhao, H. (2010). *Energy-saving direct ethanol production from viscosity reduction mash of sweet potato at very high gravity (VHG)*. Fuel Processing Technology 91, pp. 1845-1850.

3

Sorghum as a Multifunctional Crop for the Production of Fuel Ethanol: Current Status and Future Trends

Sergio O. Serna-Saldívar*, Cristina Chuck-Hernández, Esther Pérez-Carrillo and Erick Heredia-Olea
Departamento de Biotecnología e Ingeniería de Alimentos, Centro de Biotecnología. Tecnológico de Monterrey, Monterrey, N. L. México

1. Introduction

Nowadays, there is a growing interest for alternative energy sources because of the reduction of fossil fuel production. Ethanol used as automotive fuel has increased at least six times in the current century. According to the Renewable Fuels Association, in 2010 the USA bio-refineries generated 13 billion gallons of fuel ethanol and the year before worldwide production reached 19 billion. This noteworthy increment is in its majority based on maize and sugar cane as raw materials (Berg, 2004; Renewable Fuels Association, 2010). The use of these feedstocks has triggered concerns related to food security especially today when the world population has reached 7 billion people.

The relatively sudden rise in food prices during 2008, 2010 and 2011 has been attributed mainly to the use of maize for bioethanol even when other factors like droughts or changes in global consumption patterns have also played a major role (World Food Program, 2008). Food price projections indicate that this situation will worsen, breaking the downward trend registered in food prices in the last thirty years (The Economist, 2007).

Even if there was not a food-fuel controversy especially due to the current conversion of millions of tons of maize for bioethanol, the use of only this crop cannot support the ambitious objectives of renewable fuel legislation in countries like the United States of America, where a target of 36 billion gallons of liquid biofuels have been established for 2022. In order to meet this requirement all the 333 million tons of maize yearly produced by USA should be channelled to biorefineries. This production represents 2 and 16 times the maize harvested in countries like China and Mexico respectively, which in turn are two of the five top world producers.

Environmental factors have been also pushing for the quest of new crops dedicated exclusively for liquid automotive fuel in order to reduce the use of prime farming land, irrigation water and other resources. A dedicated energy crop ideally must meet several requirements such as: high biomass yield and growth rate, perennial, with reduced input necessities, fully adapted to the geographic regions where will be planted, easy to manipulate via genetic improvement, non-invasive, tolerant to stress and with a good carbon sequestration rate among others (Jessup, 2009). At the present time, energy crops are mainly represented by perennial grasses as switchgrass (*Panicum virgatum* L.), energy cane (*Saccharum* spp), sweet and forage sorghum (*Sorghum bicolor*), miscanthus (*Miscanthus* spp.) as well as other short-rotation forest resources (willow -*Salix* spp- and poplar -*Populus* spp) (Jessup, 2009; McCutchen et al., 2008).

The development of new and improved enzymes, bioprocesses and feedstocks could lead to cost reduction from an estimated of 0.69 cents to below 0.51 cents/L that nowadays is the benchmark established for starchy raw materials (Kim & Day, 2011). Besides the development of dedicated crops for energy, one of the best approaches for cost reduction and optimal use of resources is the use of flexible facilities allowing the integration of different streams of same or different feedstocks. Flexibility, balance, diversification and regionalization are indeed keywords in the development of solutions to meet future world energy demands.

In tropical, subtropical, and arid regions from the United States, Mexico, China, India, Southern Africa, and other developing countries, where agronomic harsh conditions prevail, one of the most promising crops for fuel is sorghum (*Sorghum bicolor* (L.) Moench) (Reddy et al., 2005; Zhang et al., 2010). This is a high efficient photosynthetic crop that reached a worldwide production of 56 million tons of grain in 2009 (FAOSTAT, 2011), just behind maize, wheat, rice and barley. Almost 30% of this production is harvested in North America where sorghum is mainly used for feed. Sorghum is a C4 plant, highly resistant to biotic and abiotic factors as insects, drought, salinity, and soil alkalinity. Furthermore, this crop has one of the best rates of carbon assimilation (50 $g/m^2/day$) which in turn allows a fast growth and a better rate of net CO_2 use (Prasad et al., 2007). Sorghum requires one third of the water with respect to sugar cane and 80 to 90% compared to maize (Almodares & Hadi, 2009; Wu et al., 2010b). Thus, sorghum is considered as one of the most drought resistant crops. Furthermore, sorghum requires approximately one third of the fertilizer required by sugar cane (Kim & Day, 2011) and its growth cycle is between 3 to 5 months allowing two or three crops per year instead of one commonly obtained with sugarcane. Besides environmental advantages, sorghum is one of the more acquiescent plants to genetic modification because is highly variable in terms of genetic resources and germplasm. This facilitates plant breeding and development of new cultivars adapted to different regions around the globe (Zhang et al., 2010).

Sorghum can be classified in four broad groups: grain, sweet, forage and high biomass. All belong basically to the same species and virtually there are no biological or taxonomic differences (Wang et al., 2009). Grain sorghum is used mainly as food, feed and for starch production. In the United States only a small percentage of fuel ethanol (around 2-3%) is obtained from grain sorghum (Renewable Fuels Association, 2010; Turhollow et al., 2010; Zhao et al., 2008), but in 2009 about 30% of the U.S. grain sorghum crop was used for ethanol production (Blake, 2010).

On the other hand, forage sorghum is characterized as a high biomass crop. This capacity has been boosted by intensive research programs worldwide, focused in the design of new varieties tailored for ethanol production (Rooney et al., 2007). The main product obtained from sweet sorghums is the fermentable sugar rich juice that is produced and accumulated in the stalks in a similar fashion as sugar cane. The extracted sweet juice is mainly composed of sucrose, glucose, and fructose, and thus can be directly fermented into ethanol with efficiencies of more than 90% (Wu et al., 2010b). According to Almodares & Hadi (2009) sorghum yields a better energy output/input ratio compared to other feedstocks such as sugar cane, sugar beet, maize and wheat. Altogether with the juice, the residue or bagasse can be also converted to ethanol or used for other traditional applications.

In summary, sorghum is a crop well adapted to adverse climatic conditions which at this time is one of the growing concerns in agronomic projections. This is mainly due to the change of rain patterns and climate, greenhouse effect and the steadily rise of world temperature. Given all these advantages of sorghum as a potential source of biofuels, the objective of this chapter is to explore its potential, as an integrated crop for fuel production in terms of yield and technologies available for processing. The chapter especially focuses on optimum technologies to produce bioethanol from sweet sorghums, starchy grains and biomass from dedicated crops.

2. Botanical features and agronomic characteristics

Sorghum is a member of *Poaceae* family, a high-efficient photosynthetic crop, well adapted to tropical and arid climates. As a result, sorghum is extremely efficient in the use of water, carbon dioxide, nutrients and solar light (Kundiyana, 1996; Serna-Saldívar, 2010). This crop is considered one of the most drought resistant, making it one of the most successful in semi-desert regions from Africa and Asia (Woods, 2000). This resistance is due mainly to its photosynthetic C4 metabolism that allows sorghum to accumulate CO_2 during the night, to lower the photorespiration rate in presence of light, to reduce the loss of water across the stoma and the waste of carbon (Keeley & Rundel, 2003).

The leaves of sorghum and maize are similar but in the case of sorghum they are covered by a waxy coat that protects the plant from prolonged droughts. The sorghum grain is grouped in panicles and the plant height ranges from 120 to 400 cm depending on type of cultivar and growing conditions. An advantage of sorghum compared to maize is that it has a comparatively lower seed requirement because only 10 to 15 kg/ha are used compared with 40 kg/ha required by other cereals (Kundiyana, 1996). In some regions is possible to produce multiple crops per year, either from seed (replanting) or from ratoon (Saballos, 2008; Turhollow et al., 2010).

3. Chemical composition

3.1 Juice from sweet sorghum

The mature stems of sweet sorghum contain about 73% moisture and the solids are divided in structural and non-structural carbohydrates. Approximately 13% are non-structural carbohydrates composed of sucrose, glucose and fructose, in variable amounts according to cultivar, harvesting season, maturity stage, among other agronomic factors (Mamma et al., 1996; Phowchinda et al., 1997). Anglani (1998) suggests a classification of sweet sorghums based on proportion of soluble sugars in the juice. The first group with a high content of sucrose (sugary type) and the second with more monosaccharides (glucose and fructose) compared to other soluble carbohydrates (syrup type). Smith et al. (1987) in their evaluation of six sweet sorghum varieties throughout four years in nine different locations did not find significant differences in sugar content or composition. The typical composition indicates that around 70% was sucrose and the rest glucose and fructose in equal parts. In stem dry basis, Woods (2000) reported fermentable sugars content between 41 to 44% in Keller and Wray varieties with 80 and 63% represented by sucrose and the rest by glucose and fructose. A fiber variety analyzed by the same author (H173) reached only 20% fermentable sugars based on the dry stem weight; sucrose, glucose and fructose were found in equivalent amounts (around 7% for each sugar). Compared to sugar cane, the main difference is that the sucrose content in cane is significantly higher compared to glucose and fructose (90, 4 and 6%respectively) and the total content sugar is 49% of the dry stem weight. In general terms, composition of simple sugars in sweet sorghum juice is 53-85, 9-33 and 6-21% for sucrose, glucose and fructose, respectively (Gnansounou et al., 2005; Mamma et al., 1996; Phowchinda et al., 1997; Prasad et al., 2007).

Beyond the proportion of soluble sugars in sweet sorghum plants, the yield of total sugars per harvested area is a better guide in the analysis for fuel production. Woods (2000) reported for sweet sorghum cultivars (Keller, Wray and H173) an average of 7, 10 and 4 ton of fermentable sugars/ha respectively, significantly lower compared to the 17 ton/ha for sugarcane indicated by the same author. The varieties studied by Davila-Gomez et al. (2011) yielded an average of 1.85 to 2.03 ton of sugar/ha, whereas Smith et al. (1987) in a extensive study performed in several locations of continental United States and Hawaii, obtained from 4.5 to 10.6 ton/ha. In other varieties evaluated in China, the best yields reached 18 ton/ha (Zhang et al., 2010).

Sugars in sweet sorghum are very sensitive to microbial contamination especially after crushing stalks for juice production. In data reported by Davila-Gomez et al. (2011), the percentage of sugars, as °Brix before fermentation, was lower (11 to 24% lower) than the obtained immediately after harvest in summer time, when temperatures easily reached 32°C in Northeast Mexico. The microbial contamination was the most obvious explanation of this phenomenon. Besides, the sucrose proportion in the fermented juices was lower in relation to glucose and fructose (0 to 10% of total). This can be related to invertase activity of contaminating wild yeasts that hydrolyzed sucrose into glucose and fructose. These monomers are quickly metabolized by means of facilitated diffusion into the yeast cell. Wu et al. (2010b), working with cultivars with 16 to 18% of fermentable sugars, found that as much as 20% of substrate can be lost in 3 days at 25°C. This loss corresponds to approximately 700 L ethanol/ha when a yield of 50 ton of sorghum stems/ha is considered. Daeschel et al. (1981) reported that juices can be preserved during 14 days at 4°C without detectable changes or deterioration (sour odor and foaming). These authors also reported that the dominant spoilage microorganisms were *Leuconostoc mesenteroides* and *Lactobacillus plantarum* at 25 and 32°C, respectively and recommended to process the juice within five hours after extraction.

3.2 Sorghum grain

Sorghum grain is a naked caryopsis composed of three major anatomical parts: pericarp, germ, and endosperm. The pericarp is composed of epicarp, mesorcarp and endocarp (cross and tube cells). Among cereals, sorghum is the only one that can contain significant amounts of starch granules in the mesocarp cells. The starch-devoid germ is rich in fat, soluble sugars and proteins (albumins and globulins) whereas the endosperm is divided into the single layered aleurone and the starchy endosperm cells positioned in the corneous and floury or chalky regions of the endosperm. The endosperm constitutes the largest fraction of the kernel and where almost all the starch is contained. Similar to maize, sorghum contains 60 to 70% of starch. The endosperm texture and hardness are highly related to the performance of the grain during several stages of ethanol production. In general terms, composition of sorghum is similar to maize with a few small but significant differences mainly in protein and fat concentrations. Sorghum for instance, has an average 1% less fat and 1.5 to 2.0% more crude protein compared to maize. Both sorghum and maize have more than 50% of this protein as prolamins named kafirins and zeins, respectively. In sorghum, approximately half of the prolamin fraction is bound. In contrast, approximately 70% of the maize prolamins are free or alcohol-soluble. There are some sorghum varieties that contain significant amounts of condensed tannins in the testa. These sorghums are classed as type III and have a lower nutritional value compared to other sorghums and maize. This is due to the presence of tannins that bind proteins and inactivate enzymes. As a result, high tannin sorghums may have reduced ethanol yields (Serna-Saldivar, 2010).

One of the most noteworthy differences between sorghum and maize is its starch granule-protein matrix interaction that negatively affects the susceptibility of both proteins and starch to enzyme hydrolyses. These structural differences affect protein digestibility and the speed of dextrins and glucose production during liquefaction and saccharification and thereafter the efficiency of yeast fermentation. Kafirins, despite the high sequence homology with zeins, tend to be less digestible especially after wet-cooking indicating the change in conformational structure attributed to formation of disulphide bonds. This is due to its high hydrophobicity which also makes possible the formation of additional protein aggregates that enhance the formation of more covalent bonds compared to zeins (Wong et al., 2009). Prolamins in the kernel are concentrated in protein bodies arranged among starch granules. The protein body composition in maize and sorghum is also similar, with *alpha* kafirin in the inner core surrounded by *beta* and *gamma* kafirins. The difference with maize is that during wet thermal processes the external part of protein body seems to form a net that makes difficult to access the *alpha* portion that is in turn more digestible than the *beta* and *gamma* counterparts. This phenomenon affects starch digestibility because in sorghum is 15 to 25% less digestible compared to maize. Taylor & Belton (2002) indicate that in sorghum, a complex rather than a simple obstruction mechanism between kafirins and starch is more likely to occur. This is the main reason why sorghum has lower susceptibility to hydrolysis and fermentation and yields less fuel ethanol compared to maize. Besides the starch-protein relationship, some other factors such as mash viscosity, amount of phenolic compounds, ratio of amylose to amylopectin and formation of amylose-lipid complex in the mash, limit the rate of enzymatic hydrolysis and fermentation efficiency during bioethanol production. For instance, starch in amylose-lipid complex cannot be converted into fermentable sugars, reducing conversion rate and final ethanol yield (Wang et al., 2008).

3.3 Sorghum bagasse and straw

As stated in section 3.1, besides water-soluble sugars (sucrose, glucose and fructose), sorghum is composed by structural cell wall carbohydrates primarily cellulose and hemicellulose, which in turn can be hydrolyzed and used as substrate for ethanol production (Sipos et al., 2009).

Sorghum bagasse is the residual fraction obtained after juice extraction from sweet sorghum whereas sorghum straw is the remaining material usually left on the field after threshing. The composition and proportion of fibrous-structural fractions in sorghum is widely reported and varies according to intrinsic and extrinsic factors such as cultivar type, maturity and climatic conditions. An average of 15% of the total weight corresponds to the fibrous portion within a range from 12 to 17% (Woods, 2000).

In sweet sorghum bagasse, average content of cellulose, hemicelluloses and lignin is 34-44%, 27-25%, and 18-20% respectively (Ballesteros et al., 2003; Kim & Day, 2011; Sipos et al., 2009).

Table 1 summarizes chemical composition of sweet sorghum bagasse and straw compared to energy-dedicated sugar cane, maize, wheat and rice counterparts.

Feedstock	Fiber(%)	Cellulose(%)	Hemicellulose (%)	Lignin (%)	Ash (%)
Sweet sorghum	13.0	44.6	27.1	20.7	0.4
Sweet sorghum [2]	-	25.0	22.0	4.0	-
Sweet sorghum bagasse[3]	-	41.3	24.6	14.0	3.7
Sorghum straw	-	32.4	27.0	7.0	0.7
Sugar cane	13.5	41.6	25.1	20.3	4.8
Energy cane	26.7	43.3	23.8	21.7	0.8
Corn stover	-	40.0	28.0	21.0	7.0
Wheat straw	-	38.0	32.0	19.0	8.0
Rice straw	-	36.0	28.0	14.0	20.0

[1] Modified from Kim & Day (2011) and Reddy & Yang (2005). All data expressed in dry weight basis. Percentage of fiber is based in 100% of original material and cellulose, hemicellulose, lignin and ash are percentages of the total fiber; [2]Wray variety (Woods, 2000); [3] Data yet not published from sweet sorghum bagasse harvested in Central Mexico and manually pressed for juice extraction.

Table 1. Fiber composition of different ethanol feedstock [1]

4. Ethanol fuel from sweet sorghum juice

Sweet sorghum juice can be used for syrup, molasses, sugar and ethanol production with average fermentation efficiencies from 85 to 90% (Almodares & Hadi, 2009; Prasad et al., 2007; Wang et al., 2009; Wu et al., 2010b). The sweet sorghum juice is not commonly used for crystallized sugar production because of the presence of significant amounts of inverted sugars (glucose and fructose) that makes difficult crystallization in large-scale processes. However, the sweet sorghum juice, rich in fermentable sugars, has an excellent potential for yeast fermentation (Turhollow et al., 2010; Woods, 2000).

The sweet sorghum juice is obtained through a mechanical operation with a roller mill composed by a set of cylinders, similar to the ones employed by the sugar cane mills. Water is added during the last stage of the crushing process with the aim to augment the solubilization of residual sugars associated to the bagasse. The sweet sorghum juice yields around 50% in relation to the initial weight of the stems (Wu et al., 2010b). However, these authors describe an extraction process by pressing, which results in lower yields compared to roller mills. Furthermore, pressing is a batch process which is difficult to optimize for industrial conditions.

Approximately 90% of fermentable sugars from sorghum stalks can be obtained after conventional roller-milling, yielding an extraction ratio of 0.7 in relation to the initial plant weight (Almodares & Hadi, 2009). Gnansounou et al. (2005) reported extraction ratios ranging from 0.59 to 0.65 for the sweet sorghum cultivars Kelley, Wray, Río and Tianza. On the other hand, Kundiyana (1996) observed that extraction percentages varied between 47 to 58%, close to values observed by our research group in central Mexico (unpublished data).

After extraction, the sweet sorghum juice is fermented, distilled and the ethanol finally dehydrated (Fig. 1). This is the simplest way to produce fuel ethanol because the grain and fiber processes require the hydrolysis of starch and fiber components into fermentable sugars. These steps are considered expensive, take time and expend energy and other additional resources (i.e. enzymes, chemical reagents, etc.) (Fig. 2 and 3). Despite these benefits, some challenges must be solved in order to efficiently convert the sweet sorghum crop into fuel ethanol. The main setbacks are the relatively higher rate of sugar degradation at ambient temperature and the low nitrogen content for yeast growth (Mei et al., 2009; Wu et al., 2010b). Thus, the logistics of just in time harvesting and the storage of the feedstock in facilities that retard decomposition and degradation of fermentable carbohydrates should be considered and stressed. In relation to nitrogen availability, this disadvantage can be overcome with the supplementation of urea, ammonia or yeast extract in order to avoid sluggish fermentation.

Besides sugar and nitrogen content, fermentation performance of sweet sorghum juice can also be affected with processing parameters and bioreactor configuration. Nuanpeng et al. (2011) observed in a repeated-batch study that very high gravity (VHG) fermentation is a good alternative to produce high ethanol concentrations from sweet sorghum juice when an adequate level of yeast cell concentration, nitrogen, and agitation are used. On the other hand, Laopaiboon et al. (2007) reported better results in fed-batch fermentation compared to batch configuration, in terms of ethanol concentration and product yield but not in productivity (measured as grams of ethanol generated/L/hr). These findings indicate the need to optimize parameters as feeding and withdrawn rate in order to optimize yields.

Fig. 1. Flowchart for ethanol production from sweet sorghum juice; [1]Water 73%, sugars (sucrose, glucose and fructose) 13.0%; [2]Water 84%, sugars (sucrose, glucose and fructose) 14.2%. Data from: Almodares & Hadi (2009) and Gnansounou et al. (2005).

The microorganism used, as indicated in the next sections, is also a factor that is worthwhile exploring. In the case of sweet sorghum juice, fermentation with different yeast strains has been evaluated and productivity varies significantly, but most of the strains showed an efficiency of more than 90% (Wu et al., 2010b). Liu et al. (2008) reported the use of immobilized yeast in a fluidized bed reactor that shortened process time and increased conversion efficiency. These results can be optimized when parameters as temperature, agitation rate, particles stuffing rate and pH are modified. Liu & Shen (2008) found that fermentation with immobilized yeast at 37°C, 200 rpm, 25% particles stuffing rate and pH of 5.0 in shaking flasks and 5 L bioreactor corresponds to the optimal conditions derived from an orthogonal experimental design.

5. Ethanol fuel from sorghum grain

5.1 Conventional dry grind

The five basic steps in the conventional dry-grind ethanol process are milling, liquefaction, saccharification, fermentation and ethanol distillation/dehydration (Fig. 2). Mashing goes throughout the entire process beginning with mixing the grain meal with water (and possibly backset stillage) to obtain a mash ready for fermentation. Mashing is a wet-cooking process to turn the gelatinized starch into fermentable sugars first with the use of thermostable *alpha*-amylase and then with amyloglucosidase (Zhao et al., 2008; Solomon et al., 2007; Wu et al., 2007). Starch is the substrate for grain fuel ethanol. Unlike maize, the starch content of sorghum is not the best indicator of ethanol yield obtained by the dry-grind process because this carbohydrate greatly differs in availability or susceptibility to amylases.

The comparatively higher protein content of sorghum compared to maize should be advantageous because the protein is partially degraded into free amino nitrogen compounds during biocatalysis. These compounds are a source of nitrogen for yeast nutrition. However, the relatively lower protein digestibility and nature of the endosperm proteins associated to sorghum counteracts its higher protein concentration. As a result, sorghum mashes almost always contain less free amino nitrogen compared to maize mashes. The use of proteases during or after liquefaction is a good alternative to increase free amino nitrogen in sorghum mashes (Perez-Carrillo & Serna-Saldivar, 2007). Protein digestibility in wet-cooked sorghum is relatively lower compared to other cereals, mainly because of the cross-linking of prolamins. This phenomenon reduces the availability of nitrogenous compound in sorghum mashes needed to support yeast metabolism during fermentation.

Yeast cannot use proteins as source of nitrogen, instead it utilizes amino acids and short peptides (di or tri), indicating the importance of protein fragmentation altogether with starch hydrolysis in mashing. Beyond yeast nutrimental quandary, there are also issues related to starch digestibility that affects the performance of amylolytic enzymes during liquefaction and saccharification. This trend is also related to proteins because of the interaction between protein and starch that in sorghum reduces the susceptibility of this polysaccharide in both native and gelatinized conditions. Sorghum starch has higher gelatinization temperature compared to maize and more prolamin containing bodies within the endosperm, differences that can restrict gelatinization of starch granules (Zhao et al., 2008).

It has been reported that ethanol yields from sorghum decreases as protein content increases; however, at the same protein level, ethanol fermentation efficiency can vary as much as 8%. The difference is higher than typical experimental variations which indicate that additional factors to protein affects starch-conversion rate. In a work reported by Wang et al. (2008), nine sorghum genotypes were selected and used to study the effect of protein availability on efficiency of ethanol fermentation. The results showed a strong positive linear relationship between protein digestibility and fermentation efficiency, indicating the influence, and at the same time, the usefulness of this sorghum grain features as predictor of ethanol yield (Rooney et al., 2007; Wang et al., 2008; Wu et al., 2007; Wu et al., 2010a).

In Fig. 2 a typical process of dry-grind ethanol production is depicted. An average yield of 390 L of ethanol from 1 ton of sorghum can be obtained, but yields as high as 400 L/ton with fermentation efficiencies of more than 90% has been achieved and reported (Chuck-Hernandez et al., 2009; Pérez-Carrillo & Serna-Saldivar, 2007). The Dried Distillers Grains with Solubles (DDGS) obtained in these processes contribute to the economics of biorefineries. The wet distillers grains can be dried to 12% moisture with the aim to produce a shelf-stable byproduct.

Its nutritional composition (39 and 49% of protein and carbohydrates respectively) makes it an excellent option for livestock feed, especially for ruminants.

5.2 Use of biotechnology to improve ethanol yields

5.2.1 Genetic modified sorghum

Nowadays, advances in transformation and genetic modification in plants make the development of special sorghum cultivars one of the best tactics to overcome the various known factors that reduce ethanol yields. Previous research works have concluded that fermentation efficiencies and ethanol yields are influenced by genotype and chemical composition (Wu et al., 2007, 2008; Zhao et al., 2008). These investigations have determined important traits that enhance or reduce yields. Starch, protein and tannins are the principal components related to ethanol production from sorghum grain and these characteristic can be associated to genotype and also, in the case of starch and protein, to environmental factors as sowing season and location (Wu et al., 2008). Starch composition, specifically the amylose:amylopectin ratio, is related to fermentation efficiency. Raw materials with less amylose are more efficiently converted into ethanol (Wu et al., 2006). The improvement is related to digestibility of starch, reported as higher in waxy types (Rooney & Pflugfelder, 1986). Wu et al. (2006) also attributed the increased efficiency to the lower content of amylose-lipid complexes in mashes.

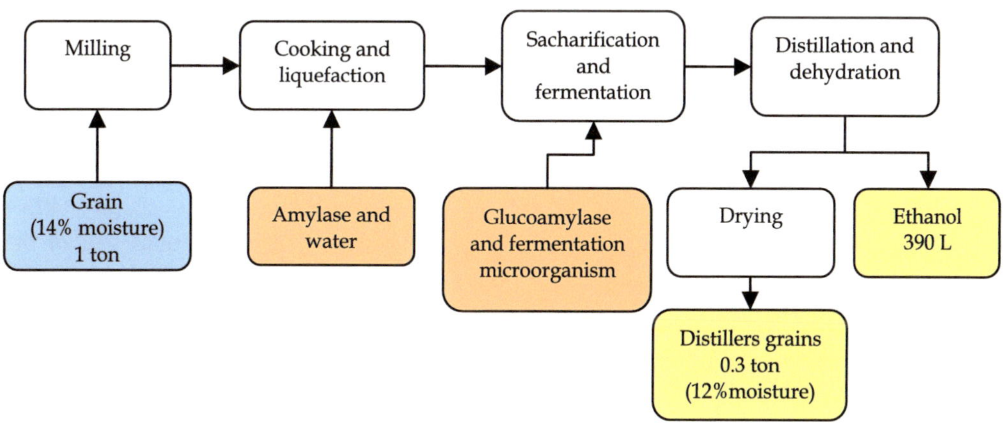

Fig. 2. Flowchart for ethanol production from sorghum grain. Data from: Serna-Saldívar (2010).

The DSC thermograms of starches from waxy sorghum and waxy maize are essentially the same: both display a single, smooth endothermic peak, with approximately the same onset, peak, and ending temperatures in the range of 60-80°C. However, in normal sorghum a second peak appears around 85 to 105°C corresponding to an amylose-lipid complex that reduces the availability of starch. Waxy starches are thereby easily gelatinized and hydrolyzed, giving high conversion efficiencies (Wu et al., 2007). Thus, the waxy characteristics improved the susceptibility of the endosperm matrix for low-energy gelatinization, enzymatic hydrolysis and total ethanol production (Wu et al., 2010a).
In the case of proteins, Wu et al. (2010a) indicate that high-lysine, high-protein-digestibility (HD) sorghum lines have been developed. These genotypes have several potential advantages for their use as feedstocks in biorefineries. First, the starch granules swells and

pastes more easily at lower temperatures; second, the proteins have improved feed value with higher bioavailability even for monogastrics. Interestingly, these high-lysine genotypes can contain 60% more of this essential amino acid compared to regular counterparts and similar content compared to quality protein maize (QPM) genotypes (Wu et al., 2010a). The enhanced protein digestibility of these lines is attributed to an improved kafirin digestibility as a result of the unique, abnormal and highly invaginated protein bodies. Segregated progeny with HD population lack the kafirin protein body matrix that surround starch granules and restrict swelling and pasting.

While modification in starch and protein digestibility affects ethanol production, one of the most important traits in starch conversion is total starch harvested per area. The primary goal of sorghum breeding programs has been and continues to be the development of high-yielding, drought-tolerant and pest-resistant hybrids. This effort will continue and additional gains in yield can be expected which will result in higher ethanol production from each hectare dedicated to sorghum (Rooney et al., 2007).

5.2.2 Exogenous enzymes

As explained before, protein digestibility is related to ethanol production and this digestibility in turn is related to the tendency of sorghum proteins to form web-like structures during mashing which reduces the possibility of enzymes to access starch. Protein solubility should decrease with the increase of protein cross-linking; thus, this parameter can be used as a quality indicator in sorghum biorefineries (Zhao et al., 2008).

The utilization of proteases before conventional starch liquefaction can be used as an alternative method to improve rate of starch hydrolysis and yield hydrolyzates with high FAN concentration (Perez-Carrillo & Serna-Saldivar, 2007).

Perez-Carrillo et al. (2008) proposed the use of protease before starch gelatinization and liquefaction of both decorticated and whole sorghum meals. The use of decortication to remove the sorghum outer layers and the exogenous protease had a positive synergic effect in terms of ethanol yield and energy savings because mashes required about half of the fermentation time compared to conventionally processed sorghum. Decorticated meals with more starch were more susceptible to *alpha*-amylase during liquefaction and produced more ethanol during fermentation (Alvarez et al., 2010). This technology produced similar ethanol yields compared to soft yellow dent maize and 44% more ethanol compared to the whole sorghum control treatment. The other advantage of mechanical decortication is that the bran, separated beforehand, is shelf-stable and can be directly channeled for production of animal feeds and consequently the yield of wet distilled grains from decorticated sorghum is significantly lower compared to the obtained after processing whole sorghum meals. Thus, if dried distilled grains are produced, the biorefinery plant will spend less energy when processing decorticated sorghum.

5.2.3 Germination and sprouting

Germinated or sprouted regular and high-tannin sorghums have improved ethanol yields compared to the unmalted kernels. Yan et al. (2009, 2010) reported a reduction in fermentation time and reported higher yields when sprouted sorghum was processed. The improved yield and efficiency is attributed to the action of intrinsic enzymes in starch, proteins and cell walls. Thus, the use of purposely malted or field sprouted sorghums can be advantageous for fuel ethanol biorefineries. Nevertheless, the industries should consider that malting requires important inputs in terms of water, labor, energy for drying and logistics.

5.2.4 Very High Gravity (VHG) fermentation

Very High Gravity (VHG) mashes are used for fuel ethanol production at industrial scale. Among the benefits include an increased productivity, a reduced capital cost, a higher ethanol concentration in the fermented mash (from 7-10% to 15-18% -v/v- or more), and a decrease in water requirements. The most concentrated ethanol in fermented mashes also reduces distillation requirements, being an important issue because after feedstock, energy is the biggest production input, representing 30% of total ethanol cost (Pradeep et al., 2010; Wang et al., 2007). This economic consideration indicates the importance of the substrate concentration at the beginning of the process. The use of mashes with higher sugar concentration influences the decision of which fermentation microorganism will be selected and used.

Yeast osmotolerance is determined by genetics and by the carbohydrate level present in mashes, fermentation temperature, osmotic pressure/water activity and substrate concentration. Osmotolerant yeast fermenting in batch conditions can produce and tolerate levels of 16 to 17% (v/v) alcohol (Casey & Ingledew, 1986). According to the same authors, higher alcohol beers can be produced if oxygenation and nitrogen sources are supplemented to worts. Predeep et al. (2010) reported a maximum ethanol concentration of 15.6% (v/v) converted with about 86.6% efficiency when finger millet mashes were fermented with *Saccharomyces bayanus*. Fermentation temperature is also an important factor affecting productivity, and generally speaking, at higher temperatures the time required to finish fermentation is decreased. Jones & Ingledew (1994) reported an increment in fermentation efficiency when dissolved solids concentration increased from 14 to 36.5 g/100 mL and also observed that the use of urea accelerated the rate of reaction and decreased time required to complete fermentation.

Working with VHG sweet sorghum juice rather than with ground sorghum grain, Wu et al. (2010b) reported an increase in glycerol (0.3 to 0.6%) and residual sugars (0.2 to 5.1%) when sugar in juices increased from 20 to 30%. A reduction in fermentation efficiency (93 to 72%) was also observed after 72 hours fermentation. Authors recommend the use of juices with no more than 20% soluble sugars in order to obtain the highest efficiency.

In general terms, yeasts can exhibit osmotic inhibition starting at 15% sugar, and this inhibition is higher in glucose followed by other carbohydrates such as sucrose and maltose. Sumari et al. (2010) stated that very few types of yeasts were known to tolerate sugar concentration above 40% and normally at this concentration their growth is sluggish. For this reason, the screening for osmotolerance and the development of new strains is necessary for industrial purposes. Sumari et al. (2010), using a molecular genetic approach, characterized a set of yeasts isolated from African brews and wines. One strain was able to ferment a medium with sucrose concentration of 1000 g/L. The phylogenetic analysis with rDNA clustered this microorganism away from the typical osmotolerant yeast. This indicates the opportunity to explore and look for new strains in nature.

Besides yeast, other microorganisms, as bacteria, are especially designed for ethanol fermentation. *Escherichia coli* is the typical modified microorganism for ethanol production because of the wide spectrum of metabolized carbohydrates, its well-known genetic makeup and the easiness of manipulation. *Zymomonas mobilis*, a rod shaped, gram negative, non-spore forming bacteria is naturally ethanologenic and compared to yeast, has higher rates of glucose uptake. *Z. mobilis* has also a higher ethanol production, increased yield and tolerance, making it a good option to use in VHG fermentation. Kesava et al. (1995), working with *Z. mobilis,* reported 95% conversion rates after 35 hours fermentation and ethanol yields of approximately 70 g/L when fermenting mashes containing 150 g glucose/L. The bacterium was able to ferment mashes containing 200 g glucose/L in a step-fed system. Perez-Carrillo et al. (2011) observed that *Z. mobilis* had lower nitrogen requirements compared to *S. cerevisiae* when fermenting mashes adjusted to 20° Plato. This bacterium has potential and possible advantages for commercial use in biorefineries.

5.3 Physico-mechanical technologies to Improve ethanol yield

Several approaches to increase ethanol yield from sorghum involve physical or mechanical treatments, v.gr: reduction of particle size, decortication or steam flaking. The aim of these treatments is to reduce physical barriers to hydrolytic enzymes in order to yield more fermentable sugars in shorter reaction times.

5.3.1 Particle size

Particle size of ground sorghum meals also plays an important role in the starch-to-ethanol conversion process. Wang et al. (2008) observed that fermentation efficiencies of finely ground samples were approximately 5% higher compared to coarsely ground counterparts. This effect is a consequence of differences in gelatinization temperature and accessibility of starch to hydrolyzing enzymes. Wang et al. (2008) reported that gelatinization temperatures of larger or coarser particles are 5-10°C higher compared to finer particles.

The conversion of meals with smaller particles enhanced digestibility due to an improvement in the relative surface-contact area. Mahasukhonthachat et al. (2010) indicate that starch digestion proceeded by diffusion mechanisms is based on an inverse square dependence of rate coefficient on average particle size.

5.3.2 Decortication

According to Rooney & Serna-Saldivar (2000) pericarp, testa, aleurone and mainly peripheral endosperm are grain tissues directly related to the lower nutrient digestibility of sorghum. These layers can be removed through decortication or pearling, an abrasive process used on a regular basis for production of refined flours or grits (Serna-Saldivar, 2010). Commercial mills are typically batch type and are equipped with a set of abrasive disks or carborundum stones to mechanically remove from 10 to 30% of the grain weight. The resulting mixture of bran and decorticated sorghum is separated via air aspiration or sifting (Serna-Saldivar, 2010). The classified pearled grain is then conventionally milled into a meal or flour. This technology requires little capital investment or alteration of existing facilities (Wang et al., 1999). The mechanical removal of the sorghum outer layers increases starch concentration and decreases fiber, fat and phenolics. The ground decorticated sorghum kernels are more susceptible to thermoresistant *alpha*-amylase hydrolysis (Perez-Carrillo & Serna-Saldivar, 2007). Furthermore, the removal of the sorghum outer layers allows greater starch loading and results in improved ethanol yields.

5.3.3 Steam-flaking

Other proposed alternative to process sorghum before dry-milling is steam-flaking. This technology, widely used in feedlots, disrupts the endosperm structure with the injection of live steam in a period of 15 to 30 min, followed by flaking through grooved rolls. Before flaking, moisture of sorghum is increased to at least 21% and a conditioning or surfactant agent as lecithin is added in order to obtain whole flakes and reduce processing losses (Serna-Saldivar, 2010). After drying and cooling, sorghum flakes can be milled using traditional processes. The pregelatinized starch associated to the ground and steamed flaked sorghum had higher susceptibility during liquefaction and produced more ethanol during fermentation. Compared to the whole sorghum counterpart the steam-flaked sorghum yielded approximately 40% more ethanol (Chuck-Hernandez et al., 2009). Currently, the cost of steam flaking one ton of sorghum is approximately $7.5 US dollars.

5.3.4 Supercritical Fluid Extrusion (SCFX)

Extrusion has been widely used for the processing of cereal grains because this thermoplastic technology is continuous and saves unit operations and energy. In extrusion, the materials are subjected to heating, mixing, and shearing, resulting in physical and chemical changes during its passage through the extruder. The major advantages of extrusion include: improvement of starch digestibility and reduction of its molecular weight, production of free sugars and dextrins, changes in the native structure of both starch granules and proteins and reduced viscosity of fermentation broths. Therefore, extrusion could be an effective process to improve the bioconversion rate of sorghum starch (Zhan et al., 2006).

An innovative processing technology patented by researchers of Cornell University combines extrusion process and supercritical-fluid technology. The main difference between supercritical-fluid (SCFX) and conventional extrusion is the injection of supercritical carbon dioxide, which replaces water as blowing agent for expansion. The injection of supercritical-fluid carbon dioxide breaks the intimate bonds between starch granules and protein matrix and results in the improvement of starch availability (Zhan et al., 2006). These researchers suggested that SCFX produces molecular degradation of starch during extrusion of sorghum. This process also increased about 8% the protein digestibility, the measurable starch content, the free sugar concentration and gelatinized starch and other parameters that increased ethanol yield (+5%) and boosted fermentation efficiency compared to the non-extruded counterpart. The SCFX cooking also affected the crude fiber, chemical fraction that after microscope examination showed disruption and fissures. These authors describe the sorghum extrudates with "porous structure". Thus, this thermoplastic procedure was indeed effective as pretreatment to improve bioconversion of sorghum into ethanol.

6. Ethanol from sorghum bagasse and straw

6.1 Raw material conditioning

After extraction of juice or grain harvesting, the lignocellulosic residue is chopped, milled, and dried at 50-60 °C to reduce the moisture content to about 10 to 15% (Herrera et al., 2003; Sipos et al., 2009). There are many options to reduce particle size; the most commonly used are hammer or rotary mills. Grinding can be used on both dry and wet materials, and the cost is one of the lowest compared with others methods used for milling biomass. The grinder reduces the particle size to a fine powder by mechanical shearing and this operation can also be made with rotating and stationary abrasive stones (Mizuno et al., 2009).

6.2 Fiber extraction

One of the most significant problems in ethanol production from lignocellulose is production cost (Mizuno et al., 2009) because the fiber conversion requires of high energy investments in order to obtain high concentrations of fermentable sugars from the insoluble polymers (Kurian et al., 2010; Mamma et al., 1996). A pre-hydrolysis step releases both the hemicellulosic and cellulosic fractions of the fiber (Herrera et al., 2003). The main processes related to the pretreatment of sorghum biomass for ethanol production are the acid and/or enzyme-catalyzed hydrolyses (Mamma et al., 1996; Sipos et al., 2009). Generally, the acid hydrolysis precedes the enzymatic in order to optimize production of C6 and C5 fermentable sugars (Sipos et al., 2009).

6.3 Pretreatments used for sorghum bagasse

The extraction of structural carbohydrates from bagasse cell walls is highly related to the effectiveness of pretreatments. Nowadays there are many proposed treatments for cellulose and hemicellulose extraction, but only few have been commercially implemented. In the following sections some of the proposed technologies for sorghum biomass are discussed.

6.3.1 Steam explosion

The ground sorghum bagasse is rehydrated with steam at atmospheric pressure and impregnated with low amounts (up to 3% w/w) of sulfur dioxide (SO_2) in plastic bags for 20-30 minutes in order to improve enzymatic saccharification (Sipos et al., 2009; Stenberg et al., 1998; Öhgren et al., 2005). The impregnated bagasse is introduced into a reactor and the temperature is maintained by injection of saturated steam, varying in a range of 170-210°C (Sipos et al., 2009; Stenberg et al., 1998; Öhgren et al., 2005). After 2 to 10 minutes, the blow-down valve is opened and the hydrolyzate is released into a cyclone (Stenberg et al., 1998). Sipos et al. (2009) achieved an extraction of 89% to 92% of cellulose with steam explosion, up to 18 g glucose, 23 g xylose and 5.5 g arabinose/L hydrolyzate. Ballesteros et al. (2003) used steam explosion pretreatment without sulfur dioxide and obtained around 50% of solids recovery and only 20% solubilization of the cellulose. Hemicellulose sugars were extensively solubilized because the raw material had originally 25% xylose and after the treatment only 2% remained on the fibrous residue.

6.3.2 Dilute acid hydrolysis

Acid hydrolysis, the most common fiber pretreatment method (Ban et al., 2008), generates significant amounts of sugars from hemicellulose. Besides it is a process relatively cheap (Gnansounou et al., 2005). Sulfuric, hydrochloric, hydrofluoric or acetic acids have been tested as catalysts (Herrera et al., 2003). The process consists on the addition of diluted aqueous acid solution (0.1 to 10 % w/v) to the ground raw material and hydrolyzing in an autoclave. A solid residue, rich in cellulose and lignin, is formed after acid hydrolysis and subsequently treated with enzymes in order to increase the amounts of fermentable sugars (Tellez-Luis et al., 2002). Kurian et al. (2010) achieved extract with 92 g/L of total sugars from sweet sorghum bagasse treated with sulfuric acid at a concentration of 5 g/kg and treated at 140°C for 30 minutes. Ban et al. (2008) treated the same raw material at a solid-liquid mass ratio of 10% with 80 g phosphoric acid/L at 120°C for 80 minutes. These authors reported 302 g reducing sugars/kg with this pretreatment.

6.3.3 Alkali pretreatment

Unlike other pretreatments, the use of strong alkali delignifies biomass by disrupting the ester bonds of cross-linked lignin and xylans, resulting in cellulose and hemicellulose enriched fraction. Alkali pretreatment processes generally utilize lower temperatures, pressures and residence times compared to other technologies (McIntosh & Vancov, 2010).

The main compounds used as pretreatment agents in alkali processes are: sodium hydroxide, ammonia and lime, because of their comparatively lower cost and the possibility of chemical and water recycling (McIntosh & Vancov, 2010). Usually two temperature conditions are used for hydrolysis: mild (60°C) or high (121°C).

6.4 Enzymatic extraction

There are several enzymes generally used to convert cellulose and hemicellulose into soluble sugars. They are a mixture of pectinases, cellulases and hemicellulases (Lin et al., 2011; Reddy & Yang, 2005). Cellulose can be hydrolyzed by the synergistic action of endo-acting enzymes knows as endoglucanases, and exo-acting enzymes, known as exoglucanases (Lin et al., 2011). Today it is common to employ enzyme complexes consisting of seven or more degrading enzymes that act synergistically. The enzyme mixture is added before or after chemical or mechanical treatments (Reddy & Yang, 2005). Enzymes appear to be the best prospects for continued improvements because can reduce production costs (Gnansounou et al., 2005).

Sipos et al. (2009) observed that the separation of the solid and the liquid phases after chemical pretreatment is beneficial to the whole process because the xylose-rich liquid fraction can be fermented into ethanol through the pentose pathway or as substrate for microbial cellulase production or transformed into other various valuable products. On the other hand, the solid fraction can be further hydrolyzed and fermented into ethanol. The use of alkali treatment before enzyme hydrolysis generated 540 g glucose/kg raw material, equivalent to a 90% conversion of available cellulose to monomeric sugars. On the other hand, 235 g xylose/kg was released after pretreatment of sorghum straw (McIntosh & Vancov, 2010). These hydrolysates were obtained with an enzyme complex containing endoglucanase, exoglucanase, xylanase, beta-glucosidase and cellulase.

6.5 Hydrolysis by-products or fermentation inhibitors

The fiber chemical hydrolysis process can produce a large number of sugar degradation products which are known to inhibit bacteria and yeast and thus the conversion of fermentable sugars into bioethanol (Ban et al., 2008). The most important inhibitors are furfural, 5-hydroxymethylfurfural and acetic acid. After the acid hydrolysis, it is necessary to adjust the pH with alkalis in order to obtain the adequate conditions for the subsequent step of fermentation. Lime or calcium hydroxide is commonly added to increase the pH to 9-10. This alkali treatment precipitates inhibitors in the form of insoluble salts and therefore acts as detoxifying treatment (Kurian et al., 2010).

6.6 Fermentation

Hydrolyzates obtained from sorghum fiber are solutions rich in both hexoses and pentoses (Kurian et al., 2010). Production of ethanol from these mashes is possible only with the use of osmotolerant and pentose fermenting yeast or bacterial strains (Table 2).

Ballesteros et al. (2003) obtained 16.2 g ethanol/L when hydrolyzates obtained from sweet sorghum bagasse were fermented with *Kluyveromyces marxianus*. On the other hand, Kurian et al. (2010) working with *Pichia stipitis* obtained 38.7 g ethanol/L with a theoretical conversion of 82.5%. In Fig. 3, a flowchart of ethanol production from sorghum bagasse is depicted. A yield of 158 L ethanol/ton biomass (wet basis) can be obtained after a sulfuric acid hydrolysis. The process yielded 110 kg of lignin and other non-fermentable materials. Almodares & Hadi (2009) and Gnansounou et al. (2005) reported that the cellulase used in Simultaneous

Microorganism	Characteristics
Clostridium acetobutilicum	Useful in fermentation of xylose to acetone and butanol; bioethanol produced in low yield
Clostridium thermocellum	Capable of converting cellulose directly to ethanol and acetic acid. Bioethanol concentrations are generally less than 5 g/l. Cellulase is strong inhibition encountered by cellobiose accumulation
Escherichia coli	Native strains ferment xylose to a mixture of bioethanol, succinic, and acetic acids but lack ethanol tolerance; genetically engineered strains predominantly produce bioethanol
Klebsiella oxytoca	Native strains rapidly ferment xylose and cellobiose; engineered to ferment cellulose and produce bioethanol predominantly
Klebsiella planticola ATCC 33531	Carried gene from *Zymomonas mobilis* encoding pyruvate decarboxylase. Conjugated strain tolerated up to 4% ethanol
Lactobacillus pentoaceticus	Consumes xylose and arabinose. Slowly uses glucose and cellobiose. Acetic acid is produced along with lactic in 1:1 ratio
Lactobacillus casei	Ferments lactose, particularly useful for bioconversion of whey
Lactobacillus xylosus	Uses cellobiose if nutrients are supplied: uses glucose, D-xylose and L-arabinose
Lactobacillus pentosus	Homolactic fermentation. Some strains produce lactic acid from sulfite waste liquors
Lactobacillus plantarum	Consumes cellobiose more rapidly than glucose, xylose, or arabinose. Appears to depolymerize pectins; produces lactic acid from agricultural residues
Pachysolen tannophilus *Saccharomyces cerevisiae* ATCC 24 860	Co-culture of *S. cerevisiae* and strains resulted in the best ethanol yield
Pichia stipits NRRL Y-7124, Y-11 544, Y-11 545	NRRL strain Y-7124 utilized over 95% xylose based on 150 g/L initial concentration. Produced 52 g/L of ethanol with a yield of 0.39 g ethanol per g xylose
Pichia stipits NRLL Y-7124 (floculating strain)	Maximum cell concentration of 50 g/L. Ethanol production rate of 10.7 g/L.h with more than 80% xylose conversion. Ethanol and xylitol yield of 0.4 and 0.03 g/ g xylose
Saccharomyces cerevisiae CBS 1200 *Candida shehatae* ATCC 24 860	Co-culture of two yeast strains utilized both glucose and xylose. Yields of 100 and 27% on glucose and xylose, respectively

[1] With data from: Balat et al. (2008) and Lee (1997).

Table 2. Native and engineered microorganisms capable of fermenting xylose to bioethanol[1]

Saccharification and Fermentation (SSF) can be added directly or from material previously deviated from pretreatment and inoculated along with *Trichoderma reesei* or other fungi such as *Neurospora crassa* and *Fusarium oxysporum.* These microorganisms were capable of directly fermenting cellulose (Mamma et al., 1996). *F. oxysporum* was used in a SSF along with *S. cerevisiae,* yielding 5.2 to 8.4 g ethanol per 100 g of fresh sorghum. The efficiency was calculated based on soluble sugars and not in total polysaccharides (Mamma et al., 1996).

7. Estimated ethanol yields

Fig. 1 to 3 summarizes and compares average ethanol yields from sorghum grain, sweet juice and biomass. Ethanol yields vary according to variety, geography, soil fertility and temperature.

Sweet sorghums usually yield from 50 up to 120 tons of stalks after the first cut. This feedstock contains 73% moisture, 13% soluble sugars, 5.3% cellulose, 3.7% hemicelluloses and 2.7% lignin. The stalks yield up to 70% sweet juice and 15.33 ton/ha of spent bagasse (Almodares & Hadi, 2009; Prasad et al., 2007).

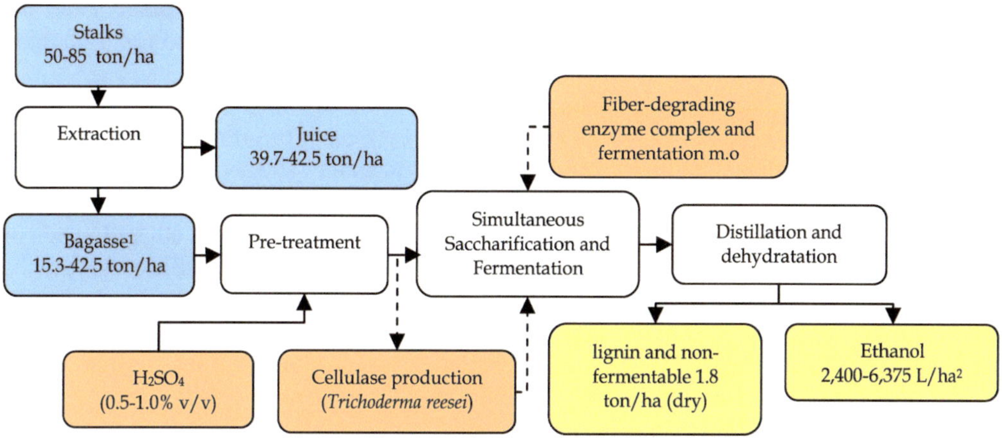

Fig. 3. Flowchart for ethanol production from sweet sorghum bagasse. [1] Average composition of sweet sorghum bagasse: Water 54%, simple fermentable sugars 5.4%; Cellulose 17%; Hemicellulose 12%; Lignin 11.7% [2] Practical yield from fermentation with *I. orientalis*: 3,865 L/ha. From: Almodares & Hadi (2009); Gnansounou et al. (2005).

Water added during extraction is considered part of the sweet juice yield (Fig. 1) and the sweet juice commonly contains around 14% soluble sugars. This substrate allows the production of 3,450 L ethanol/ha with a fermentation efficiency of 95%, similar to the result reported Kim & Day (2011) (3,296 L/ha). These last researchers did not consider losses that negatively affect fermentation efficiencies. Almodares & Hadi (2009), on the other hand, reported a yield of 3,000 L ethanol/ha directly when processing juice extracted from varieties that yielded from 39 to 128 ton stalks/ha. Although Wu et al. (2010b) did not report ethanol yields per hectare, the calculated ethanol production from the amount of total fermentable sugars extracted from a high yielding M81E cultivar planted at two different locations and bioconverted with a 95% of fermentation efficiency was in the range of 4,750 to 5,220 L/ha. These potential ethanol yields are equivalent to the bioconversion of 12 to 13 tons of maize kernels.

Experimental data obtained from sweet sorghums cultivated in Central Mexico indicated that these materials are capable of yielding 6.38 tons of sugar/ha/cut. Consequently, when are adequately bioconverted have the potential of producing 4,132 L ethanol (unpublished data). Regarding to the lignocellulosic fraction, if 15.33 ton of bagasse/ha is obtained containing 29% cellulose and hemicellulose and 5.4% of remaining unextracted soluble sugars, up to 2,400 L of ethanol can be obtained (Fig. 3). This yield represents almost half of the 4,058 L/ha described by Kim & Day (2011) as theoretical ethanol.

In central Mexico, 42.5 ton of bagasse/ha with 50% fermentable sugars are commonly obtained. This biomass is capable of yielding 6,375 L ethanol with perfect conversion efficiency. However, experimental data where the acid-treated biomass was fermented with *Issatchenkia orientalis* indicated only 60% fermentation efficiency (3,865 L/ha) (unpublished

data). These results indicate that there are still many areas for potential improvements especially when processing spent biomass.
Almodares & Hadi (2009) reported that a yield up to 2 ton of grain/ha can be expected from sweet sorghum. If this material is milled, hydrolyzed and fermented, a final ethanol yield of 780 L can be expected. Nevertheless, the sweet sorghum grain during optimum harvesting is not fully matured and generally collected along the vegetative parts of the plant. Thus, the immature sweet sorghum kernels are usually processed with the bagasse and not fermented using grain technologies.
The biomass production per cultivated surface (Fig. 3) is the key and most important factor that affects ethanol yields indicating the importance of both plant breeding for the generation of new improved cultivars and the agronomic conditions mainly affected by soil fertility and water availability. The new biomass cultivars should adapt to marginal lands in order to minimize competition with basic grain production. The potential to obtain ethanol yields of 6630, 7000 and 10000 L/ha (with 95% of extraction and fermentation efficiency) can be achieved because yields of 50 to 120 tons of biomass/ha are reported. Comparatively Kim & Day (2011) indicated that the theoretical yield of maize kernels can be as high as 5,100 L/ha and up to 8,625 L/ha when the whole plant is bioconverted into ethanol (grain + corn stover).
One of the most important factors to be addressed during yield calculation is indeed the energy required for ethanol production. Biomass and starch require more energy for hydrolysis compared to sweet sorghum juice. The technologies for starchy kernels and sweet juice are matured but the conversion and estimation of energy balances when processing lignocellulosic material will be critically important for the evaluation of economic advisability.

8. Future trends

One of the most promising research priorities in agricultural production is the genetic improvement of crops with high economic relevance. In the case of sorghum for fuels there are important advances in the development of biomass, sweet and high yielding grain varieties and hybrids, but is yet one of the most important and critical research topics. The new cultivars should be adapted to marginal lands and also they must be resistant to pests, other phytopathogens and stable facing water stress.
The creation of new varieties for ethanol production is not an easy task because the relevant traits, such as plant height, total soluble solids, juice production and lignin : cellulose : hemicellulose ratio are "non additive" (Reddy et al., 2005). On the other hand and according to Turhollow et al. (2010), the genetic mapping combined with its relatively fast hybridation and field tests, can facilitate the design and development of dedicated bioenergy cultivars.
It is also of upmost importance to develop machinery to harvest sweet and biomass sorghums because the use of existing sugarcane equipments reduce yields and efficiencies. Furthermore, it is also imperative to development new agronomical and technological packages that include "just in time" harvesting.
The use of biomass sorghum represents one of the most relevant topics in research even when there are not economic and energy efficient technologies. However, there have been important advances in terms of fiber degradation to yield extracts rich in C5 and C6 fermentable sugars. The development of new and more environmental-friendly pretreatments that include the use of fiber degrading enzymes and hot water and new strains of yeast and bacteria are critical points for the economics of biomass transformation.

The new microorganisms must be designed or genetically engineered to be more efficient in terms of enhanced capacity to fully ferment C5 and C6 sugars at high temperatures (Canizo, 2009). The development of new strains of *Saccharomyces cerevisiae* designed for pentose utilization, with high tolerance to inhibitors, and with a better genomic stability has not been yet fully addressed despite the recent advances in genetic engineering. Unfortunately, there are only few industrial and commercial strains in the market.

Process wise, biorefineries should focus on designing new bioreactors, flow-patterns, new cocktails of enzymes to optimize hydrolysis, the utilization of immobilized microorganisms and the development of new distillation and ethanol dehydration technologies that favors the total energy balance.

9. References

Almodares, A. & Hadi, M.R. (2009). Production of bioethanol from sweet sorghum: A review, *African Journal of Agricultural Research*, Vol.5, No.9, (September 2009), pp. 772-780, ISSN 1991-637X

Alvarez, M., Perez-Carrillo, E. & Serna-Saldívar, S.O. (2010). Effect of decortication and protease treatment on the kinetics of liquefaction, saccharification, and ethanol production from sorghum, *Journal of Chemical Technology and Biotechnology*, Vol. 885, No. 8, (August 2010), pp. 1122-1129, ISSN 1097-4660.

Anglani, C. (1998). Sorghum carbohydrates-A review, *Plant Foods for Human Nutrition*, Vol. 52, No.1, (March 1998), pp. 77-83. ISSN 0921-9668

Balat, M., Balat, H. & Öz, C. (2008). Progress in bioethanol processing, *Progress in Energy and Combustion Science*, Vol.34, No. 5, (October 2008), pp.551-573, ISSN 0360-1285

Ballesteros, M., Oliva, J.M., Negro, M.J., Manzanares, P. & Ballesteros, I. (2003). Ethanol from lignocellulosic materials by a simultaneous saccharification and fermentation process (SFS) with *Kluyveromyces marxianus* CECT 10875, *Process Biochemistry*, Vol.39, No. 12, (October 2004), pp. 1843-1848, ISSN 1359-5113

Ban, J., Yu, J., Zhang, X. & Tan, T. (2008). Ethanol production from sweet sorghum residual, *Frontiers of Chemical Engineering in China*, Vol.2, No.4 (December 2008), pp. 452-455. ISSN 2095-0179

Berg, C. (2004). World fuel ethanol, analysis and outlook, Japan. 28.03.11, Available from http://www.meti.go.jp/report/downloadfiles/g30819b40j.pdf

Blake, C. (2010). Sorghum expansion tied to cellulosic ethanol. 30.06.11, Available from http://southeastfarmpress.com/grains/sorghum-expansion-tied-cellulosic-ethanol

Canizo, J.R. (2009). Estudio del efecto de los parámetros de operación de los pretratamientos acidos diluidos y de la modificación genética de *Kluyveromyces marxianus* sobre la producción de bioetanol a partir de material lignocelulósico (forraje Bermuda NK37, Cynodon dactylon). Maestría en Biotecnología. Tecnológico de Monterrey. Monterrey, México.

Casey, G.P. & Ingledew, W.M. (1986). Ethanol tolerance in yeasts, *Critical Reviews of Microbiology*, No. 13, Vol. 3, pp. 219-280, ISSN 1040-841X.

Chuck-Hernandez, C., Perez-Carrillo, E. & Serna-Saldivar, S.O. (2009). Production of bioethanol from steam-flaked sorghum and maize, *Journal of Cereal Science*, Vol. 50, No. 1 (July 2009), pp. 131-137, ISSN 0733-5210.

Daeschel, M. A., Mundt, J. O. & McCarty, I. E. (1981). Microbial changes in sweet sorghum (*Sorghum bicolor*) juices, *Applied Environmental Microbiology*, Vol. 42, No. 2, (August 1981), pp. 381-382. ISSN 0099-2240.

Davila-Gomez, F.J., Chuck-Hernandez, C., Perez-Carrillo, E., Rooney, W.L. & Serna-Saldivar, S.O. (2011). Evaluation of bioethanol production from five different varieties of sweet and forage sorghums (*Sorghum bicolor* (L) Moench), *Industrial Crops and Products*, Vol. 33, No. 3 (May 2011), pp.611-616, ISSN 0926-6690.

FAOSTAT. (2011). Cereal production, In: FAOSTAT, 28.03.11, Available from http://faostat.fao.org/

Gnansounou, E., Dauriat, A. & Wyman, C.E. (2005). Refining sweet sorghum to ethanol and sugar: economic trade-offs in the context of North China. *Bioresource Technology*. Vol.96, No.9, (June 2005), pp. 985-1002, ISSN 0960-8524

Herrera, A., Téllez-Luis, S.J., Ramírez, J.A. & Vázquez, M. (2003). Production of xylose from sorghum straw using hydrochloric acid, *Journal of Cereal Science*, Vol.37, No. 3 (May2003), pp. 267-274, ISSN 0733-5210.

Jessup, R. (2009). Development and status of dedicated energy crops in the United States, *In Vitro Cellular & Developmental Biology*, Vol. 45, No. 3, (June 2009), pp. 282-290, ISSN 1054-5476.

Jones, A.M. & Ingledew, W.M. (1994). Fuel alcohol production: optimization of temperature for efficient Very-High-Gravity fermentation, *Applied and Environmental Microbiology*, Vol. 60, No. 3, (March 1994), pp. 1048-1051, ISSN 0099-2240.

Keeley, J.E. & Rundel, P.W. (2003). Evolution of CAM and C4 carbon-concentrating mechanisms, *International Journal of Plant Science*, Vol. 164, No. 3 (Suppl.), (January 2003), pp. S55-S77, ISSN 1058-5893.

Kesava, S.S., Rakshit, S.K. & Panda, T. (1995). Production of ethanol by *Zymomonas mobilis*: the effect of batch step-feeding of glucose and relevant growth factors, *Process Biochemistry*, Vol.30, No. 1, pp. 41-47, ISSN 1359-5113.

Kim, M. & Day, D. (2011). Composition of sugar cane, energy cane, and sweet sorghum suitable for ethanol production at Louisiana sugar mills, Journal of Industrial Microbiology & Biotechnology, Vol. 38, No. 7, (August 2010), pp 803-807, ISSN: 1367-5435.

Kundiyana, D.K. (July 1996). Sorganol: In-field production of ethanol from sweet sorghum, 28.03.11, Available from http://digital.library.okstate.edu/etd/umi-okstate-1974.pdf

Kurian, J.K., Minu, A.K., Banerji, A. & Kishore, V.V.N. (2010). Bioconversion of hemicellulose hidrolysate of sweet sorghum bagasse to ethanol by using *Pichia stipitis* NCIM 3497 and *Debaryomyces hansenii* sp., *Bioresources*, Vol.5, No.4 (November 2010), pp. 2404-2416.

Laopaiboon, L., Thanonkeo, P., Jaisil P. & Laopiboon, P. (2007) Ethanol production from sweet sorghum juice in batch and fed-batch fermentations by *Saccharomyces cerevisiae*, *World Journal of Microbiology and Biotechnology*, Vol. 23, No.10 (October 2007), pp. 1497-1501, ISSN 0959-3993.

Lee, J. (1997) Biological conversión of lignocellulosic biomass to ethanol, *Journal of Biotechnoloy*, Vol. 56, No. 1 (July 1997), pp. 1-24. ISSN 0168-1656.

Lin, Z.X., Zhang, H.M., Ji, X.J., Chen, J.W. & Huang, H. (2011). Hydrolytic enzyme of cellulose for complex formulation applied research, *Applied Biochemical Biotechnology*, Vol.164, No.1 (January 2011), pp. 23-33, ISSN 0273-2289.

Liu, R. & Shen, F. (2008). Impacts of main factors on bioethanol fermentation from stalk juice of sweet sorghum by immobilized *Saccharomyces cerevisiae* (CICC 1308), *Bioresource Technology*, Vol. 99, No. 4, (March 2008), pp. 847-854, ISSN 0960-8524.

Liu, R., Li, J. & Shen, F. (2008). Refining bioethanol from stalk juice of sweet sorghum by immobilized yeast fermentation, *Renewable Energy*, Vol. 33, No. 5, (May 2008), pp. 1130-1135, ISSN 0960-1481.

Mahasukhonthachat, K., Sopade, P.A. & Gidley, M.J. (2010). Kinetics of starch digestion in sorghum as affected by particle size, *Journal of Food Engineering*, Vol.96, No. 1 (January 2010), pp. 18-28, ISSN 0260-8774.

Mamma, D., Koullas, D., Fountoukidis, G., Kekos, D., Macris, B.J. & Koukios, E. (1996). Bioethanol from sweet sorghum: simultaneous saccharification and fermentation of carbohydrates by a mixed microbial culture, *Process Biochemistry*, Vol. 31, No.4, (May 1996), pp. 377-381, ISSN 1359-5113.

McCutchen, B.F., Avant, R.V. & Baltensperger, D. (2008). High-tonnage dedicated energy crops: the potential of sorghum and energy cane, Proceedings of the twentieth annual conference of the National Agricultural Biotechnology Council. pp. 119-122. Columbus, OH, USA. June 3–5, 2008.

McIntosh, S. & Vancov, T. (2010). Enhanced enzyme saccharification of *Sorghum bicolor* straw using dilute alkali pretreatment, *Bioresource Technology*, Vol.101, No. 17 (September 2010), pp. 6718–6727, ISSN 0960-8524.

Mei, X., Liu, R., Shen, F. & Wu, H. (2009). Optimization of fermentation conditions for the production of ethanol from stalk juice of sweet sorghum by immobilized yeast using response surface methodology, *Energy & Fuels*, Vol.23, No.1, (January 2009), pp. 487-491, ISSN 0887-0624.

Mizuno, R., Ichinose, H., Honda, M., Takabatake, K., Sotome, I., Takai, T., Maehara, T., Okadome, H., Isobe, S., Gau, M. & Kaneki, S. (2009). Use of whole crop sorghums as a raw material in consolidated bioprocessing bioethanol production using *Flammulina velutipes, Bioscience, Biotechnology, and Biochemistry*, Vol.73, No.7 (July 2009), pp. 1671-1673, ISSN 0916-8451.

Nuanpeng, S., Laopaiboon, L., Srinophakun, P., Klanrit, P., Jaisil, P. & Laopaiboon, P. (2011). Ethanol production from sweet sorghum juice under very high gravity conditions: batch, repeated-batch and scale up fermentation, *Electronic Journal of Biotechnology*, Vol. 14, No. 1 (January 2011), http://dx.doi.org/10.2225/vol14-issue1-fulltext-2, ISSN 0717-3458.

Öhgren, K., Galbe, M. & Zacchi, G. (2005). Optimization of steam pretreatment of SO_2-impregnated corn stover for fuel ethanol production, in: Davison, B. H., Evans, B. R., Finkelstein, M., McMillan, J. D. (eds), Twenty-Sixth Symposium on Biotechnology for Fuels and Chemicals. Humana Press, USA, ISBN 978-1-59259-991-2.

Perez-Carrillo, E. & Serna-Saldivar, S.O. (2007). Effect of protease treatment before hydrolysis with *alpha*-amylase on the rate of starch and protein hydrolysis of maize, whole sorghum and decorticated sorghum, *Cereal Chemistry*, Vol. 84, No. 6 (November/December, 2007), pp. 607-613, ISSN 0009-0352.

Pérez-Carrillo, E., Serna-Saldívar, S.O., Alvarez, M.M. & Cortes-Callejas, M.L. (2008). Effect of sorghum decortication and use of protease before liquefaction with thermoresistant *alpha*-amylase on efficiency of bioethanol production, *Cereal Chemistry*, Vol. 85, No.6 (November/December, 2008), pp. 792-798, ISSN 0009-0352.

Pérez-Carrillo, E., Cortes-Callejas, M.L., Sabillón-Galeas, L.E., Montalvo-Villarreal, J.L., Canizo, J.R., Moreno-Zepeda, M.G. & Serna-Saldívar, S.O. (2011). Detrimental effect of increasing sugar concentrations on ethanol production from maize or decorticated sorghum mashes fermented with *Saccharomyces cerevisiae* or *Zymomonas mobilis*, *Biotechnology Letters*, Vol. 33, No. 2 (February, 2011), pp. 301-307), ISSN 0141-5492.

Phowchinda, O., Delia-Dupuy, M.L. & Strehaiano, P. (November 1997). Alcoholic fermentation from sweet sorghum: some operating problems, 28.03.11, Available from http://www.energy-based.nrct.go.th/Article/Ts-3%20alcoholic%20fermentation%20from%20sweet%20sorghum%20some%20operating%20problems.pdf

Pradeep, P., Goud, G.K. & Reddy, O. V. S. (2010). Optimization of very high gravity (VHG) finger millet (ragi) medium for ethanolic fermentation by yeast, *Chiang Mai Journal of Science*, Vol. 37, No. 1, (July 2009), pp. 116-123, ISSN 0125-2526.

Prasad, S., Singh, A., Jain, N. & Joshi, H.C. (2007). Ethanol production from sweet sorghum syrup for utilization as automotive fuel in India, *Energy & Fuels*, Vol. 21, No. 4, (May 2007), pp. 2415-2420, ISSN 0887-0624.

Reddy, B.V.S., Ramesh, S., Reddy, P.S., Ramaiah, B., Salimath, P.M. & Kachapur, R. (2005). Sweet sorghum - a potential alternate raw material for bio-ethanol and bio-energy. International Crops Research Institute for the Semi-Arid Tropics. 28.03.11, Available from http://www.icrisat.org/Biopower/BVSReddySweetSorghumPotentialAlternative.pdf

Reddy, N. & Yang, Y. (2005). Biofibers from agricultural by products for industrial applications, *TRENDS in Biotechnology*, Vol.23, No.1 (January 2005), 22-27, ISSN 0167-7799.

Renewable Fuels Association. (2010). The Industry-Statistics. 28.03.11, Available from http://www.ethanolrfa.org

Rooney, L. & Serna-Saldívar, S. (2000). Sorghum, in: Kulp, K., Ponte, J. (eds.), Handbook of Cereal Science and Technology. Marcel Dekker, New York, NY, ISBN 0824782941.

Rooney, L.W. & Pflugfelder, R.L. (1986). Factors affecting starch digestibility with special emphasis on sorghum and corn, *Journal of Animal Science*, Vol. 63, No. 6 (June 1986) pp. 1607-1623, ISSN 0021-8812 .

Rooney, W., Blumenthal, J., Bean, B. & Mullet, J.E. (2007). Designing sorghum as a dedicated bioenergy feedstock, *Biofuels, Bioproducts and Biorefining*, Vol. 1, No. 2, (September 2007), pp. 147-157, ISSN 1932-1031.

Saballos, A. (2008). Development and utilization of sorghum as a bioenergy crop. Chapter 8, in W. Vermerris (ed.). *Genetic Improvement of Bioenergy Crops*. Springer. USA, ISBN 0387708049.

Serna-Saldivar, S. (2010). Cereal Grains: Properties, Processing, and Nutritional Attributes CRC Press, ISBN 9781439815601

Sipos, B., Réczey, J., Somorai, Z., Kádár, Z., Dienes, D. & Réczey, K. (2009). Sweet sorghum as feedstock for ethanol production: enzymatic hydrolysis of steam-pretreated bagasse, *Applied Biochemistry and Biotechnology*, Vol.153, No. 1-3 (May 2009) pp. 151–162, ISSN 0273-2289.

Smith, G. A., Bagby, M. O., Lewellan, R. T., Doney, D. L., Moore, P. H., Hills, F. J., Campbell, L. G., Hogaboam, G. J., Coe, G. E. & Freeman, K. (1987). Evaluation of sweet sorghum for fermentable sugar production potential, *Crop Science*, Vol. 27, No.4, (July-August 1987), pp. 788-793, ISSN 0931-2250.

Solomon, B.D., Barnes, J.R. & Halvorsen, K.E. (2007). Grain and cellulosic ethanol: history, economics, and energy policy, *Biomass and Bioenergy*, Vol. 31, No. 6 (June 2007), pp. 416-425, ISSN 0961-9534.

Stenberg, K., Tengborg, C., Galbe, M. & Zacchi, G. (1998). Optimisation of steam pretreatment of SO_2-impregnated mixed softwoods for ethanol production, *Journal of Chemical Technology & Biotechnology*, Vol. 71, No. 4 (April 1998), pp. 299-308 ISSN 1097-4660.

Sumari, D., Hosea, K. M. M. & Magingo, F. S. S. (2010). Genetic characterization of osmotolerant fermentative *Saccharomyces* yeasts from Tanzania suitable for industrial very high gravity fermentation, *African Journal of Microbiology Research*, Vol. 4, No. 11, (June 2010), pp. 1064-1070, ISSN 1996-0808.

Taylor, J.R.N. & Belton, P.S. (2002). Sorghum, In: Pseudocereals and less common cereals, Belton, P. & J. Taylor, pp. 25-91, Springer, Germany, ISBN 3540429395-

Tellez-Luis, S.J., Ramírez, J.A. & Vázques, M. (2002). Mathematical modelling of hemicellulosic sugar production from sorghum straw, *Journal of Food Engineering*, Vol.52, No. 3 (May 2002), pp. 285-291, ISSN 0260-8774.

The Economist. (2007). Food prices. Cheap no more. 28.03.11, Available from http://www.economist.com/displaystory.cfm?story_id=10250420

Turhollow, A.F., Webb, E.G. & Downing, M.E. (June 2010). Review of Sorghum Production Practices: Applications for Bioenergy, 28.03.11, Available from http://info.ornl.gov/sites/publications/files/Pub22854.pdf.

Wang, D., Bean, S., McLaren, J., Seib, P., Madl, R., Tuinstra, M., Shi, Y., Lenz, M., Wu, X. & Zhao, R. (2008). Grains sorghum is a viable feedstock for ethanol production, *Journal of Industrial Microbiology and Biotechnology*, Vol. 35, No. 5 (May 2008), pp. 313-320, ISSN 1367-5435.

Wang, F.Q., Gao, C.J., Yang, C.Y. & Xu P. (2007). Optimization of an ethanol production medium in very high gravity fermentation, *Biotechnology Letters*, Vol. 29, No.2 (February 2007), pp. 233-236, ISSN 1573-6776.

Wang, M.L., Zhu, C., Barkley, N.A., Chen, Z., Erpelding, J.E., Murray, S.C., Tuinstra, M.R., Tesso, T., Pederson, G.A. & Yu, J. (2009). Genetic diversity and population structure analysis of accessions in the US historic sweet sorghum collection, *Theoretical and Applied Genetics*, Vol. 120, No.1, (December 2009), pp. 13-23, ISSN 0040-5752.

Wang, S., Ingledew, W.M., Thomas, K.C., Sosulski, K. & Sosulski, F.W. (1999). Optimization of fermentation temperature and mash specific gravity of fuel alcohol production, *Cereal Chemistry*, Vol. 76, No. 1 (January/February 1999), pp. 82-86, ISSN 0009-0352.

Wong, J.H., Lau, T., Cai, N., Singh, J., Pedersen, J.F., Vensel, W.H., Hurkman, W.J., Wilson, J.D., Lemaux, P.G. & Buchanan, B.B. (2009). Digestibility of protein and starch from sorghum (*Sorghum bicolor*) is linked to biochemical and structural features of grain endosperm, *Journal of Cereal Science*, Vol. 49, No. 1 (January 2009), pp. 73-82, ISSN 0733-5210.

Woods, J. (2000). Integrating Sweet Sorghum and Sugarcane for Bioenergy: Modelling The Potential for Electricity and Ethanol Production in SE Zimbabwe. Thesis for the degree of Doctor of Philosophy. King's College London. University of London, UK.

World Food Program. (2008). Rising food prices: impact on the hungry. In: Database of Press Releases related to Africa. 28.03.11, Available from http://appablog.wordpress.com/2008/03/14/rising-food-prices-impact-on-the-hungry/

Wu, X., Jampala, B., Robbins, A., Hays, D., Yan, S., Xu, F., Rooney, W., Peterson, G., Shi, Y. & Wang, D. (2010a). Ethanol fermentation performance of grain sorghum (*Sorghum bicolor*) with modified endosperm matrices, *Journal of Agricultural and Food Chemistry*, Vol. 58, No. 17 (September 2010), pp. 9556-9562, ISSN 0021-8561.

Wu, X., Staggenborg, S., Propheter, J.L., Rooney, W.L., Yu, J. & Wang, D. (2010b). Features of sweet sorghum juice and their performance in ethanol fermentation, *Industrial Crops and Products*, Vol. 31, No. 1(January 2010), pp. 164-170, ISSN 0926-6690.

Wu, X., Zhao, R., Bean, S.R., Seib, P.A., McLaren, J.S., Madl, R.L., Tuinstra, M., Lenz, M.C. & Wang, D. (2007). Factors impacting ethanol production from grain sorghum in the dry-grind process, *Cereal Chemistry*, Vol.84, No.2 (March/April 2007), pp. 130-136, ISSN 0009-0352.

Wu, X., Zhao, R., Liu, L., Bean, S., Seib, P.A., McLaren, J., Madl, R., Tuinstra, M., Lenz, M. & Wang, D. (2008). Effects of growing location and irrigation on attributes and ethanol yields of selected grain sorghums, *Cereal Chemistry*, Vol.85, No.4 (July/August 2008), pp. 495-501, ISSN 0009-0352.

Wu, X., Zhao, R., Wang, D., Bean, S.R., Seib, P.A., Tuinstra, M.R., Campbell, M. & O'Brien, A. (2006). Effects of amylose, corn protein, and corn fiber contents on production of ethanol from starch-rich media, *Cereal Chemistry*, Vol.83, No.5 (September/October 2006), pp. 569-575, ISSN 0009-0352.

Yan, S., Wu, X., Dahlberg, J., Bean, S.R., MacRitchie, F., Wilson, J.D. & Wang, D. (2010). Properties of field-sprouted sorghum and its performance in ethanol production, *Journal of Cereal Science*, Vol.51, No.3 (May 2010), pp. 374-380, ISSN 0733-5210.

Yan, S., Wu, X., MacRitchie, F. & Wang, D. (2009). Germination-improved ethanol fermentation performance of high-tannin sorghum in a laboratory dry-grind process, *Cereal Chemisty*, Vol.86, No. 6 (November/December 2009), pp. 597-600, ISSN 0009-0352.

Zhan, X., Wang, D., Bean, S.R., Mo, X., Sun, X.S. & Boyle D. 2006. Ethanol production from supercritical-fluid-extrusion cooked sorghum, *Industrial Crops Products,* Vol. 23, No.3 (May 2006), pp. 304-310, ISSN 0926-6690.

Zhang, C., Xie, G., Li, S., Ge, L. & He, T. (2010). The productive potentials of sweet sorghum ethanol in China, *Applied Energy*, Vol. 87, No.7, (July 2010), pp. 2360-2368, ISSN 0306-2619.

Zhao, R., Bean, S.R., Ioerger, B.P., Wang, D. & Boyle, D.L. (2008). Impact of mashing on sorghum proteins and its relationship to ethanol fermentation, *Journal of Agricultural and Food Chemistry*, Vol. 56, No. 3, (January 2008), pp. 946-953, ISSN 0021-8561.

4

Simultaneous Production of Sugar and Ethanol from Sugarcane in China, the Development, Research and Prospect Aspects

Lei Liang, Riyi Xu, Qiwei Li, Xiangyang Huang,
Yuxing An, Yuanping Zhang and Yishan Guo
Bio-engineering Institute, Guangdong Academy of Industrial Technology
Guangdong Key Laboratory of Sugarcane Improvement and Biorefinery,
Guangzhou Sugarcane Industry Research Institute
P. R. China

1. Introduction

With the ever growing concern on the speed at which fossil fuel reserves are being used up and the damage that burning them does to the environment, the development of sustainable fuels has become an increasingly attractive topic (Wyman & Hinman, 1990; Lynd & Wang, 2004; Herrera, 2004; Tanaka, 2006; Chandel et al., 2007; Dien et al., 2006; Marèlne Cot, et al., 2007). The interest partially caused by environment concern, especially global warming due to emission of Greenhouse Gas (GHG). Other factors include the rise of oil prices due to its unrenewability, interest in diversifying the energy matrix, security of energy supply and, in some cases, rural development (Walter et al., 2008). The bioethanol such as sugarcane ethanol is an important part of energy substitutes (Wheals et al., 1999). This chapter was focused on the development and trends of the sugarcane ethanol in China. Based on the analysis of the challenge and the chance during the development of the sugarcane ethanol in China, it introduced a novel process which is suitable for China, and mainly talked about simultaneous production of sugar and ethanol from sugarcane, the development of sugarcane varieties ,ethanol production technology, and prospect aspects. We hope it will provide references for evaluation the feasibility of sugarcane ethanol in China, and will be helpful to the fuel ethanol development in China.

2. Sugarcane for bioethanol - A new highlight of sugar industry development

The technology of producing fuel ethanol using sugarcane, which has a characteristic of high rate of energy conversion, wide adaptability, and strong resistance, etc, has received extensive attention (Watanabe, 2009). Brazil, Australia and other countries have made breakthroughs in the sugarcane improvement, ethanol fermentation process and its application (Goldemberg et al., 2008; International Energy Agency (IEA), 2004). Brazil is the world's largest sugar producer and exporter of fuel ethanol, which is expected that annual output of 65 billion liters by 2020(Walter et al., 2008). Energy security and environmental stress force China to seek and develop biofuels as a substitute of fossil energy. Meanwhile, China has also introduced policies that encourage the development of fuel ethanol using sugarcane and other non-food crop, to ease pressure on energy demand. Recently, the study and the industrial-scale production of biofuels, particularly, fuel ethanol and biodiesel, have progressed remarkably in China as a result of government preferential policies and funding supports (Zhong et al., 2010).

Fig. 1. Highlight of sugarcane for bioethanol

3. Benefits of sugarcane for ethanol

The reasons why we choose ethanol from sugarcane as the most promising biofuels are illustrated below. Firstly, the balance of GHG emissions of sugarcane ethanol is the best among all biofuels currently produced (Macedo et al., 2008; Cerri et al., 2009; Oliveira et al., 2005). As reviewed in several studies, bioethanol based on sugarcane can achieve greenhouse gas reductions of more than 80% compared to fossil fuel use (Macedo et al., 2008). Figure 2 (BNDES, 2008) showed correspond to the consumption of ethanol produced from maize (USA), from wheat (Canada and Europe) and from sugarcane (produced in Brazil and consumed in Brazil or in Europe). Sugarcane ethanol is much better than ethanol from maize and wheat (a maximum of 35%) in case of the avoided emissions.

Secondly, as we known, cropland is very limited for planting in China. So it is very important that the land use is keeping in a high efficient level. Ethanol from sugarcane is the most productive among different crops. The fortunate experience of ethanol use in Brazil may also be coupled with a superior sucrose yield and a higher potential of biomass production of sugarcane - an average of 87 tons per hectare in South Central Brazil - than observed in other crops. As shown in figure 3, only beets can be compared with sugarcane in terms of ethanol production per cultivated hectare. However, the industrial process of ethanol production from beets depends on an external power input (electricity and fuel) while sugarcane electricity is provided by bagasse burning at the mill. (BNDES, 2008).

Ethanol produced from sugarcane is the biofuel with the best energy balance (see table1). This can be illustrated as the ratio between renewable products and the energy input as fossil fuel for Brazilian sugarcane ethanol is 9.3 (compared with 1.2-1.4 in the case of ethanol produced from American maize, and approximately 2.0 in the case of ethanol produced from European wheat). Apart from these above, other environmental impacts of the sugarcane sector, such as water consumption, contamination of soils and water shields due to the use of fertilizers and chemicals, and loss of biodiversity, are less important in comparison to other crops (Watanabe, 2009). Above in all, Sugarcane is by far the best alternative from the economical, energy and environmental point of view, for bio-fuel production.

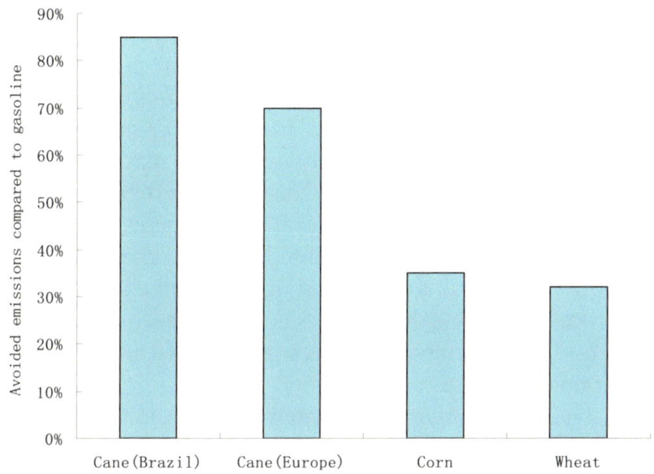

Fig. 2. Avoided GHG emissions in comparison with full life-cycle of gasoline

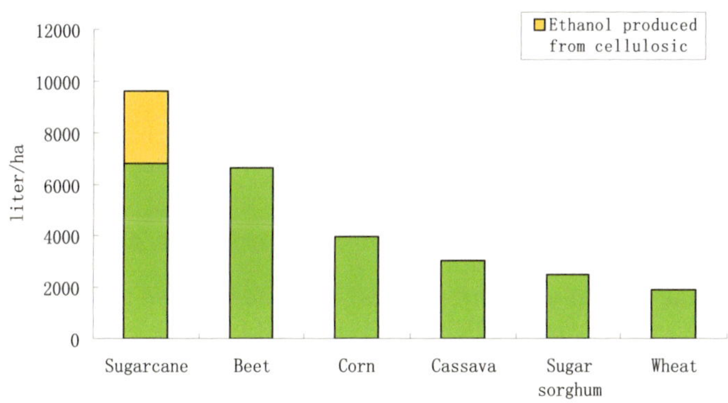

Fig. 3. Average ethanol productivity per area for different crops. Source: BENDES(2008)

Feedstock	Energy ratio
Sugarcane	9.3
Lignocellulosic residues	8.3~8.4
Cassava	1.6~1.7
Beet	1.2~1.8
Wheat	0.9~1.1
Corn	0.6~2.0

Table 1. Comparison of different feedstock for biofuel production. Source:BNDES(2008)

4. The challenge and perspectives to develop sugarcane ethanol in China

Sugarcane is mainly planted in southern China, such as Guangxi, Yunnan, Guangdong, Hainan et al, Its total planting areas were about 20 million acres in 2010 statistically, and Guangxi contribute about 60 percent of the total. Lands suitability for sugarcane is limited. It is very difficult to expand the land for sugarcane production because of the industrialization in China. An additional challenge is the harvesting. High investment requirements and difficulties with mechanization on, for example steep land, increase the risks of the implementation of mechanized harvest. About over 90 percent of the China sugarcane area was still manually harvested. Expansion of sugarcane areas will be affected by the cost/benefit of manual labor. Under the driving of the market opportunities, national policies giving incentives to the sugarcane agri-business, the further expansion of sugarcane areas forecasted for China is expected to about 2 million acres, which mustn't reduce the availability of arable land for the cultivation of food and feed crops.

There are risks of environmental degradation in different stages of sugarcane ethanol production and processing. Negative impacts have been caused by the lack of implementation of best management practices and ineffective legislation and control. Nevertheless, further improvements are necessary.

A major concern of developing sugarcane ethanol in China is the threat to sugar security. Rapid expansion of bioethanol production could potentially reduce the availability of sugar production, causing a reduction in its supply and increase of sugar price. In recent years, the sugar productions are stably at about 12 million tons, the max exceeded 14.84 million tons in 2008. While the total demand for sugar is about 12 million tons in China. With the combination of the further expansion of about 2 million acres sugarcane areas, and applying the advanced technology, for example: genetically modified sugarcane and improved cultivation techniques, yields can be increased from 5 tons to about 6-7 tons . So the sugar productions in China are expected to over 16 million tons. Based on these estimates, without affecting the supply of sugar, the current potential of sugarcane ethanol production reached over 2 million tons.

5. Simultaneous production of sugar and ethanol from sugarcane

As the major raw material, most of sugarcanes are refined into sugar in China now. Also the international sugar price is running in high level, and it needs to balance the domestic sugar supply and demand through imports, so it is impossible to produce large amounts of ethanol by sugarcane. However, it is unfavorable to sugar price stability and its healthy development if only refining sugar. To achieve more economic benefits, a viable option is to explore the "Simultaneous production of sugar and ethanol " mode. In recent years, we have made some progress on the sugarcane breeding, ethanol production technologies and process optimization for simultaneous production of sugar and ethanol.

5.1 Material distribution

At present, sugar is produced following the three stage boiling technology or the three and a half stage boiling technology. It takes a long time and high energy consumption to boil the B sugar and C sugar. The value the by-product is low. There are high costs and weak adaptability to the market.

Generally, it is advantage to regulate sugar production and ethanol production according to market demand the flexibility while applying the "Simultaneous production of sugar and ethanol" mode. It is necessary to distribute the raw material fluxes rationally. However, less literature is related to juice and syrup distribution for simultaneous production of sugar and ethanol. In this paper, material fluxes balance calculation is carried out according to Brazil experience and the parameters of three and a half stage boiling process. The feed syrup is 60 Bx, the purity is 87%, and the feed syrup fluxes are 100 tons. The sugar combined fuel ethanol process is showed as Figure 4:

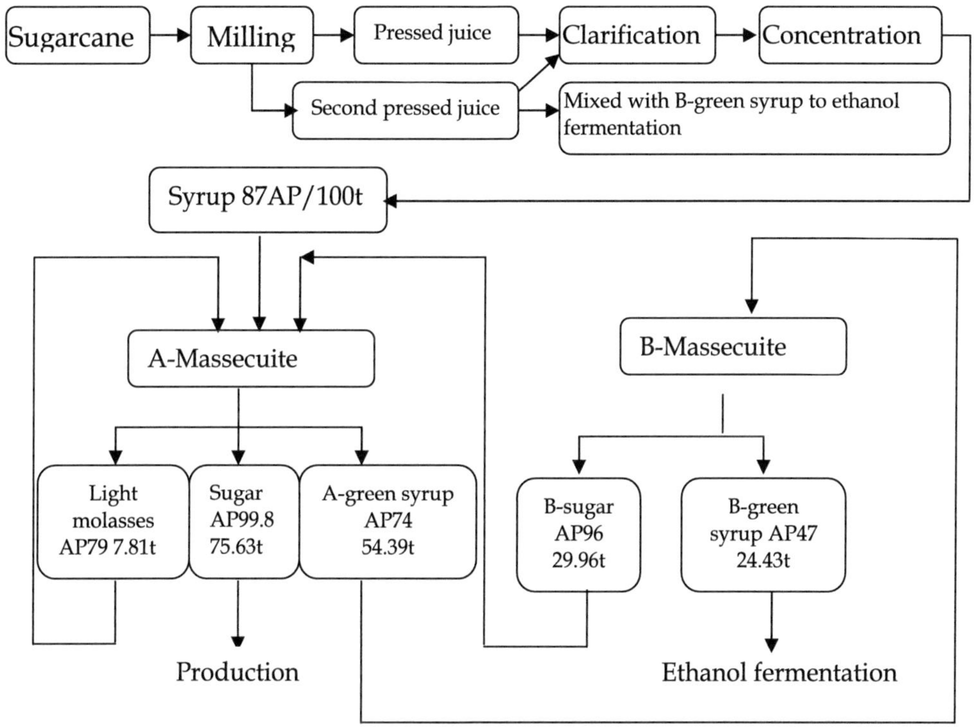

Fig. 4. Sugar combined ethanol process and its material balance

5.2 Sugarcane for simultaneous production of sugar and ethanol

In China, biotechnology research and genetic improvement have led to the development of strains which are more resistant to disease, bacteria, and pests, and also have the capacity to respond to different environments, thus allowing the expansion of sugarcane cultivation. The leading sugar enterprise in charge for applied research on agriculture, together with research developed by state institutes and universities. Efforts have been concentrated in taking advantage of its genetic diversity and high photosynthetic efficiency characteristic, high separation sugarcane population was generated via distant hybridization technology. To obtain the new material of sugarcane for ethanol, we took total biomass, total fermentable sugars as targets and adopted advanced photosynthetic efficiency living early-generation determination technology, molecular markers and cell engineering technology combined with conventional breeding. Then, in order to optimize the selection of energy sugarcane, we took a series of pilot test and technical and economic indexes of evaluation. By 2010, more than 10 sugarcane varieties for simultaneous production of sugar and ethanol are cultivated in China, such as "00-236", "FN91-4710","FN94-0403", FN95-1702","G94-116",

"Y93-159", "Y94-128", "G-22" et al.. Although potential benefits are high, there is still a lack of understanding of the potential impacts of genetically modified organisms on environmental parameters.

Fig. 5. Sugarcane for simultaneous production of sugar and ethanol

5.3 Ethanol production technologies for simultaneous production of sugar and ethanol

5.3.1 Genome shuffling of *Saccharomyces cerevisae* for multiple-stress resistant yeast to produce bioethanol

In the fermentation process, sugars are transformed into ethanol by addition of microorganism. Ethanol production from sugars has been commercially dominated by the yeast *S. cereviseae* (Tanaka, 2006). Practically, yeast cells are often exposed in multiple stress environments· Therefore, it is helpful to fermentation efficiency and economic benefits to breed the yeast strains with tolerance against the multiple-stress such as temperature, ethanol, osmotic pressure, and so on (Cakar et al., 2005). Yeast strain improvement strategies are numerous and often complementary to each other, a summary of the main technologies is shown in Table 2. The choice among them is based on three factors: (1) the genetic nature of traits (monogenic or polygenic), (2) the knowledge of the genes involved (rational or blind approaches) (3) the aim of the genetic manipulation (Giudici et al ., 2005; Gasch et al., 2000).

Genetics of Dpt Strategies			Aims
Rational approaches (for known genes)	Monogenic	Single target mutagenesis or cassette mutagenesis	Silencing of one genetic Function
		Metabolic engineering	Inserting a new function, modulating a function already present
	Polygenic	Multiple target mutagenesis	Silencing of many genetic functions
		Metabolic engineering (for a small number of genes)	Inserting more functions, modulating more already present functions
Blind approaches (for unknown genes)	Monogenic	Random mutagenesis	Silencing of a genetic function
	Polygenic	Metagenomic techniques	Inserting genes cluster
		Sexual recombination	Improving Dpt, obtaining a combination of Dpts
		Genome shuffling	Improving Dpt, obtaining a combination of Dpts

Table 2. Summary of the main genetic improvement strategies. Dpt Desired phenotype

It is difficult to improve the multi-tolerance of the yeast by rational genetic engineering technology before its mechanism completely clarified. Nevertheless, for quantitative traits, the number of responsible genes QTLs is so great that a "gene-by-gene" engineering strategy is impossible to perform. In these cases, blind strategies, such as genome shuffling (Zhang et al., 2002), could be applied in order to obtain quickly strains with recombinant traits. Genome shuffling is an accelerated evolutionary approach that, on the base of the recursive multiparental protoplast fusion, permits obtaining the desired complex phenotype more rapidly than the normal breeding methods (Figure 6). Genome shuffling technology can bring a rapidly improvement of breeding a hybrid with whole-genome random reorganization. After the initial strains in various long term evolution experiments (Figure 7), we successfully applied the genome shuffling technology that combines the advantage of multi- parental recursive fusion with the recombination of entire genomes normally associated with conventional mutant breeding to selecting the multiple-stress resistant yeast (Figure 8).

5.3.2 Continuous fermentation

Traditionally, ethanol has been produced batch wise. However, high labor costs and the low productivity offered by the batch process have led many commercial operators to consider the continuous fermentation. Continuous fermentation can be performed in different kind of bioreactors - stirred tank reactors or plug flow reactors. Continuous fermentation often gives a higher productivity, offers ease of control and is less labor intensive than batch fermentation (Cheng et al., 2007). However contamination is more serious in this operation (Skinner & Leathers, 2004). In the fuel ethanol industry, control of bacterial contamination is achieved by acidification and using antibiotics such as penicillin G, streptomycin, tetracycline (Aquarone E,1960; Day et al., 1954), virginiamycin(Hamdy et al., 1996; Hynes et

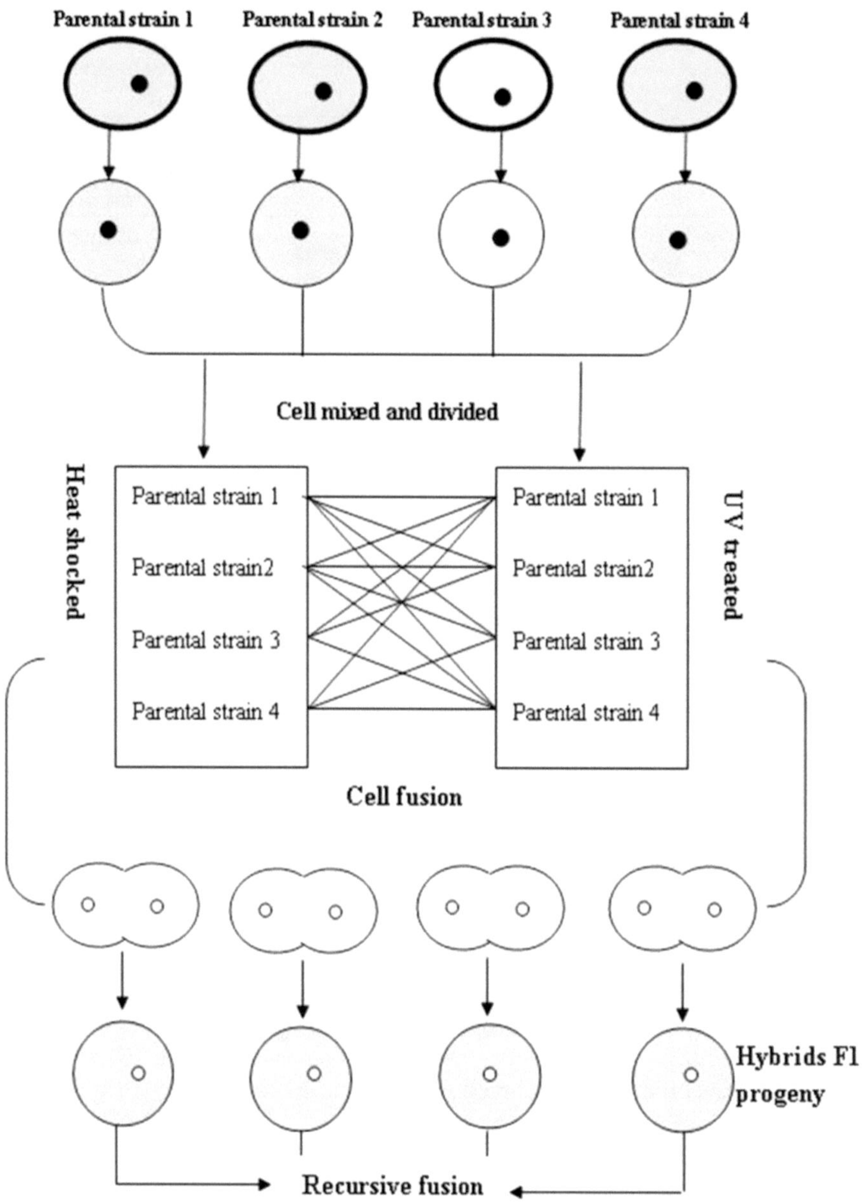

Fig. 6. Protoplast fusion of the genome shuffling process

al., 1997; Islam et al., 1999), monensin(Stroppa et al., 2000), or mixtures thereof. Fig 9 shows the process of continuous fermentation of molasses and sugarcane juice to produce ethanol. A high cell density of microbes in the continuous fermenter is locked in the exponential phase, which allows high productivity and overall short processing of 6 - 12 h as compared to the conventional batch fermentation (30 - 60 h). This results in substantial savings in labor and minimizes investment costs by achieving a given production level with a much smaller plant.

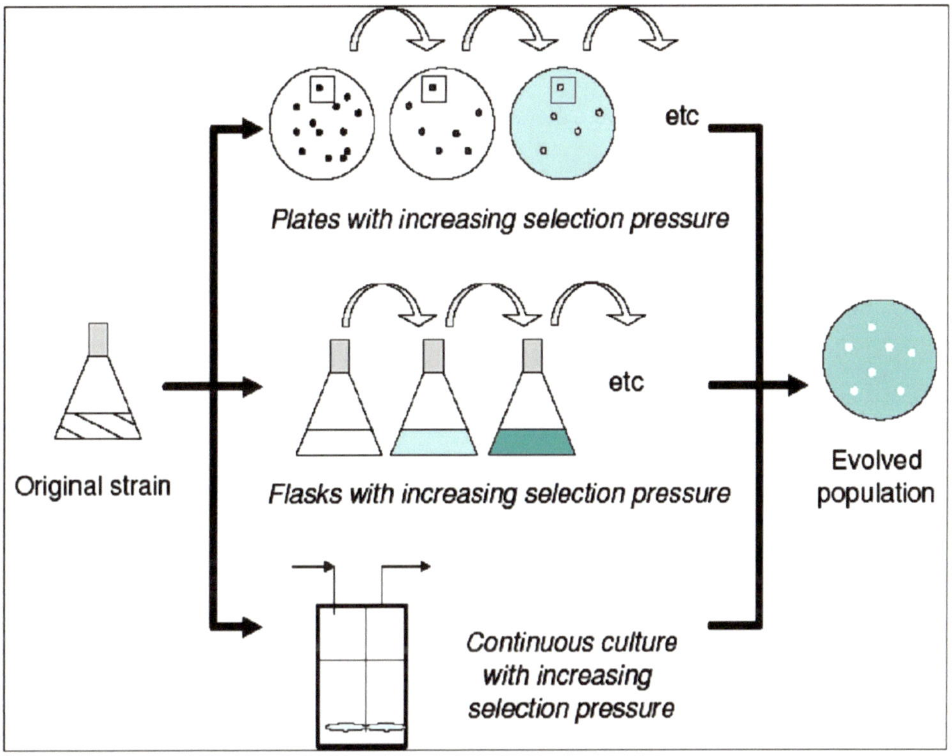

Fig. 7. Approach for evolutionary engineering

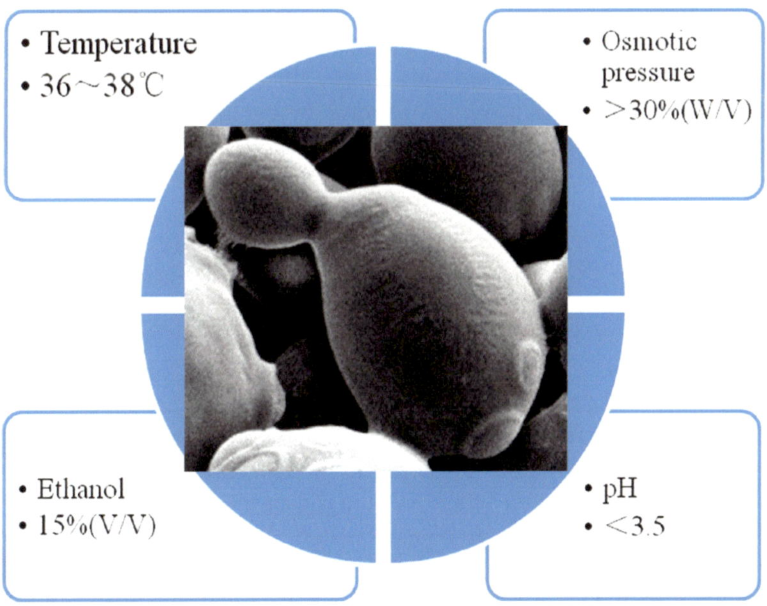

Fig. 8. Multiple-stress Resistant Yeast

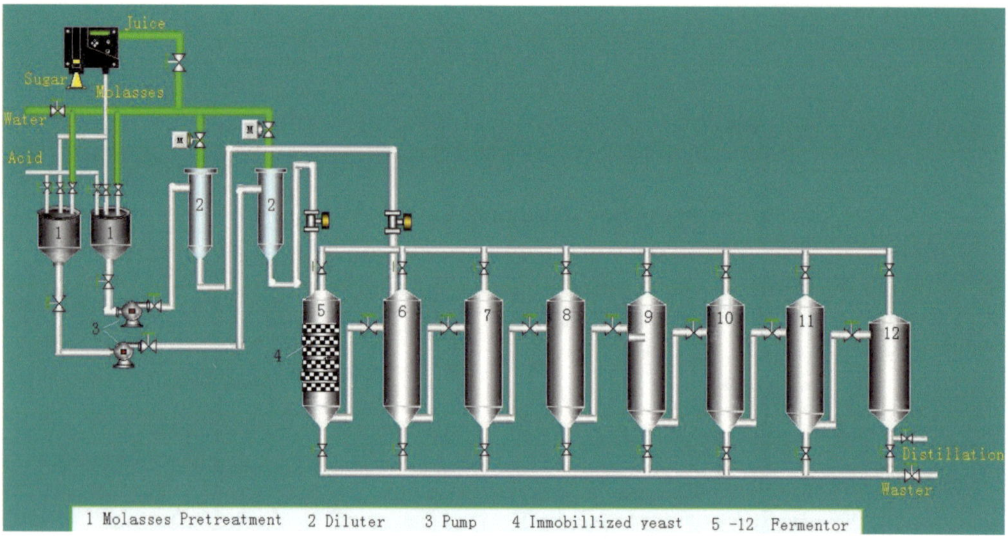

Fig. 9. Continuous fermentation of molasses and sugarcane juice to produce ethanol

5.3.3 Sugarcane pieces as yeast supports for alcohol production from sugarcane juice and molasses

A limitation to continuous fermentation is the difficulty of maintaining high cell concentration in the fermenter. The use of immobilized cells circumvents this difficulty. Immobilization by adhesion to a surface (electrostatic or covalent), entrapment in polymeric matrices or retention by membranes has been successful for ethanol production (Godia et al., 1987). The applications of immobilized cells have made a significant advance in fuel ethanol production technology. Immobilized cells offer rapid fermentation rates with high productivity - that is, large fermenter volumes of mash put through per day, without risk of cell washout. In continuous fermentation, the direct immobilization of intact cells helps to retain cells during transfer of broth into collecting vessel. Moreover, the loss of intracellular enzyme activity can be kept to a minimum level by avoiding the removal of cells from downstream products (Najafpour, 1990). Immobilization of microbial cells for fermentation has been developed to eliminate inhibition caused by high concentration of substrate and product and also to enhance ethanol productivity and yield. Neelakantam (2004) demonstrated that a high yeast inoculation at the start of the sugarcane juice fermentation allows the yeast outgrow the contaminant bacteria and inhibit its growth and metabolism.

Varies immobilization supports for variety of products have been reported such as polyvinyl alcohol (PVA, see Fig10), alginates (Kiran Sree, 2000; Corton et al., 2000), Apple pieces (Kourkoutas et al., 2006), orange peel (S.plessas, 2007), and delignified cellulosic residues (Kopsahelis, 2006; Bardi & Koutinas, 1994). We applied sugarcane pieces as yeast supports for alcohol production from sugarcane juice and molasses(Fig 11).The results(Liang et al.,2008) showed ethanol concentrations (about 77g/l or 89.76g/l in average value) , and ethanol productivities (about 62.76 g/l.d or 59.55g/l .d in average value)were high and stable, and residual sugar concentrations were low in all fermentations(0.3-3.6g/l)with conversions ranging from 97.7-99.8%, showing efficiency(90.2-94.2%) and operational stability of the biocatalyst for ethanol fermentation. the results presented in this paper (see table 3), according to initial concentration of sugars in the must, showed that the

Fig. 10. Yeast immobilized in Polyvinyl Alcohol

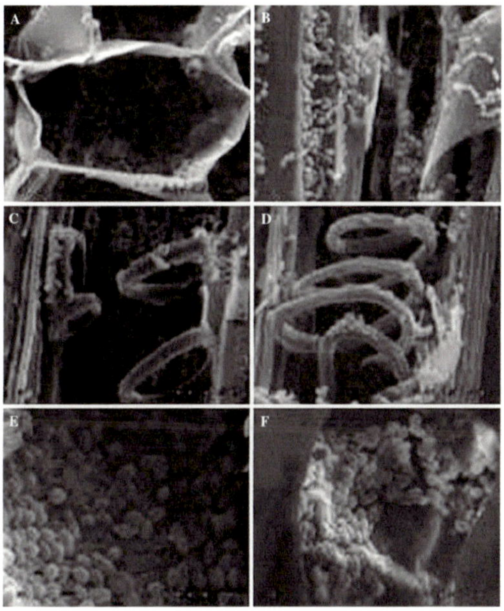

Fig. 11. Scanning electron micrographs of the middle part of the support after yeast immobilization.

sugarcane supported biocatalyst was equally efficient to that described in the literature for ethanol fermentation. Sugarcane pieces were found suitable as support for yeast cell immobilization in fuel ethanol industry. The sugarcane immobilized biocatalysts showed high fermentation activity. The immobilized yeast would dominate in the fermentation broth due to its high populations and lower fermentation time, that in relation with low price of the support and its abundance in nature, reuse availability make this biocatalyst attractive in the ethanol production as well as in wine making and beer production. After a long period of using, spent immobilized supports can be used as protein-enriched(SCP production) animal feeds.

Carrier	Medium	Initial sugar (g/l)	Ferm.time (h)	Residual sugar (g/l)	Ethanol (g/l)	Ethanol productivity (g/l.d)	Conversion (%)
Apple pieces (Y. Kourkoutas et al.,2001)	Grape must	206	80	30.8	85	26	85
Dried figs (Bekatorou et al., 2002)	Glucose	120	45	1.4	45.0	24.0	98
Spent grains (Kopsahelis et al.,2006)	molasses	187	30	8.8	51.4	42.7	95.3
Orange peel (S.plessas et al.,2007)	Glucose	125	9	4	51.4	128.3	96.8
	molasses	128	14	2	58.9	100.1	98.4
	Raisin extract	124	12	2.3	55.3	110.4	98.1
Sugarcane pieces present study	Molasses	154	27	2.3	77.12	62.76	98.5
	Sugarcane juice	176	32	0.85	89.76	59.55	99.5

Table 3. Fermentation parameters (average value) obtained in batch fermentation with *Saccharomyces cerevisiae,* immobilized on various carriers, at 30°C

5.3.4 Ethanol purification and water recovery

Distillation and molecular - sieve absorption are used to recover ethanol from the raw fermentation beer. The flow sheet of this section is presented in Figure 12 and figure 13. Distillation itself is a two-way progress include heating and cooling. That could be possible to save much steam and cooling water if we take good advantage of the heat exchange in the system. Due to its energy-saving, so far negative pressure distillation system has been popular in China. Take molasses alcohol as an example, compare to air distillation system, negative pressure distillation system could save approximately 2t steam per ton 95% (v/v) alcohol. The system showed in figure contains 3 columns, which is .fractioning column 1, fractioning column 2, and separating methanol column respectively. Making use of the different boiling points the alcohol in the fermented wine is separated from the main resting solid components. The remaining product is hydrated ethanol with a concentration of 95% (v/v). Further dehydration is normally done by molecular-sieve absorption, up to the specified 99.7°GL in order to produce anhydrous ethanol which is used for blending with pure gasoline to obtain the country's E10 mandatory blend. The fermented mash which contains 10~13 %(v/v) alcohol is preheated by the alcohol gas from the top of the first column and gas is cooled simultaneously. Then the gas stream is cooled by 3 heat exchangers, the cooler is water. Subsequently the liquid distillate which contains 30% (v/v) alcohol is feeding on the middle tray of column 2. Wastewater of column 1 is heated by the alcohol gas from the top of column 2 in the reboiler, meanwhile the steam flash evaporated in the vacuum bottom. The waste goes to anaerobic jar and then aeration tank. Cooled alcohol is pumped back to the top trays of column 2. Fusel oil is extracted from the middle trays of the column 2. Liquid distillate contains 95 %(v/v) alcohol and exceeded methanol amount. In order to decrease the concentration of aldehyde and methanol, one more column is needed. The 96%v/v alcohol with 4% water is feeding on the molecular-sieve absorption system.Finally 99.5%v/v ethanol which could be added to the gas to make gasohol is achieved.

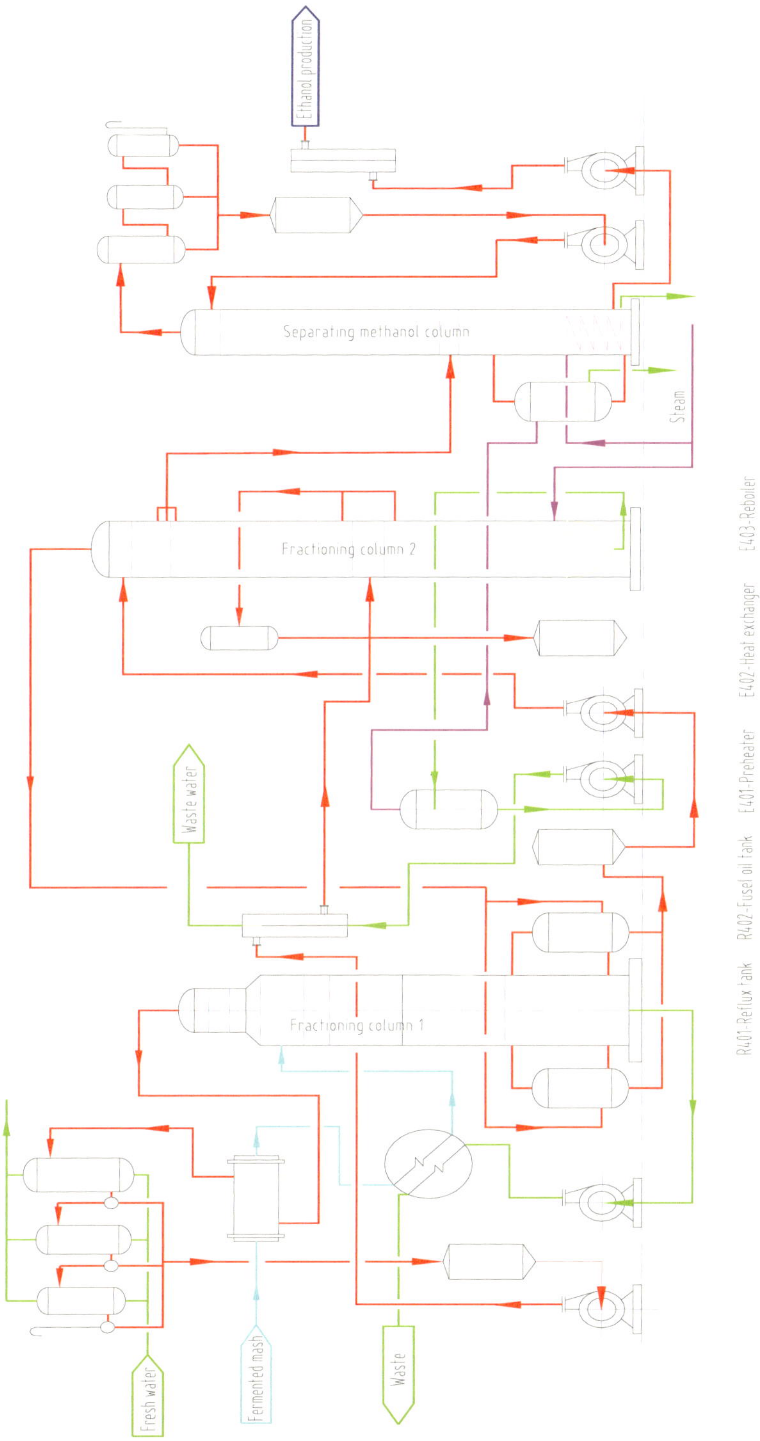

Fig. 12. Ethanol separation and dehydration.

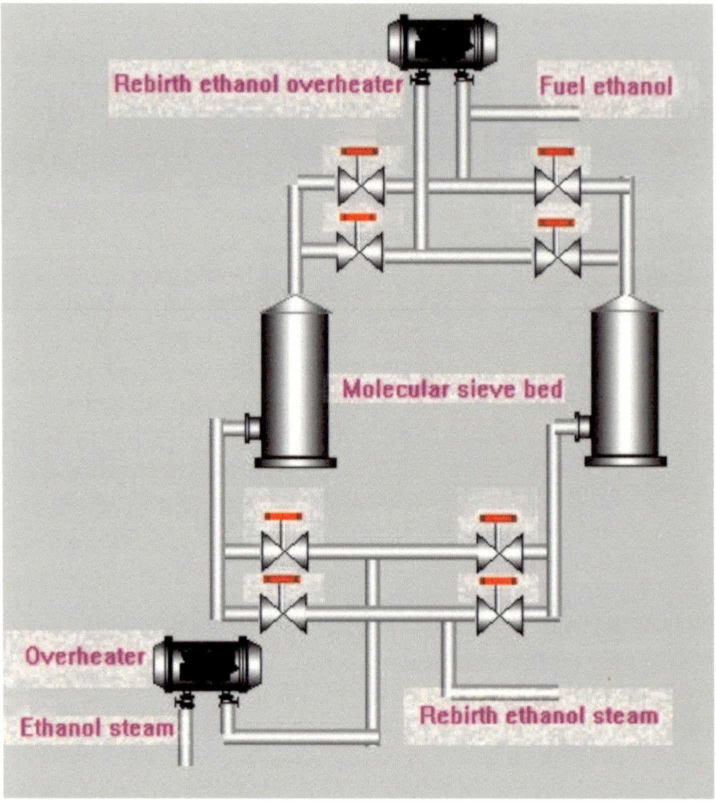

Fig. 13. Ethanol dehydration with molecular sieve bed

5.4 Economic analysis for simultaneous production of sugar and ethanol from sugarcane

Based on the economic analysis, the profits of the three different modes in 5,000 tons sugarcane pressed plants are showed in table 4. There are high costs of fermentation and distillation for sugarcane directly for ethanol fermentation due to low concentration of sugarcane juice, and about 15 tons waste water need treatment. It is also uneconomic to produce fuel ethanol using concentrated juice because of high energy costs. Therefore, that sugarcane is used directly for fuel ethanol production does not reflect its best economic benefits and flexible market response capacity.

In traditional opinion, people prefer to produce sugar as possible as they can rather than use more molasses to produce ethanol. They think that it is uneconomic to produce 1 ton ethanol with nearly 2 tons sugar consumption. In fact, we can achieve the maximized economic benefits applying "the simultaneous production of sugar and ethanol " mode, in which we boil the A-syrup that have the good characteristic of low energy consumption, to produce the top-grade white sugar production. B-green syrup and second pressed juice are mixed to produce the fuel ethanol. Costs of the ethanol production can be greatly reduced.

According to the calculations, it will bring more economic benefits while employ "the simultaneous production of sugar and ethanol" mode.

Project	Sugar product only	sugar combined fuel ethanol	Ethanol product only
Sugarcane milled (t/d)	100	100	100
Fuel ethanol (t)	-	1	7
Sugar (t)	12	11.5	-
Molasses (t)	3	-	-
Sugar product costs (RMB/t sugar)	4000	3970	-
Fuel ethanol product costs (RMB)	-	1000	5500
Total costs (RMB)	48000	46655	38500
Fuel ethanol product incomes (RMB)	-	8000	56000
Sugar product incomes (RMB)	72000	69000	-
Molasses incomes (RMB)	2700	-	-
Total incomes (RMB)	74700	77000	56000
profits (RMB)	26700	30345	17500

Table 4. The profits of the three different modes

6. Conclusions

Various technologies have been identified for immediate increases in the efficiency and sustainability of current and future sugarcane ethanol. In conclusion, recycle utilization design are seems to be suitable for sugarcane bioethanol development, for example, recycling of byproducts of sugarcane in the fields reduces chemical fertilizers application rates, reducing water consumption with closure of water-processing circuits and the use of bagasse to generate electricity or to manufacture bagasse polymer composites (Xu et al., 2010), improving the energy balance of ethanol production; as well as in production and

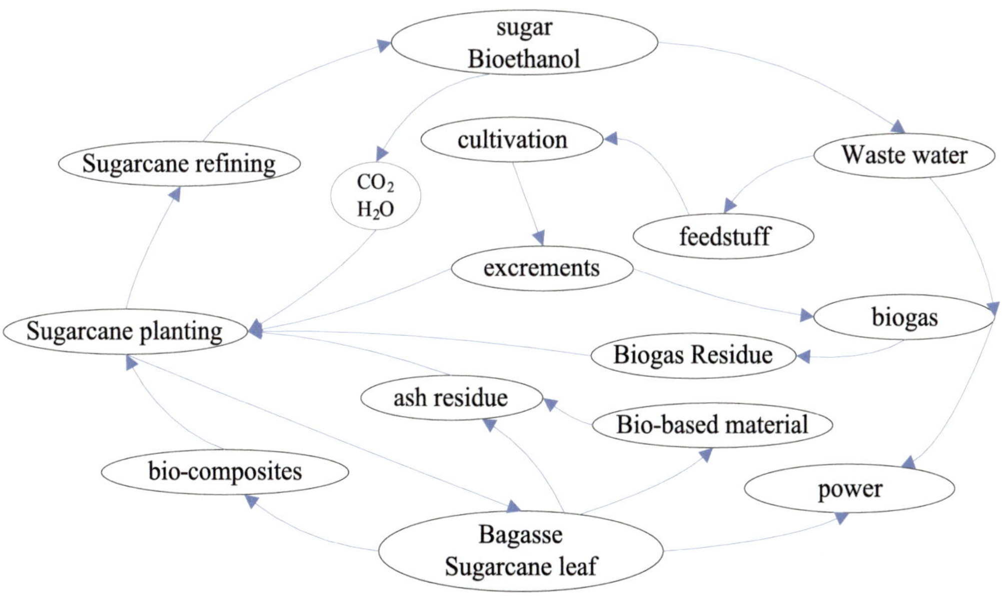

Fig. 14. Recycle utilization design for sugarcane bioethanol development

harvesting processes. At present, we think bagasse is not preferable for directly bioethanol production due to their high bioconversion costs. Adequate developed technology is available to achieve sustainable sugarcane production and bioethanol. However, the adoption of new technologies requires a favorable economic and political environment that facilitates investments in clean technologies. Pollution problems require strict enforcement of legislation and inspection of agricultural and industrial activities.

Developing the sugarcane ethanol provides a novel option for utilization of the sugar industry, and it will be also helpful to the fuel ethanol development in China.

7. Acknowledgements

This research work was supported by Ministry of Science and Technology of China, (NCSTE-2006-JKZX-023), Guangdong Science and Technology Department (2010A010500005) and the Natural Science Foundation of Guangdong Province (10451031601006220). We specially thank Guangdong Key Laboratory of Sugarcane Improvement and Biorefinery, Guangzhou Sugarcane Industry Research Institute for supporting this research.

8. References

Alexander, A.G. (1984). Energy cane as a multiple products alternative. Proceedings Pacific Basin Biofuels Workshop, Honolulu

Alexander, A.G. (1997). Production of energy sugarcane. *Sugar Journal*, 1, 5-79

Aquarone, E. (1960). Penicillin and tetracycline as contamination control agents in alcoholic fermentation of sugarcane molasses. *Appl Microbiol*, 8, 263–268

Bardi, E. P. & Koutinas, A. A. (1994). Immobilization of yeast on delignified cellulosic material for room and low-temperature wine making. *Journal of Agricultural and Food Chemistry*, 42, 221–226.

BNDES. Banco Nacional de Desenvolvimento Econômico e Social: Sugarcane-based bioethanol: energy for sustainable development / coordination BNDES and CGEE - Rio de Janeiro: BNDES, 2008304 p.

BNDES; CGEE (Orgs.). (2008). Sugarcane-based bioethanol: energy for sustainable development. Rio de Janeiro: BNDES, 316 p.

Cakar, Z. P., Seker, U. O., Tamerler, C. et al. (2005). Evolutionary engineering of multiple-stress resistant *Saccharomyces cerevisiae*. *FEMS Yeast Research*, 5, 569–578

Cerri, C.C., Maia, S.M.F., Galdos, M.V., Cerri, C.E.P., Feigl, B.J. & Bernoux, M.k. (2009) Brazilian greenhouse gas emissions: the importance of agriculture and livestoc. *Scientia Agricola*. 66, 6,831–843.

Chandel, A. K., Chan E.S., Rudravaram, R. et al. (2007). Economics and environmental impact of bioethanol production technologies: an appraisal. *Biotechnology and Molecular Biology Review*. 2, 1, 14-32

Cheng, J.F., Liu, J.H., Shao, H.B. & Qiu, Y.M. (2007). Continuous secondary fermentation of beer by yeast immobilized on the foam ceramic. *Research journal of biotechnology*, 2, 3, 40

Corton, E., Piuri, M., Battaglini, F.& Ruzal, S.M. (2000). Characterization of *Lactobacillus* carbohydrate fermentation activity using immobilized cells technique. *Biotechnology Progress*,16,1,59-63

Day, W.H., Serjak, W.C., Stratton, J.R. & Stone, L. (1954). Antibiotics as contamination control agents in grain alcohol fermentations. *J Agric Food Chem*, 2, 252–258

Dien, B.S., Jung, H.J.G., Vogel, K.P., Casler, M.D., Lamb, J.A.F.S., Iten, L., Mitchell, R.B. & Sarath, G. (2006). Chemical composition and response to dilute acid pretreatment and enzymatic saccharification of alfalfa, reed canary grass and switch grass. *Biomass and Bioenergy*, 30, 10, 880-891

Gasch, A. P., Spellman, P. T., Kao, C.M. et al. (2000) Genomic expression programs in the response of yeast cells to environmental changes. *Mol Biol Cell*, 11, 4241-4257

Goldemberg, J., Coelho, S.T. & Guardabassi, P. (2008). The sustainability of ethanol production from sugarcane. *Energy Policy*, 36, 2086– 2097

Godia, F., Casas, C. & Sola, C. (1987). A survey of continuous ethanol fermentation systems using immobilized cells. *Process Biochem*, 22, 43–48

Hamdy, M.K., Toledo, R.T., Shieh, C.J., Pfannenstiel, M.A. & Wang, R. (1996). Effects of virginiamycin on fermentation rate by yeast. *Biomass Bioenerg*,11,1–9

Herrera, S. (2004). Industrial biotechnology- a chance at redemption. *Nature Biotechnol.* 22, 671-675

Hynes, S.H., Kjarsgaard, D.M., Thomas, K.C. & Ingledew, W.M. (1997). Use of virginiamycin to control the growth of lactic acid bacteria during alcoholic fermentation. *J Ind Microbiol Biotechnol*, 18, 284–291

Giamalva, M. J., Clarke S. J. & Stein, J.M. (1984). Sugarcane hybrids of biomass. Biomass, 6, 61-68

Giudici, P., Solieri, L., Andrea, M., Pulvirenti et al. (2005). Strategies and perspectives for genetic improvement of wine yeasts. *Appl Microbiol Biotechnol*, 66, 622–628

International Energy Agency. (2004). Biofuels for transport—an international perspective. Paris: International Energy Agency.

Islam, M., Toledo, R. & Hamdy, M.K. (1999). Stability of virginiamycin and penicillin during alcohol fermentation. *Biomass Bioenergy*, 17, 369–376

Kiran Sree, N., Sridhar, M. & Venkateswar Rao, L.(2000). High alcohol production by repeated batch fermentation using an immobilized osmotolerant Saccharomyces cerevisiae. *Journal of Industrial Microbiology & Biotechnology*, 24, 222–226

Kopsahelis, N. (2006). Comparative study of spent grains and delignified spent grains as yeast supports for alcohol production from molasses. *Bioresour.Technol*, doi:10.1016/j.biortech.2006.03.030

Kourkoutas, Y., Kanellski, M. & Koutinas, A.A. (2006). Apple pieces as immobilization support of various microorganisms. *LWT*, 39, 980-986

Lynd, L.R. & Wang, M.Q. (2004). A product-nonspecific framework for evaluating the potential of biomass-based products to displace fossil fuels. *J. Ind. Ecol*, 7, 17-32

Liang, L., Zhang, Y., Liang S., et al. (2008). Study of sugarcane pieces as yeast supports for ethanol production from sugarcane juice and molasses. *Journal of Industrial Microbiology and Biotechnology*, 35,1605-1613

Macedo, I.C., Seabra, J.E.A. & Silva, J.E.A.R. (2008). Greenhouse gases emissions in the produciton and use of ethanol from sugarcane: the 2005/2006 averages and prediciton for 2020. *Biomass and Bioenergy*, 2008, 32,582–595

Marèlne Cot, M., Loret, M.O., Francois, J. et al. (2007). Physiological behavior of *saccharomyces cerevisiae* in aerated fed-batch fermentation for high level production of bioethanol. *FEMS Yeast Res*, 7,22-32

Najafpour, G.D. (1990). Immobilization of microbial cells for production of organic acids. *J Sci Islam Repub Iran,* 1, 172–176

Neelakantam V. Narendranath Ronan Power. (2004). Effect of yeast inoculation rate on the metabolism of contaminating lactobacilli during fermentation of corn mash. *J Ind Microbiol Biotechnol* ,31, 581–58

Oliveira de, M.E.D., Vaughan, B.E. & Edward. (2005) Ethanol as Fuel: Energy, Carbon Dioxide Balances, and Ecological Footprint. *BioScience,*55, 57

Rothkopf, G. (2007). A Blueprint for Green Energy in the Americas. Inter-American Development Bank Retrieved 2008-08-22. See chapters Introduction (pp. 339–444) and Pillar I: Innovation (pp. 445–482)

Skinner, K.A. & Leathers, T.D. (2004). Bacterial contaminants of fuel ethanol production. *J Ind Microbiol Biotechnol*, 31, 401–408

S.plessas, A. B. (2007). Use of Saccharomyces Cerevisiae Cells Immobilized on Orange Peel as Biocatalyst for Alcoholic Fermentation. *Bioresource Technology,* 98, 860-865

Stroppa, C.T., Andrietta, M.G.S., Andrietta, S.R., Steckelberg, C. & Serra, G.E. (2000). Use of penicillin and monensin to control bacterial contamination of Brazilian alcohol fermentations. *Int Sugar J*, 102,78–82

Tanaka, L. (2006). Ethanol fermentation from biomass resources: Current state and prospects. *Appl. Microbiol. Biotechnol*, 69, 627-642

Walter, A., Dolzan, P., Quilodrán, O. et al. (2008).A sustainability analysis of the brazilian ethanol. Campinas

Watanabe, M. (2009). Ethanol Production in Brazil: Bridging its Economic and Environmental Aspects. International Association for Energy Economics. Brazil

Wheals, E.A., Basso, L.C., Alves, D.M.G. & Amorim, H.V. (1999). Fuel ethanol after 25 years. *Trends Biotechnol*, 17, 482–487

Wyman, C.E. & Hinman, N.D. (1990). Ethanol. Fundamentals of production from renewable feedstocks and use as transportation fuel. *Appl Biochem. Biotechnol*, 24/25, 735-75.

Xu, Y., Wu, Q., Lei, Y. & Yao, F. (2010). Creep behavior of bagasse fiber reinforced polymer composites. *Bioresource Technology*, 101, 3280-3286

Zhang, Y.X., Perry, K., Vinci, V. A. et al. (2002). Genome Shuffling Leads to Rapid Phenotypic Improvement in Bacteria. *Nature*, 415, 644-646

Zhang, M.Q., Chen, R.K., Luo, J, et al. (2000) Analyses for inheritance and combining ability of photochemical activities measured by chlorophyll fluorescence in the segregating generation of sugarcane. *Field Crops Res*, 65, 31-39

Zhong, C., Cao, Y.X., Li, B.Z. & Yuan, Y.J. (2010). Biofuels in China: past, present and future. *Biofuels Bioproducts and Biorefining*, 4, 3, 326–342

Part 2

Second Generation Bioethanol Production (Lignocellulosic Raw-Material)

5

Hydrolysis of Lignocellulosic Biomass: Current Status of Processes and Technologies and Future Perspectives

Alessandra Verardi[1], Isabella De Bari[2*],
Emanuele Ricca[1] and Vincenza Calabrò[1]
[1]*Department of Engineering Modeling, University of Calabria, Rende (CS)*
[2]*ENEA Italian National Agency for New Technologies, Energy and the Sustainable Economical Development, Rotondella (MT)*
Italy

1. Introduction

Bioethanol can be produced from several different biomass feedstocks: sucrose rich feedstocks (e.g. sugar-cane), starchy materials (e.g. corn grain), and **lignocellulosic biomass.** This last category, including biomass such as corn stover and wheat straw, woody residues from forest thinning and paper, is promising especially in those countries with limited lands availability. In fact, residues are often widely available and do not compete with food production in terms of land destination. The process converting the biomass biopolymers to fermentable sugars is called **hydrolysis**. There are two major categories of methods employed. The first and older method uses acids as catalysts, while the second uses enzymes called **cellulases**. Feedstock **pretreatment** has been recognized as a necessary upstream process to remove lignin and enhance the porosity of the lignocellulosic materials prior to the enzymatic process (Zhu & Pan, 2010; Kumar et al., 2009).

Cellulases are proteins that have been conventionally divided into three major groups: **endoglucanase**, which attacks low cristallinity regions in the cellulose fibers by endoaction, creating free chain-ends; **exoglucanases** or cellobiohydrolases which hydrolyze the 1, 4-glycocidyl linkages to form cellobiose; and **β-glucosidase** which converts cello-oligosaccharides and disaccharide cellobiose into glucose residues. In addition to the three major groups of cellulose enzymes, there are also a number of other enzymes that attack hemicelluloses, such as **glucoronide**, **acetylesterase**, **xylanase**, **β-xylosidase**, **galactomannase** and **glucomannase**. These enzymes work together synergistically to attack cellulose and hemicellulose. Cellulases are produced by various **bacteria and fungi** that can have cellulolytic mechanisms significantly different.

The use of enzymes in the hydrolysis of cellulose is more effective than the use of inorganic catalysts, because enzymes are highly specific and can work at mild process conditions. In spite of these advantages, the use of **enzymes in industrial processes** is still limited by

---

* Corresponding Author

several factors: most enzymes are relatively unstable at high temperatures, the costs of enzyme isolation and purification are high and it is quite difficult to recover them from the reaction mixtures. Currently, extensive research is being carried out on cellulases with **improved thermostability**. These enzymes have high specific activity and increased flexibility. For these reasons they could work at low dosages and the higher working temperatures could speed up the hydrolysis reaction time. As consequence, the overall process costs could be reduced. Thermostable enzymes could play an important role in assisting the liquefaction of concentrated biomass suspensions necessary to achieve ethanol concentrations in the range 4-5 wt%.

The immobilization of enzymes has also been proposed to remove some limitations in the enzymatic process (Hong et al., 2008). The main advantage is an easier recovery and reuse of the catalysts for more reaction loops. Also, enzyme immobilization frequently results in improved thermostability or resistance to shear inactivation and so, in general, it can help to extend the enzymes lifetime.

This chapter contains an overview of the lignocellulosic hydrolysis process. Several process issues will be deepened: cellulase enzyme systems and hydrolysis mechanisms of cellulose; commercial mixtures; currents limits in the cellulose hydrolysis; innovative bioprocesses and improved biocatalysts.

2. Structure of lignocellulose biomass

Lignocellulosic biomass is typically nonedible plant material, including dedicated crops of wood and grass, and agro-forest residues. Lignocellulosics are mainly composed of **cellulose, hemicellulose, and lignin**.

Cellulose is a homopolysaccharide composed of β-D-pyranose units, linked by β-1, 4-glycosidic bonds. Cellobiose is the smallest repetitive unit and it is formed by two glucose monomers. The long-chain cellulose polymers are packed together **into microfibrils** by hydrogen and van der Waals bonds. Hemicellulose and lignin cover the microfibils (Fig.1). Hemicellulose is a mixture of polysaccharides, including pentoses, hexoses and uronic acids. Lignin is the most complex natural polymer consisting of a predominant building block of phenylpropane units. More specifically, p-coumaryl alcohol, coniferyl alcohol and sinapyl alcohol are the most commonly encountered alcohols (Harmesen et al., 2010). Lignocellulosic materials also contain small amounts of pectin, proteins, extractives (i.e. no- structural sugars, nitrogenous material, chlorophyll and waxes) and ash (Kumar et al., 2009).

The composition of the biomass constituents can vary greatly among various sources (Table 1). Accurate measurements of the biomass constituents, mainly lignin and carbohydrates, are of prime importance because they assist tailored process designs for the maximum recovery of energy and products from the raw materials.

Since 1900, researchers have developed several methods to measure the lignin and carbohydrates content of lignocellulosic biomass. Globally recognized Organizations, such as American Society for Testing and Materials (ASTM), Technical Association of the Pulp and Paper Industry (TAPPI) and National Renewable energy and Laboratory (NREL) have developed methods to determine the chemical composition of biomass, based on modifications of the two main procedures developed by Ritter (Ritter et al., 1932) and by Seaman (Saeman et al., 1954), (Table 2).

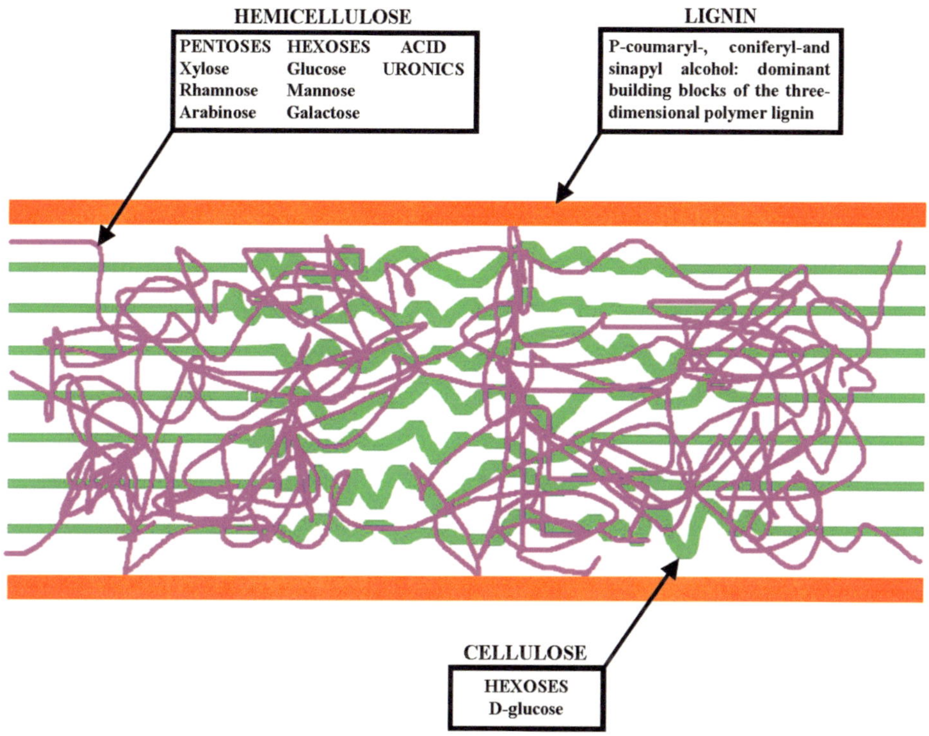

Fig. 1. Lignocellulosic materials: composition of major compounds (Kumar, 2009)

Lignocellulosic materials	Cellulose (%)	Hemicellulose (%)	Lignin (%)
Coastal bermudagrass	25	35.7	6.4
Corn Cobs	45	35	15
Cotton seed hairs	80-95	5-20	0
Grasses	25-40	35-50	10-30
Hardwoods steam	40-55	24-40	18-25
Leaves	15-20	80-85	0
Newspaper	40-55	25-40	18-30
Nut shells	25-30	25-30	30-40
Paper	85-99	0	0-15
Primary wastewater solids	8-15	NA	24-29
Softwoods stems	45-50	25-35	25-35
Solid cattle manure	1.6-4.7	1.4-3.3	2.7-5.7
Sorted refuse	60	20	20
Swine waste	6.0	28	NA
Switchgrass	45	31.4	12.0
Waste papers from chemical pulps	60-70	10-20	5-10
Wheat straw	30	50	15

Table 1. Composition of some common sources of biomass (Sun and Cheng, 2002)

TAPPI		ASTM		NREL
Method	Title	Method	Title	Title
T 13 os 54; Later T222 om-06	Lignin in Wood (original) Acid- Insoluble Lignin in Wood and Pulp (later)	D 1106-96 (2007)	Standard Test Method for Cromatographic Analysis of Chemically Refined Cellulose (1996)	Determination of Structural Carbohydrates and Lignin in Biomass
T249 cm-00	Carboydrate Composition of Extractive -Free Wood and Wood Pulp by Gas-Liquid Chromatography	ASTM D1915-63 (1989) withdrawn, replaced by D5896	Standard Test Method for Chromatographic Analysis of Chemically Refined Cellulose (1996)	
		AST D5896-96	Standard Test Method for Carbohydrate Distribution of Cellulosic Material	
		E1721	Standard Test Method for Determination of Acid-Insoluble Residue in Biomass	
		E1758	Determination of Carbohydrates in Biomass by High Performance Liquid Chromatography	

Table 2. Methods provided by globally recognized organizations for the chemical composition of biomass (Sluiter et al., 2010)

3. Products from lignocellulosic biomass

Lignocellulosic biomass is a potential source of several bio-based products according to the **biorefinery** approach. Currently, the products made from bioresources represent only a minor fraction of the chemical industry production. However, the interest in the bio-based products has increased because of the rapidly rising barrel costs and an increasing concern about the depletion of the fossil resources in the near future (Hatti-Kaul et al., 2007). The goal of the biorefinery approach is the generation of energy and chemicals from different biomass feedstocks, through the combination of different technologies (FitzPatrick et al., 2010).

The biorefinery scheme involves a multi-step biomass processing. The first step concerns the feedstock pretreatment through physical, biological, and chemical methods. The outputs from this step are **platform (macro) molecules or streams** that can be used for further processing (Cherubini & Ulgiati, 2010). Recently, a detailed report has been published by

DOE describing the value added chemicals that can be produced from biomass (Werpy, 2004). Figure 2 displays a general biorefinery scheme for the production of specialty polymers, fuel, or composite materials (FitzPatrick et al., 2010).
Besides ethanol, several other products can be obtained following the hydrolysis of the carbohydrates in the lignocellulosic materials. For instance, xylan/xylose contained in hemicelluloses can be thermally transformed into furans (2-furfuraldeyde, hydroxymethil furfural), short chain organic acids (formic, acetic, and propionic acids), and cheto compounds (hydroxy-1-propanone, hydroxy-1-butanone) (Güllü, 2010; Bozell & Petersen, 2010).

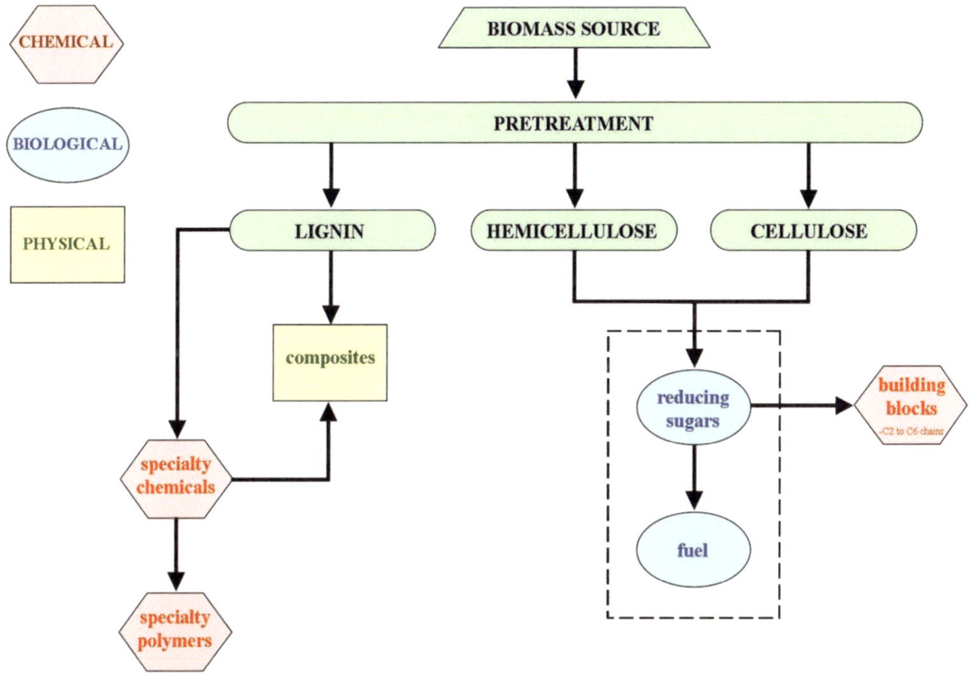

Fig. 2. Scheme of a lignocellulosic biorefinery. The shape of each step describes the type of process used, chemical, biological, and physical (legend) (FitzPatrick et al., 2010)

Furfural can be further processed to form some building blocks of innovative polymeric materials (i.e. 2, 5-furandicarboxylic acid). In addition, levulinic acid could be formed by the degradation of hydroxymethil furfural (Demirabas, 2008). Another product prepared either by fermentation or by catalytic hydrogenation of xylose is xylitol (Bozell & Petersen, 2010). Furthermore, through the chemical reduction of glucose it is possible to obtain several products, such as sorbitol (Bozell & Petersen, 2010). The residual lignin can be an intermediate product to be used for the synthesis of phenol, benzene, toluene, xylene, and other aromatics. Similarly to furfural, lignin could react to form some polymeric materials (i.e. polyurethanes) (Demirabas, 2008).

4. Production for ethanol from lignocellulosic biomass

Ethanol is the most common renewable fuel recognized as a potential alternative to petroleum-derived transportation fuels. It can be produced from lignocellulosic materials in various ways characterized by common steps: hydrolysis of cellulose and hemicellulose to monomeric sugars, fermentation and product recovery (fig 3). The main differences lie in the hydrolysis phase, which can be performed by dilute acid, concentrated acid or enzymatically (Galbe & Zacchi, 2002).

4.1 Acid hydrolysis

The main advantage of the acid hydrolysis is that acids can penetrate lignin without any preliminary pretreatment of biomass, thus breaking down the cellulose and hemicellulose polymers to form individual sugar molecules. Several types of acids, concentrated or diluted, can be used, such as sulphurous, sulphuric, hydrocloric, hydrofluoric, phosphoric, nitric and formic acid (Galbe & Zacchi, 2002). Sulphuric and hydrochloric acids are the most commonly used catalysts for hydrolysis of lignocellulosic biomass (Lenihan et al., 2010).
The acid concentration used in the **concentrated acid hydrolysis** process is in the range of 10-30%. The process occurs at low temperatures, producing high hydrolysis yields of cellulose (i.e. 90% of theoretical glucose yield) (Iranmahboob et al., 2002).

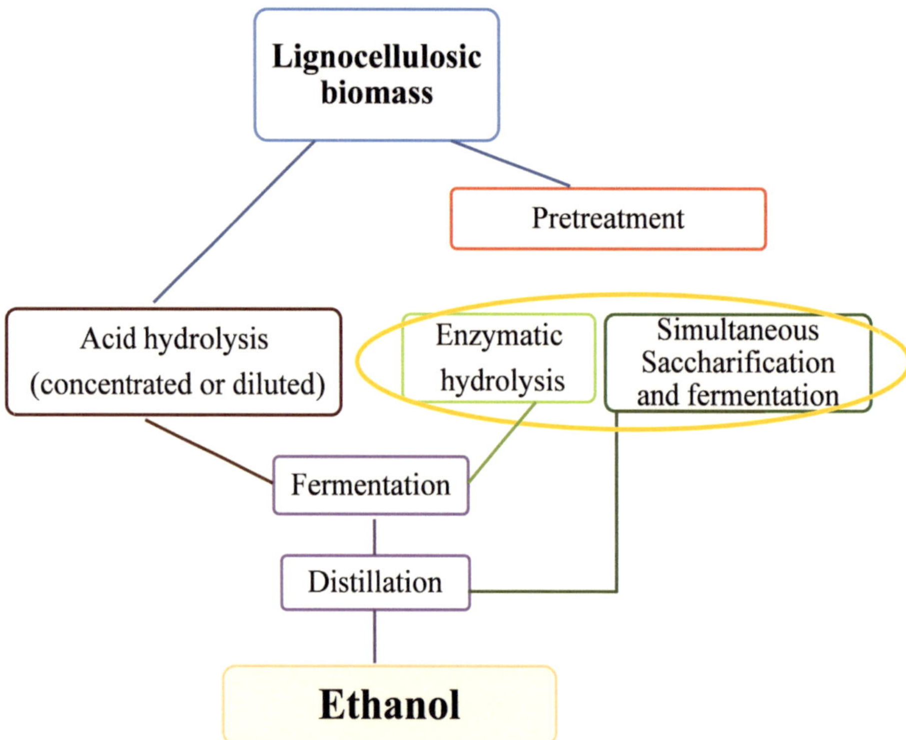

Fig. 3. Process for production ethanol from lignocellulosic biomass. The circle in the scheme indicates two alternative process routes: simultaneous hydrolysis and fermentation (SSF); separate hydrolysis and fermentation (SHF).

However, this process requires large amounts of acids causing corrosion problems to the equipments. The main advantage of the **dilute hydrolysis process** is the low amount of acid required (2-5%). However this process is carried out at high temperatures to achieve acceptable rates of cellulose conversion. The high temperature increases the rates of hemicellulose sugars decomposition thus causing the formation of toxic compounds such as furfural and 5-hydroxymethyl-furfural (HMF). These compounds inhibit yeast cells and the subsequent fermentation stage, causing a lower ethanol production rate (Larsson et al., 1999; kootstra et al., 2009). In addition, these compounds lead to reduction of fermentable sugars (Kootstra et al., 2009). In addition, high temperatures increase the equipment corrosion (Jones & Semrau, 1984).

In 1999, the BC International (BCI) of United States has marketed a technology based on two-step dilute acid hydrolysis: the first hydrolysis stage at mild conditions (170-190°C) to hydrolyze hemicellulose; the second step at more severe conditions to hydrolyze cellulose 200-230°C (Wyman, 1999).

In 1991, the Swedish Ethanol Development Foundation developed the CASH process. This is a two-stage dilute acid process that provides the impregnation of biomass with sulphur dioxide followed by a second step in which diluted hydrochloric acid is used. In 1995, this foundation has focused researches on the conversion of softwoods using sulphuric acid (Galbe & Zacchi, 2002).

4.2 Pretreatment

A pretreatment step is necessary for the enzymatic hydrolysis process. It is able to remove the lignin layer and to decristallize celllulose so that the hydrolytic enzymes can easily access the biopolymers.The pretreatment is a critical step in the cellulosic bioethanol technology because it affects the quality and the cost of the carbohydrates containing streams (Balat et al., 2008). Pretreatments methods can be classified into different categories: physical, physiochemical, chemical, biological, electrical, or a combination of these (kumar et al., 2009), (Table 3).

On the whole, the final yield of the enzymatic process depends on the combination of several factors: biomass composition, type of pretreatment, dosage and efficiency of the hydrolytic enzymes (Alvira et al., 2010).

The use of enzymes in the hydrolysis of cellulose is more advantageous than use of chemicals, because enzymes are highly specific and can work at mild process conditions. Despite these advantages, the use of enzymes in industrial applications is still limited by several factors: the costs of enzymes isolation and purification are high; the specific activity of enzyme is low compared to the corresponding starch degrading enzymes. As consequence, the process yields increase at raising the enzymatic proteins dosage and the hydrolysis time (up to 4 days) while, on the contrary, decrease at raising the solids loadings. One typical index used to evaluate the performances of the cellulase preparations during the enzymatic hydrolysis is the **conversion rate** to say the obtained glucose concentration per time required to achieve it (g glucose/L/h/). Some authors reported conversion rates of softwoods substrates (5%w/v solids loading) in the range 0.3-1.2 g/L/h (Berlin et al., 2007). In general, compromise conditions are necessary between enzymes dosages and process time to contain the process costs.

In 2001, the cost to produce cellulase enzymes was 3-5$ per gallon of ethanol (0.8-1.32$/liter ethanol), (Novozymes and NREL)[1]. In order to reduce the cost of cellulases for bioethanol production, in 2000 the National Renewable Laboratory (NREL) of USA has started collaborations with Genencor Corporation and Novozymes. In particular, in 2004, Genencor has achieved an estimated cellulase cost in the range $0.10-0.20 per gallon of ethanol (0.03-

[1] News on: *Sci Focus Direct on Catalysts*, 2005

		Operating conditions	Advantages	Disadvantages
Physical	Chipping Grinding Milling	Room temperature Energy input < 30Kw per ton biomass	Reduces cellulose critallinity	Power consumption higher than inherent biomass energy
Physio-chemical	Steam pretreatment	160-260°C (0. 69-4.83MPa) for 5-15 min	Causes hemicellulose auto hydrolysis and lignin transformation; cost-effective for hardwoods and agricultural residues	Destruction of a portion of the xylan fraction; incomplete distruption of the lignin-carboydrate matrix; generation of inhibitory compounds; less effective for softwoods
	AFEX (Ammonia fiber explosion method)	90°C for 30 min.1-2kg ammonia /kg dry biomass	Increases accessible surface area, removes lignin and hemicellulose;	Do not modify lignin neither hydrolyzes hemicellulose;
	ARP (Ammonia recycle percolation method)	150-170°C for 14 min Fluid velocity 1cm/min	Increases accessible surface area, removes lignin and hemicellulose;	Do not modify lignin neither hydrolyzes hemicellulose;
	CO_2 explosion	4kg CO_2/kg fiber at 5.62 Mpa 160 bar for 90 min at 50 °C under supercritical carbon dioxide	Do not produce inhibitors for downstream processes. Increases accessible surface area, does not cause formation of inhibitory compounds	It is not suitable for biomass with high lignin content (such as woods and nut shells) Does not modify lignin neither hydrolyze hemicelluloses
	Ozonolysis	Room temperature	Reduce lignin content; does not produce toxic residues	Expensive for the ozone required;
	Wet oxidation	148-200°C for 30 min	Efficient removal of lignin; low formation of inhibitors; low energy demand	High cost of oxygen and alkaline catalyst
Chemical	Acid hydrolysis: dilute-acid pretreatment	Type I: T>160°, continuous-flow process for low solid loading 5-10%,)- Type II: T<160°C, batch process for high solid loadings (10-40%)	Hydrolyzes hemicellulose to xylose and other sugar; alters lignin structure	Equipment corrosion; formation of toxic substances

		Operating conditions	Advantages	Disadvantages
	Alkaline hydrolysis	Low temperature; Long time high. Concentration of the base; For soybean straw: ammonia liquor (10%) for 24 h at room temperature	Removes hemicelluloses and lignin; increases accessible surface area	Residual salts in biomass
	Organosolv	150-200 °C with or without addition of catalysts (oxalic, salicylic, acetylsalicylic acid)	Hydrolyzes lignin and hemicelluloses	High costs due to the solvents recovery
Biological		Several fungi (brown-, white- and soft-rot fungi	Degrades lignin and hemicelluloses; low energy requirements	Slow hydrolysis rates
Electrical	Pulsed electrical field in the range of 5-20 kV/cm,	~2000 pulses of 8 kV/cm	Ambient conditions; disrupts plant cells; simple equipment	Process needs more research

Table 3. Methods for biomass lignocellulosic pretreatment (Kumar et al., 2009)

0.05$/liter ethanol) in NREL´s cost model (Genencor, 2004)[2]. Similarly, collaboration between Novozymes and NREL has yielded a cost reduction in the range $0.10-0.18 per gallon of ethanol (0.03-0.047$/liter ethanol), a 30-fold reduction since 2001 (Mathew et al., 2008).

Unlike the acid hydrolysis, the enzymatic hydrolysis, still has not reached the industrial scale. Only few plants are available worldwide to investigate the process (pretreatment and bioconversion) at demo scale. More recently, the steam explosion pretreatment, investigated for several years in Italy at the ENEA research Center of Trisaia (De Bari et al., 2002, 2007), is now going to be developed at industrial scale thanks to investments from the Italian Mossi & Ghisolfi Group.

5. Enzymatic hydrolysis: Cellulases

5.1 Cellulolytic capability of organisms: Difference in the cellulose-degrading strategy

Different strategies for the cellulose degradation are used by the cellulase-producing microorganisms: aerobic bacteria and fungi secrete soluble extracellular enzymes known as *non complexed cellulase system*; anaerobic cellulolytic microorganisms produce *complexed cellulase systems*, called *cellulosomes* (Sun et al., 2002). A third strategy was proposed to explain the cellulose-degrading action of two recently discovered bacteria: the aerobic *Cytophaga hutchinsonii* and the anaerobic *Fibrobacter succinogenes* (Ilmén et al., 1997).

[2] Genencor, relations, 21 October 2004, avaible from: http:/genencor.com/cms/connect/genencor/media_relations/news/archive/2004/gen_211004_en.htm

- *Non-complexed cellulase system*. One of the most fully investigated non-*complexed cellulase system* is the *Trichoderma reesei* model. *T. reesei* (teleomorph *Hypocrea jecorina*) is a saprobic fungus, known as an efficient producer of extracellular enzymes (Bayer et al., 1998). Its non-complexed cellulase system includes two cellobiohydrolases, at least seven endoglucanases, and several β-glucosidases. However, in *T. reesei* cellulases, the amount of ß-glucosidase is lower than that needed for the efficient hydrolysis of cellulose into glucose. As a result, the major product of hydrolysis is cellobiose. This is a dimer of glucose with strong inhibition toward endo- and exoglucanases so that the accumulation of cellobiose significantly slows down the hydrolysis process (Gilkes et al., 1991). By adding ß-glucosidase to cellulases from either external sources, or by using co-culture systems, the inhibitory effect of cellobiose can be significantly reduced (Ting et al., 2009).
  It has been observed that the mechanism of cellulose enzymatic hydrolysis by *T.reesei* involves three simultaneous processes (Ting et al., 2009):
  1. Chemical and physical changes in the cellulose solid phase. The chemical stage includes changes in the degree of polymerization, while the physical changes regard all the modifications in the accessible surface area. The enzymes specific function involved in this step is the *endoglucanase.*
  2. Primary hydrolysis. This process is slow and involves the release of soluble intermediates from the cellulose surface. The activity involved in this step is the *cellobiohydrolase.*
  3. Secondary hydrolysis. This process involves the further hydrolysis of the soluble fractions to lower molecular weight intermediates, and ultimately to glucose. This step is much faster than the primary hydrolysis and *β-glucosidases* play a role for the secondary hydrolysis.
- *Complexed cellulase system.* Cellulosomes are produced mainly by anaerobic bacteria, but their presence have also been described in a few anaerobic fungi from species such as *Neocallimastix*, *Piromyces*, and *Orpinomyces* (Tatsumi et al., 2006; Watanabe & Tokuda, 2010). In the domain Bacteria, organisms possessing cellulosomes are only found in the phylum *Firmicutes*, class *Clostridia*, order *Clostridiales* and in the *Lachnospiraceae* and *Clostridiaceae* families. In this latter family, bacteria with cellulosomes are found in various clusters of the genus Clostridium (McCarter & Whiters, 1994; Wilson, 2008).
  Cellulosomes are protuberances produced on the cell wall of the cellulolytic bacteria grown on cellulosic materials. These protuberances are stable enzyme complexes tightly bound to the bacteria cell wall but flexible enough to bind strongly to cellulose (Lentig & Warmoeskerken, 2001). A cellulosome contains two types of subunits: *non-catalytic subunits*, called *scaffoldins*, and *enzymatic subunits*. The scaffoldin is a functional unit of cellusome, which contain multiple copies of *cohesins* that interact selectively with domains of the *enzymatic subunits*, CBD (cellulose binding domains) and CBM (carbohydrates binding modules). These have complementary cohesins, called dockerins, which are specific for each bacterial species (Fig. 4) (Gilligan & Reese, 1954; Lynd et al., 2002; Arai et al., 2006;).
  For the bacterial cell, the biosynthesis of a cellulosome enables a specific adhesion to the substrate of interest without competition with other microorganisms. The cellulosome allows several advantages: (1) synergism of the cellulases; (2) absence of unspecific adsorption (McCarter & Whiters, 1994; Zhang & Lynd, 2004). Thanks to its intrinsic Lego-like architecture, cellulosomes may provide great potential in the biofuel industry.

The concept of cellulosome was firstly discovered in the thermophilic cellulolytic and anaerobic bacterium, *Clostridium thermocellum* (Wyman, 1996). It consists of a large number of proteins, including several cellulases and hemicellulases. Other enzymes that can be included in the cellulosome are lichenases.

- *Third cellulose-degrading strategy*. The third strategy was recently proposed to explain the cellulose-degrading behavior of two recently sequenced bacteria: *Cytophaga hutchinsonii* and *Fibrobacter succinogenes* (Ilmén, 1997). *C. hutchinsonii* is an abundant aerobic cellulolytic soil bacterium (Fägerstam & Petterson, 1984), while *F. succinogenes* is an anaerobic rumen bacterium which was isolated by the Rockville, (Maryland), and San

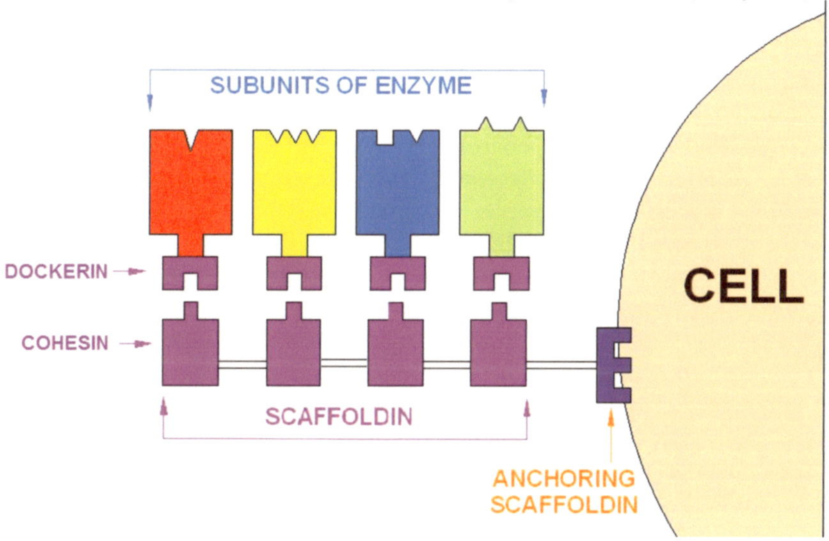

Fig. 4. Schematic representation of a cellulosoma

Diego (California) Institute of Genomic Research (TIGR) (Mansfield et al., 1998). In the aerobic *C. hutchinsonii* no genes were found to code for CBM and in the anaerobic *F. succinogenes* no genes were identified to encode dockerin and scaffoldin. Thus, a third cellulose degrading mechanism was proposed. It includes the binding of individual cellulose molecules by outer membrane proteins of the microrganisms followed by the transport into the periplasmic space where they are degraded by endoglucanases (Ilmén, 1997).

5.2 Characteristics of the commercial hydrolytic enzymes

Most cellulase enzymes are relatively unstable at high temperatures. The maximum activity for most fungal cellulases and β-glucosidase occurs at 50±5°C and a pH 4.5- 5 (Taherzadeh & Karimi, 2007; Galbe & Zacchi, 2002). Usually, they lose about 60% of their activity in the temperature range 50–60 °C and almost completely lose activity at 80°C (Gautam et al., 2010). However, the enzymes activity depends on the hydrolysis duration and on the source of the enzymes (Tengborg et al., 2001). In general, cellulases are quite difficult to use for prolonged operations.

As mentioned before, the enzyme production costs mainly depend on the productivity of the enzymes-producing microbial strain. Filamentous fungi are the major source of cellulases and mutant strains of *Trichoderma* (*T. viride, T. reesei, T. longibrachiatum*) have long been considered to be the most productive (Gusakov et al., 2007; Galbe & Zacchi, 2002). Preparations of cellulases from a single organism may not be highly efficient for the hydrolysis of different feedstocks. For example, *Thrichoderma reesei* produces endoglucanases and exoglucanases in large quantities, but its β-glucosidase activity is low, resulting in an inefficient biomass hydrolysis. For this reason, the goal of the enzymes producing companies has been to form cellulases cocktails by enzymes assembly (multienzyme mixtures) or to construct engineered microrganisms to express the desired mixtures (Mathew et al., 2008). Enzyme mixtures often derive from the co-fermentation of several micro-organisms (Ahamed & Vermette, 2008; Kabel et al., 2005; Berlin et al., 2007), (Table 4). All the commercial cellulases listed in table 4 have an optimal condition at 50°C and pH of 4.0-5.0. More recently, some enzymes producers have marked new mixtures able to work in a higher temperature ranging from 50 to 60°C (Table5).

In 2010, new enzymes were produced by two leading companies, Novozymes and Genencor, supported by the USA Department of Energy (DOE). Genencor has launched four new blends: **Accelerase®1500, Accelerase®XP, Accelerase®XC and Accelerase®BG**. Accelerase®1500 is a cellulases complex (exoglucanase, endoglucanase, hemi-cellulase and β-glucosidase) produced from a genetically modified strain of *T. reesei.* All the other Accelerase are accessory enzymes complexes: Accelerase®XP enhances both xylan and glucan conversion; Accelerase®XC contains hemicellulose and cellulase activities; Accelerase® BG is a β-glucosidase enzyme. In February 2010, Genencor has developed an enzyme complex known as **Accellerase®Duet** which is produced with a genetically modified strain of *T. reesei* and that contains not only exoglucanase, endoglucanase, β-glucosidase, but includes also xylanase. This product is capable of hydrolyzing lignocellulosic biomass into fermentable monosaccharides such as glucose and xylose (Genencos, 2010)[3]. Similarly, Novozymes has produced and commercialized two new enzymatic mixtures: **cellic Ctec**, and **cellic Htec**. **Cellic CTec** is used in combination with Cellic HTec and this mixture is capable to work with a wide variety of pretreated feedstocks, such as sugarcane bagasse, corn cob, corn fiber, and wood pulp, for the conversion of the carbohydrates in these materials into simple sugars (Novozyme, 2010)[4].

In order to meet the future challenges, innovative bioprocesses for the production of new generation of enzymes are needed. As already described, conventional cellulases work within a range of temperature around 50°C and they are typically inactivated at temperatures above 60-70 °C due to disorganization of their three dimensional structures followed by an irreversible denaturation (Viikari et al., 2007). Some opportunities of process improvement derive from the use of thermostable enzymes.

5.3 Enzymes for the cellulose liquefaction: Thermophilic enzymes

The thermophilic microrganisms can be grouped in thermophiles (growth up to 60 °C), extreme thermophiles (65-80 °C) and hyperthermophiles (85-110 °C). The unique stability of the enzymes produced by these microrganisms at elevated temperatures, extreme pH and high pressure (up to 1000 bar) makes them a valuable resource for the industrial

---

[3] Genencor, products, 14 January 2010, avaible from: http:// www.genencor.com/ wps/ wcm/ connect/ genencor/ genencor/ products and services/ business development/ biorefineries/ products/ accellerase product line en.htm

[4] Novozyme, brochure, 29 January 2010, Viable from: http:// www.bioenergy. novozymes.com/ files/ documents/ Final%20Cellic%20Product%20Brochure_ 29Jan2010.pdf

Commercial mixture	FPU (U/ml)[a]	Cellobiase (U/ml)[b]	Proteins (U/ml)[c]	Source	Supplier
Bio-feed beta L	<5	12	8	*T. longibrachiatum* *T. reesei*	Novozymes (Bagsvaerd, Denmark)
Cellubrix (Celluclast)	56	136	43	*T. longibrachiatum* *A. niger*	Novozymes
Cellulase 2000L	10	nd	7	*T. longibrachiatum* *T. reesei*	Rodhia -Danisco (Vinay, France)
Cellulyve 50L	24	nd	34	*T. longibrachiatum* *T. reesei*	Lyven (Colombelles France)
Energex L	<5	19	28	*T. longibrachiatum* *T. reesei*	Novozymes
GC220	116	215	64	*T. longibrachiatum* *T. reesei*	Genencor-Danisco (Rochester, USA)
GC440	<5	70	29	*T. longibrachiatum* *T. reesei*	Genencor
GC880	<5	86	43	*T. longibrachiatum* *T. reesei*	Genencor
Novozymes 188	<5	1,116	57	*A.niger*	Novozymes
Rohament CL	51	28	44	*T. longibrachiatum T. reesei*	Rhom-AB Enzymes (Rajamäki, Finland)
Spezyme CP	49	nd	41	*T. longibrachiatum* *T. reesei*	Genencor
Ultraflo L	<5	20	18	*T. longibrachiatum* *T. reesei*	Novozymes
Viscozyme L	<5	23	27	*T. longibrachiatum* *T. reesei*	Novozymes
Viscostar 150L	33	111	40	*T. longibrachiatum T. reesei*	Dyadic (Jupiter, Usa)

A) One FPU (filter paper unit) is the amount of enzyme that forms 1 μmol of reducing sugars/min during the hydrolysis reaction of filter paper Whatman No.1
B) One CBU (cellobiase unit) corresponds to the amount of enzyme which forms 2 μmol of glucose/min from cellobiose

Table 4. Commercial cellulases

Commercial mixture	B-glucosidase activity(U/ml)[a]	pH	Temperature (°C)	Source	Supplier
Biocellulase A	32	5	55	*A. niger*	Quest Intl. (Sarasota, Fl)
Cellulase AP 30 K	60	4.5	60	*A. niger*	Amano Enzyme Inc.

Table 5. Commercial cellulases able to work at temperature ranging from 50 to 60°C.

bioprocesses that run at harsh conditions (Demain et al., 2005). Of special interest is the thermoactivity and thermostability of these enzymes in the presence of high concentrations of organic solvents, detergents and alcohols. On the whole, thermophilic enzymes have an increased resistance to many denaturing conditions such as the use of detergents which can be often the unique efficient mean to obviate the irreversible adsorption of cellulases on the substrates. Furthermore, the utilization of high operation temperatures, which cause a decrease in viscosity and an increase in the diffusion coefficients of substrates, have a significant influence on the cellulose solubilization. It is worth noting that, differently from the mesophilic enzymes, most thermophilic cellulases did not show inhibition at high level of reaction products (e.g. cellobiose and glucose). As consequence, higher reaction rates and higher process yields are expected (Bergquist et al., 2004). The high process temperature also reduces any contamination of the fermentation medium.

Several cellulose degrading enzymes from various thermophilic organisms have been investigated. These include cellulases mainly isolated from anaerobic bacteria such as *Anaerocellum thermophilum* (Zverlov et al., 1998), *Clostridium thermocellum* (Romaniec et al., 1992), *Clostridium stercorarium* (Bronnenmeier et al., 1991; Bronnenmeier & Staudenbauer, 1990) and *Caldocellum saccharolyticum* (Te'o V et l., 1995), *Pyrococcus furiosus* (Ma & Adams, 1994), *Pyrococcus horikoshi* (Rahman et al.,1998), *Rhodothermus* strains (Hreggvidsson et al., 1996), *Thermotoga sp.*, (Ruttersmith et al., 1991), *Thermotoga marittima* (Bronnenmeier et al., 1995), *Thermotoga neapolitana* (Bok et al., 1998).

Xylanase have been detected in *Acidothermus cellulolyticus* in different *Thermus, Bacillus, Geobacillus, Alicyclobacillus* and *Sulfolobales* species (Sakon et al., 1996).

Although many cellulolytic anaerobic bacteria such as *Clostridium thermocellum* produce cellulases with high specific activity, they do not produce high enzymes quantities. Since the anaerobes show limited growth, most researches on thermostable cellulases production have been addressed to aerobic species. Several mesophilic or moderately thermophilic fungal strains are also known to produce enzymes stable and active at high temperatures. These enzymes are produced from species such as *Chaetomium thermophila* (Venturi et al., 2002), *Talaromyces emersonii* (Grassick et al., 2004), *Thermoascus aurantiacus* (Parry et al., 2002). They may be stable at temperatures around 70 °C for prolonged periods. Table 6 summarizes some of thermostable enzymes isolated from Archea, Bacteria and Fungi.

During the last decade several efforts have been devoted to develop different mixtures of selected thermostable enzymes. In 2007, mixtures of thermostable enzymes, including cellulases from *Thermoascus auranticus, Thrichoderma reseei, Acremonium thermophilum* and *Thermoascus auranticus,* have been produced by ROAL, Finland (Viikari et al., 2007). Multienzyme mixtures were also reconstituted using purified *Chrysosporium lucknowense* enzymes (Gusakov et al., 2005).

Despite the noticeable advantages of thermostable enzymes, cultivation of thermophiles and hyperthermophyles requires special and expensive media, and it is hampered by the low specific growth rates and product inhibition (Krahe et al., 1996; Schiraldi et al., 2002;Turner et al., 2007). Large scale commercial production of thermostable enzymes still remains a challenge also dependent on the optimization of their production from mesophilic microorganisms.

6. Immobilization of enzymes

Thanks to the latest breakthroughs in the research for improving the enzymes, nowadays most enzymes are produced for a commercially acceptable price. Nonetheless, the industrial

Archea					
Enzymes	**Organism**	**pH optimum**	**T optimum (°C)**	**Stability (half life)**	***Refs.***
β-glucosidase	*Pyrococcus furiousus*	5	102	13h at 110°C	*Ma & Adams, 1994*
	Pyrococcus horikoshi	6	100	15h at 90°C	*Rahman et al., 1998*
Endoglucanase	*Pyrococcus furiousus*	6	100	40h at 95°C	*Bergquist et al., 2004*
	Pyrococcus horikoshi	6-6.5	100	19h at 100°C	*Bergquist et al., 2004*
Bacteria					
Enzymes	**Organism**	**pH optimum**	**T optimum (°C)**	**Stability (half life)**	**Refs.**
Endoglucanase	*Acidothermus cellulolyticus*	5.0	83	Inactivated at 110°C	*Sakon J. et al. 1996*
	Anaerocellum thermophilum	5-6	95-100	40min at 100°C	*Zverliv et al., 1998*
	Clostridium stercorarium	6-6.5	90	Stable for several days	*Bronnenmeier K et al., 1991*
	Clostridium thermocellum	6.6	70	33% of activity remained after 50h at 60°C	*Bergquist et al., 2004*
	Clostridium thermocellum	7.0	70	50% of activity remained after 48h at 60°C	*Romaniec et al. 1992*
	Rhodothermus marinus	7.0	95	50% of activity remained after 3.5h at 100°C, 80% after 16h at 90°C	*Bergquist et al., 2004*
	Thermotoga marittima	6.0-7.5	95	2h at 95°C	*Bronnenmeier K, et al., 1995*
	Thermotoga neapolitana	6.0	95	>240min at 100°C	*Bok JD et al., 1995*
Exoglucanase	*Clostridium stercorarium*	5-6	75	3 days at 70°C	*Bronnenmeier K et al., 1990*
Fungal					
Enzymes	**Organism**	**pH optimum**	**T optimum (°C)**	**Stability (half life)**	**Refs.**

Endoglucanase	*Chaetomium termphilum*	4.0	60	60min at 60°C	*Venturi L. Et al., 2002*
	Thermoascus aurantiacus	4.5	75	98h at 70°C and 41h at 75°C	*Parry N., 2002*
Exoglucanase (CBH IA)	*Talaromyces emersonii*	3.6	78	34 min at 80°C	*Grassik A., 2004*

Table 6. Thermostable cellulases

utilization of cellulases could be even more convenient by improving their stability in long-term operations and by developing methods/processes for the downstream recovery and reuse. These objectives can be achieved by the **immobilization of the enzymes** (Cao, 2005). The main advantages of the enzyme immobilization are:

1. more convenient handling of enzymes
2. easy separation from the product
3. minimal or no protein contamination of the product
4. possible recovery and reuse of enzymes
5. enhanced stability under storage and operational conditions (e.g. towards denaturation by heat or organic solvents or by autolysis) (Sheldon, 2007).

The main methods of enzyme immobilization can be classified into four classes: support binding (carrier), entrapment, encapsulation and cross-linking.

Support binding is based on fixing the enzyme to the external or internal surface of a substrate, by physical (adsorption), ionic or covalent bonding. **Adsorption** is a simple and inexpensive method of immobilization, and does not modify the enzyme chemical structure. However, it does not produce strong bonds between enzyme and substrate and this could cause a progressive lost of the enzyme from the support. **Ionic-binding** determines a strong bond between enzyme and support. The supports may be functionalized with a variety of chemical groups to achieve the ionic interaction, including quaternary ammonium, diethylaminoethyl and carboxymethyl derivates (Brady & Joordan, 2009). **Covalent binding** is the most widely used method of immobilization. Here the amino group of lysine is typically used as point of covalent attachment (Brady & Joordan, 2009). Lysine is a very common amino-acid in proteins, often localized on the surface of proteins. It has a good reactivity and provides acceptable bonds stability (Krenkova & Forest, 2004). Supports containing *epoxy groups* are widely used in the immobilization by covalent binding. These can react with lysine and with many other nucleophilic groups on the protein surface (e.g. Cys, Hys, and Tyr). Epoxy groups also react, in a slower way, with carboxylic groups (Mateo et al., 2007). The support used in this immobilization method is typically a prefabricated carrier, such as synthetic resins, biopolymers, inorganic polymers such as silica or zeolites.

Entrapment is based on inclusion of the enzyme in a polymer network (i.e. organic polymer, silica sol-gel). Unlike the previous methods, entrapment requires the synthesis of the polymeric network in the presence of the enzyme (Sheldon, 2007). This method has the advantage of protecting the enzyme from direct contact with the environment, reducing the effects of mechanical sheer and hydrophobic solvents. However, low amount of enzymes can be immobilized (Lalonde & Margolin, 2002).

Encapsulation is a method similar to entrapment, but, in this case, the enzyme is enclosed in a membrane that acts as a physical barrier around it (Cao L., 2005). The disadvantage is that entrapping or encapsulating matrix offer a certain resistance to the substrates diffusion.

Cross-linking results in the formation of enzyme aggregates by using bifunctional reagent, like glutaraldehyde, able to bind enzymes to each other without resorting to any support. In 1996, cross-linked enzyme crystals (CLEC; St. Clair and Navia 1992) were commercialized by Altus Biologics (Margolin, 1996). However the CLEC formation requires laborious and expensive processes of protein purification and it is applicable only to crystallisable enzymes. In addition, only one kind of enzyme can be used in the CLEC formation (Brady & Joordan, 2009). In 2001 a less-expensive method, known as CLEA (cross linked enzyme aggregates) was developed in Sheldon´s Laboratory and commercialized by CLEA Technologies (Netherlands), (Sheldon et al., 2005). Recently a new method has been developed, especially suitable for lipase immobilization. It is defined Spherezymes and it is based on the formation of a water-in-oil emulsion, in which lipases and surfactant are dissolved. Following the addition of a bifunctional cross-linker, permanent spherical particles of enzyme are generated (Brady & Joordan, 2009).
The most interesting immobilization procedures are in the area of covalent binding. Supports containing epoxy groups are widely used in the immobilization by covalent binding because these generate intense multipoint covalent attachment with different nucleophiles present on the surface of the enzyme molecules (Mateo et al., 2007). One limitation of the epoxy supports is the slow reaction of immobilization. To overcome this problem, Mateo and coworkers have designed epoxy supports able to ensure a mild physical adsorption of the enzymes followed by a very fast intramolecular covalent binding with the material epoxy groups. These supports were used to immobilize and stabilize enzymes such as glutaryl acylase (Mateo et al., 2001), β-galactosidase from *Thermus sp.* (Pessela et al., 2003), and peroxidase (Abad et al., 2002). Epoxy supports, known as **Sepabeads®** are marketed by Resindion s.r.l. and quickly have begun to supersede another commercial support, known as Eupergit. This last is a microporous, epoxy-activated acrylic beads with a diameter of 100-250μ, used for a wide variety of different enzymes (Boller et al., 2002).

6.1 Immobilization of cellulases

In literature, only few papers are available on the cellulases immobilization. This is due to the fact that cellulose is not soluble and some immobilization techniques, such as enzymes entrapment, impede the interaction enzyme-substrate. Immobilization of cellulases via covalent bonds appears to be the most suitable technique. Besides the enzyme stabilization, the covalent-immobilization allows the use of supported enzymes for several cycles of reactions (Brady & Joordan, 2009; Li et al., 2007; Mateo et al., 2007; Dourado et al., 2002; Yuan et al., 1999).
In 1999, Yuan and coworkers, immobilized cellulases onto acrylamide grafted acrylonitrile copolymer membranes (PAN) by means of glutaraldehyde. They showed that the enzyme stability was increased after the immobilization process. Also, the activity of the immobilized cellulases was higher than the free cellulases at pH 3 - 5 and at temperatures above 45 °C (Yuan et al., 1999).
In 2002, cellulases from *T. reesei* were immobilized on Eudragit L-100 by researchers of the University of Minho (Portugal). They used the commercial mixture Celluclast® 1.5L supplied by Novozymes (Denmark). This method allowed to improve the stability of the enzymes without significant loss of its specific activity. The adsorption of cellulases on Eudragit lowered the enthalpy of denaturation, but affected only slightly the denaturation temperature (Dourado et al., 2002).

In 2006, Li and coworkers, immobilized cellulase enzymes by means of liposomes. These are phospholipid vesicle, ranging in size from 25 nm to 1μm. In this method, glutaraldehyde-activated liposome bound to the enzyme thus forming the liposome-cellulase complex. Following this step, the complex was immobilized on chitosan-gel. The immobilized enzyme by the liposome molecules showed efficiency higher by 10% compared to the enzyme immobilized in chitosan-gel without liposome. The immobilized cellulase-liposome complex showed a loss of activity of 20% with respect to the original value after six cycles of reaction. Therefore, liposome-binding cellulase appeared to prevent or limit the enzyme deactivation (Li et al., 2007).

In recent investigations, two commercial cellulase enzymes (Celluclast 1.5 and Novozym 188) were immobilized on epoxy Sepabeads® support (Resindion s.rl.). The preliminary data showed that 60% of loaded Celluclast proteins were adsorbed by the support and that more than 90% of these proteins remained stably linked even after repeated washings (Verardi et al., 2011).

7. Process strategies for the hydrolysis and fermentation of lignocellulosics

After the pretreatment step, the bioconversion of lignocellulosic materials includes the biopolymers hydrolysis and the sugar fermentation. These two steps can be performed separately (SHF, separate hydrolysis and fermentation) or simultaneously (SSF, simultaneous saccharification and fermentation). SSF technology is generally considered more advantageous than SHF technology, for several reasons:

- reduced number of the process steps (Koon Ong, 2004)
- reduced end product inhibition because of the rapid conversion of glucose into ethanol by yeast (Viikari et al., 2007)
- reduced contamination by unwanted microorganisms thanks to the presence of ethanol (Elumalai & Thangavelu, 2010).

However, the optimum temperature for the enzymatic hydrolysis is typically higher than that of fermentation. Therefore, in SHF process, the temperature for the enzymatic hydrolysis can be optimized independently from the fermentation temperature, whereas a compromise must be found in SSF process (Olofsson et al., 2008). Another obstacle of the SSF process is the difficulty to carry out continuous fermentation by recirculating and reusing the yeast due to the presence of the solid residues from the hydrolysis.

High solids loadings are usually required to obtain high ethanol levels in the fermentation broths (**high gravity fermentation**). In particular, solids loadings of pretreated biomass up to 30% (w/w) could be necessary to reach an ethanol concentration of 4-5 wt% that is considered a threshold level for a sustainable distillation process. However, increasing the amount of the solids content in a bioreactor, the hydrolytic performances of the enzymes mixture tends to worsen. In particular, the high initial substrate consistency causes a **viscosity increase** (Sassner et al., 2006) that is an obstacle toward the homogeneous and effective distribution of the enzymes in the bioreactor. This problem could be partly overcome by using thermostable enzymes. In particular, the hydrolysis could be carried out in two steps: a former step at elevated temperatures with thermostable hydrolytic enzymes producing the liquefaction of biomass (SHF); the latter step, aimed at completing the biomass saccharification, could be carried out at milder temperatures by using the SSF approach (Olofsson et al., 2008).

8. Innovative bioreactor geometries and process strategies

A major requirement in cost-efficient lignocellulosics-to-ethanol process is to employ reactor systems yielding the maximal conversion of the cellulose with the minimal enzyme dosage. As consequence, one of the most important parameter for the design and operation of bioreactors for lignocellulosic conversion is the effective use of the biocatalysts to obtain high specific rates of cellulose conversion (namely the yield of glucose obtained per amount of enzymes). The maximization of the product concentration, i.e. the amount of glucose obtained per liquid volume, is also an important parameter as well as the optimization of the volumetric productivity, in this case the rate of glucose formation per reactor volume.

When the hydrolysis is carried out with high dry matter contents, hence high cellulose levels, the product concentration will drive up. For this reason, some recent researches have been finalized into attempting the enzymatic biomass conversion at high-solids loads (Jørgensen et al., 2007; Tolan, 2002). The most important problem of high solid loadings is related to the fact that the viscosity of the reaction mixture is very high and the rheology of the mixture has to be well studied: normal stress might become very significant during bioconversion. In particular, mixing and mass transfer limitations, and, presumably increased inhibition by intermediates come into play. Various fed-batch strategies have been attempted with the scope of supplying the substrate without reaching excessive viscosities and unproductive enzyme binding to the substrate (Rosgaard et al., 2007a; Rudolf et al., 2005).

As said, the currently employed cellulolytic enzyme systems, that include the widely studied *T. reesei* enzymes, are significantly inhibited by the hydrolysis products cellobiose and glucose. This inhibition retards the overall conversion rate of lignocellulosics-to glucose (Gan et al., 2002; Katz and Reese, 1968). Product inhibition is particularly significant during processing at high substrate loadings mainly because the glucose concentration is higher than that obtained in diluted biomass suspensions. (Kristensen et al., 2009; Rosgaard et al., 2007a). As consequence, both the conversion rate and the glucose yields achievable in batch processing of lignocellulose are reduced (Rosgaard et al., 2007b; Tengborg et al., 2001).

General criteria in the bioreactor design and in the selection of the operating conditions could be: use of reactors or reaction regimes that allow a rapid reduction of the glucose concentration; running of the reactions at low to medium substrate concentrations in order to maintain higher conversion rates and hence obtain higher volumetric productivity of the reactor (Andrić et al. 2010, a).

The integration of the bioreactor with a separation unit (reaction–separation hybrids) has shown promising results with product inhibited or equilibrium limited enzyme-catalyzed conversions, because it is possible to remove the products as they are formed (Ahmed et al., 2001; Gan et al., 2002). In this regard, membrane (bio) reactors could be a viable process configuration. Unlike the SSF approach in which the glucose consumption is carried out by the microrganisms simultaneously available in the hydrolyzate, the use of membrane bioreactors would accomplish the same function without any compromise in the reaction temperature. A membrane (bio-) reactor is a multifunction reactor that combines the reaction with a separation, namely in this case product removal by membrane separation, in one integrated unit, i.e. in-situ removal, or alternatively in two or more separate units. The membrane bioreactors hitherto used for the separation in enzymatic processes have been mainly ultra- and nanofiltration (Pinelo et al., 2009). However, the use of this technology is limited by the bank-up of unreacted lignocellulosics (lignin and particularly recalcitrant cellulose) in large-scale and/or continuous processing (Andrić et al. 2010, b). Already in the past, some authors improved the efficiency of the continuous stirred tank bioreactor (CSTR) by incorporating separation membranes in the reactor design. In particular, Henley et al. (1980) incorporated an UF membrane (UF) or hollow-fiber cartridge (HFC) into the CSTR-UF and CSTR-HFC system, respectively (Henley et Al., 1980). Ishihara et al. (1991) accomplished a semi-continuous hydrolysis reaction by using a continuously stirred reservoir tank, connected to a suction filter unit for the removal of the lignin-rich residue and an ultra-filtration membrane unit (tubular module), through which the filtrate was pumped in order to separate the hydrolysis products from cellulases. The concentration of the lignocellulosic substrate in the reactor was maintained almost constant by the addition of fresh substrate at appropriate intervals. The filter and ultrafiltration units were operated intermittently, while the enzymes were added at the start, recovered in the UF module, and recycled back into the reactor (Ishihara et al., 1991). More recently, Yang et al. (2006) designed the removal of reducing sugars during the cellulose enzymatic hydrolysis through a system consisting in a tubular reactor, in which the substrate was retained with a porous filter at the bottom and buffer entered at the top through a distributor. The hollow-fiber ultrafiltration module with polysulfone membrane enabled the permeation and the separation of the sugars. To keep the volume constant in the tubular reactor, all the remaining buffer was recycled back from the UF membrane and the make-up buffer was continuously supplied from the reservoir (Yang et al., 2006). In some applications an additional microfiltration unit has exceptionally been used to retain the unconverted lignin-rich solid fraction due to the presence of tightly bound enzymes (Knutsen and Davis, 2004) or has been employed to remove the unconverted substrate from the reactor. These set-ups result in slightly complex process layouts for the hydrolysis.

It is evident that the optimization of the reactor designs will permit to overcome both the rheological and inhibition limit of the bioconversion and maximize the enzymatic conversion. Therefore, the reactor design become strong relevant for large-scale processing of cellulosic biomass (Lynd et al., 2008; Wyman, 2008).

9. Conclusion

In this chapter an overview of the current knowledge on the hydrolysis of lignocellulosics for bioethanol production has been presented. In the last years several important breakthroughs have been made either on the biochemical and technological sides. This is confirmed by several industrial initiatives spread over the world. Among these, in recent days, the first brick of the lignocellulosic bioethanol demo plant (40 kton/y) has been layed in Northern Italy by the Mossi and Ghisolfi Group. Some cooperation agreements were strengthen with Novozymes for improving the efficiency of the hydrolysis step. This event represents an important stage for all the Europe making the production of lignocellulosic ethanol closer to the industrialization and opening the way to new lignocellulosic biorefineries.

10. Acknowledgments

The authors wish to thank Valentino Mannarino for his support in the realization of some pictures and Alessandro Blasi for helpful discussions and effective cooperation.

11. References

Abad, J.M.; Vélez, M.; Santamaria, C.; Guisán, J., M.; Matheus, P., R.; Vázquez, L., Gazaryan, I.; Gorton, L.; Gibson, T. & Fernández, V., M. (2002). Immobilization of Peroxidase

Glycoprotein on Gold Electrodes Modified with mixed epoxy-boronic acid monolayers. *Journal of the American chemical Society*, Vol.124, No.43 (October, 2002), pp. 12845–12853.

Ahmed F, Stein A. & Lye G. J. (2001). In-situ product removal to enhance the yield of biocatalytic reactions with competing equilibria: α-glucosidase catalysed synthesis of disaccharides. *Journal of Chemical Technology and Biotechnology*, Vol. 76, No. 9, (September 2001), pp. 971–977.

Ahamed, A. & Vermette. P. (2008) Enhanced enzyme production from mixed cultures of Trichoderma reesei RUT-C30 and Aspergillus niger LMA grown as fed batch in a stirred tank bioreactor, *Biochemical Engineering Journal,* Vol.42, No.1 (October, 2008), pp.41-46.

Alvira, P., Tomás-Pejó, E., Ballesteros, M. & Negro, M., J. (2010). Pretreatment technologies for an efficient bioethanol production process based on enzymatic hydrolysis: A review, *Bioresource Technology*, Vol.101, No.13, (July, 2009), pp.4851-4861.

Andrić P., Anne S. Meyer A. S., Jensen P. A. & Johansen K. D., (2010a) Reactor design for minimizing product inhibition during enzymatic lignocellulose hydrolysis: I. Significance and mechanism of cellobiose and glucose inhibition on cellulolytic enzymes. *Biotechnology Advances*, Vol. 28, No. 3, (June 2010), pp. 308–324.

Andrić P., Anne S. Meyer A. S., Jensen P. A. & Johansen K.D.,(2010b) Reactor design for minimizing product inhibition during enzymatic lignocellulose hydrolysis II. Quantification of inhibition and suitability of membrane reactors, *Biotechnology Advances*, Vol. 28, No. 3, (June 2010), pp.407–425.

Arai, T., Kosugi, A., Chan, H., Koukiekolo, R., Yukawa, H., Inui, M., Doi, R.,H. (2006). Properties of cellulosomal family 9 cellulases from *Clostridium cellulovorans*. Applied Microbiology and Biotechnology, Vol. 71, No.5 (August, 2006), pp. 654-660.

Balat.,M., Balat, H. & Oz, C. (2008). Progress in bioethanol processing, *Progress and Energy Combustion science,* Vol. 34, No.5, (October 2008), pp. 551-573.

Bayer, E.,A.; Chanzy, H.; Lamed, R.; Shoham, Y. (1998). Cellulose, cellulases and cellulosomes. *Current Opinion in structural Biology*, Vol.8, No.5 (October, 1998), pp. 548-557.

Begum, M., F. & Absar, N., (2009). Purification and Characterization of Intracellular Cellulase from Aspergillus oryzae ITCC-4857.01, *Mycobiology*, Vol.35, No.2 (May, 2009), pp. 121-127.

Bergquist, P.,L.; Te'o, V.,S.,J.; Gibbs, M.,D.; Curah, N.,C. & Nevalainen, K.,M.,H (2004). Recombinant enzymes from thermophilic micro-organisms expressed in fungal hosts. *Biochemical Society Transaction*, Vol. 32, No.2 (April, 2004), pp.293–297

Berlin, A.; Maximenco, V.; Gilkes, N. & Saddeler J. (2007). Optimazion of enzyme complexes for lignocelluloses hydrolysis, *Biotechnology and Bioengineering*, Vol.97 , No.2, (June, 2007), pp.287-296.

Bok, J. D.; Yernool, D., A. & Eveleigh, D., E. (1998). Purification, characterization, and molecular analysis of thermostable cellulases CelA and CelB from *Thermotoga neapolitana*. *Applied and Environmental Microbiology*, Vol.64, No.12 (December, 1998), pp. 4774–4781.

Boller,T.; Meier, C. & Menzler ,S. (2002), Eupergit Oxirane acrylic beads: How to Make Enzymes fit for Biocatalysis. *Organic Process Research & Development* , Vol.6, No. 4 (February, 2002), pp. 509-519

Bozell., J.,J. & Petersen, G.,R. (2010). Technology development for the production of biobased products from biorefinery carbohydrates-The US Department of Energy's "Top10" revisited, *Green Chemistry*, Vol. 12., No.4, (Jun, 2010), pp.525-728.

Brady D. & Jordaan J., (2009). Advances in enzyme immobilisation, *Biotechnology Letters*, Vol. 31, No. 11,(July 2009) pp. 1639-1650.

Bronnenmeier, K. & Staudenbauer, W. (1990) Cellulose hydrolysis by a highly thermostable endo-1,4-glucanase (Avicelase I) from *Clostridium stercorarium*. *Enzyme and Microbial Technology*, Vol. 12, No.6 (June, 1990), pp. 431–436.

Bronnenmeier, K.; Kern, A.; Liebl, W.; Staudenbauer, W. (1995) Purification of *Thermotoga maritima* enzymes for the degradation of cellulosic materials. *Applied and Environmental Microbiology*, Vol.61, No.4 (April, 1995), pp. 339–1407

Bronnenmeier, K.; Rücknagel, K. & Staudenbauer, W. (1991) Purification and properties of a novel type exo-1,4-β-glucanase (Avicelase II) from the cellulolytic thermophile Clostridium stercorarium. *European Journal Biochemistry*, Vol.200, No.2 (September, 1991), pp.379–385.

Cao, L. (2005). *Carrier-bound Immobilized Enzymes, Principles, Applications and Design, Book* , Wiley-VCH Verlag GmbH & Co.KGaA, Weinheim, ISBN-13: 978-3-527-31232-0; ISBN-10: 4-527-31232-3

Cherubini F., Ulgiati S. (2010). Crop residues as raw materials for biorefinery systems –A LCA case study. *Applied Energy*, Vol. 87, No.1 (January, 2010), pp. 47-57

De Bari, I.; Nanna, F. & Braccio, G. (2007). SO_2-catalyzed steam fractionation of aspen chips for bioethahnol production: Optimization of the catalyst impregnation. *Industrial & Engineering Chemistry Research*, Vol.46, No.23 (October, 2007) pp. 7711-7720.

De Bari, I.; Viola, E.; Barisano, D.; Cardinale, M.; Nanna, F.; Zimbardi, F.; Cardinale, G. & Braccio, G. (2002). Ethanol Production at Flask and Pilot Scale from Concentrated Slurries of Steam-Exploded Aspen. *Industrial & Engineering Chemistry Research*, Vol. 41, No.7 (March, 2002), pp. 1745-1753.

Demain, A.; Newcomb, M.; Wu, J.,H.,D. (2005) Cellulase, clostridia and ethanol. *Microbiology and Molecular Biology Reviews*, Vol. 69 No.1 (March, 2005), pp. 124–154

Demirabas, A. (2008). Products from Lignocellulosic Materials via Degradation Processs, *Energy Sources Part A*, Vol.30, No.1,(January, 2008) pp.27-37.

Dourado F., Bastos M., Mota M., Gama F.M. (2002) Studies on the propertie of Celluclast/Eudragit L-100 conjugate. *Journal of Biotechnology*, Vol. 99, No. 2, (October 2002), pp. 121-131.

Elumalai, S. & Thangavelu, V. (2010). Simultaneous Sacccharification and fermentation (SSF) of pretreated sugarcane bagasse using Cellulase and Saccharomyces cerevisiae-Kinetics and modeling. *Chemical Engineering Research Bullettin* Vol. 14, No. 1 (April 2010), pp. 29-35.

Fägerstam, L.,G. & Pettesson, L.,G. (1984). The primary structure of a 1,4–β-glucan cellobiohydrolases of Trichoderma reesei QM9414. *FEBS Letters*, Vol.119, No.2 (February, 1984), pp.97–101.

FitzPatrick, M. ; Champagne, P. ; Cunningham, M., F. & Whitney, R., A. (2010). A biorefinery processing perspective : Treatment of lignocellulosic materials for the production of value-addued products, *Bioresource Technology*, Vol.101, No.23, (December, 2010), pp. 8915-8922.

Galbe, M. & Zacchi, G. (2002). A review of the production of ethanol from softwood. *Applied Microbiology and Biotechnology* Vol.59, No.6, (July, 2002), pp. 618-628.

Gan Q, Allen S.J., Taylor G. (2002). Design and operation of an integrated membrane reactor for enzymatic cellulose hydrolysis. *Biochemical Engineering Journal*, Vol. 12, No. 3, pp.223–229.

Gautam, S., P; Bundela,P., S.; Pandey, A., K.; Khan, J.; Awasthi, M., K. & Sarsaiya, S.(2010). Optimization for the Production of Cellulase Enzyme from Municipal Solid Waste Residue by Two Novel Cellulolytic Fungi, *Biotechnology Research International*, Vol 2011, (December, 2010), pp. 1-8.

Gilkes, N.,R.; Henrissat, B.; Kilburn, D.,G.; Miller, R.,C., Jr. &Warren, R.,A.,J. (1991). Domains in microbial β-1,4-glycanases: sequence conservation, function, and enzyme families. *Microbiological Reviews*, Vol. 55, No.2, pp. 303–315.

Gilligan, W. & Reese, E.,T.(1954). Evidence for multiple components in microbial cellulases. *Canadian Journal of Microbiology* , Vol.1,No.2, (October,1954), pp. 90-107.

Grassick, A.; Murray, P.; Thompson, R.; Collins, C.; Byrnes, L.; Birrane, G.; Higgins, T. & Tuohy, M. (2004), Three-dimensional structure of a thermostable native cellobiohydrolase, CBH IB, and molecular characterization of the cel7 gene from the filamentous fungus, *Talaromyces emersonii*. *European Journal of Biochemistry*, Vol. 271, No. 22 (November, 2004), pp. 4495–4506.

Güllü, D. (2010). Effect of catalyst on yiels of liquid products from biomass via pyrolysis. *Energy sources*, Vol. 25, No.8, (June, 2010), pp.753-756.

Gusakov, A.,V.; Sinitsyn, A.,P.; Salanovich, T.,N.; Bukhtojarov, F.,E.; Markov, A.,V.; Ustinov, B., B.; van Zeijl, C.; Punt, P. & Burlingame, R. (2005). Purification, cloning and characterisation of two forms of thermostable and highly active cellobiohydrolase I (Cel7A) produced by the industrial strain of Chrysosporium lucknowense. *Enzyme and Microbial Technology*, Vol.36, No.1, (January, 2005), pp.57–69

Harmsen, P.,F.,H. ; Huijen, W.,J.,J.; Bermúdez López, L.,M. & Bakker, R.,R.,C (2010). Literature Review of Physical and chemical Pretreatment Processes for Lignocellulosic Biomass, in *Wageningen UR, Food & Biobased Research*, (September 2010), pp. 1-54., ISBN 9789085857570

Hatti-Kaul, R. ; Törnvall, U. ; Gustaffson, L.& Börjesson, P.(2007). Industrial biotechnology for the production of bio-based Chemicals.A cradle-to-perspective. *Trends in Biotechnology*, Vol.25, No.23 (March, 2007), pp.119-124.

Henley R. G., Yang R. Y. K., Greenfield P. F. (1980). Enzymatic saccharification of cellulose in membrane reactors. *Enzyme Microbial Technology*, Vol. 2, No. 3, (July 1980), pp. 206–208.

Hong, J.; Xu, D.; Gong, P.; Yu, J.; Ma, H. & Yao, S. (2008). Covalent-bonded immobilization of enzyme on hydrophilic polymer covering magnetic nanogels. *Microporous and mesoporous materials*, Vol.109,Nos. 1-3 (March, 2008), pp. 470-477.

Ilmén, M.; Saloheimo, A.; Onnel, M.; Pentillä, M.,E. (1997). Regulation of Cellulase Gene Expression in the Filamentous Fungus Trichoderma reesei. *Applied and Environmental Microbiology*,Vol.63,No.4 (April,1997), pp.1298-1306

Iranmahboob, J.; Nadim, F.; Monemi, S. (2002). Optimizing acid-hydrolysis: a critical step for production of ethanol from mixed wood chips. *Biomass and Bioenergy*, Vol.22, No.5, (May, 2002), pp.401-404.

Ishihara M, Uemura S, Hayashi N, Shimizu K., (1991). Semicontinuous enzymatic hydrolysis of lignocelluloses. *Biotechnology and Bioengineering*, Vol. 37, No. 10, (April 1991) pp.948–54.

Jones, J.,L. & Semrau, K., T. (1984). Wood Hydrolysis for ethanol production-previous experience and the economics of selected processes. Biomass, Vol.5, No.2 (August, 1983), pp.109-135

Jørgensen H, Vibe-Pedersen J, Larsen J, Felby C. (2007). Liquefaction of lignocellulose at highsolids concentrations. *Biotechnology and Bioengineering*. Vol. 96, Issue 5, (April 2007), pp-862–870.

Kabel, M.; van der Maarel, M.,J.,E.,C.; Klip, G.; Voragen, A,.G. & Schols H. (2005). Standard assays do not predict the efficiency of commercial cellulose preparations towards plants materials. *Biotechnology and Bioengineering*, Vol.93, No.1 (January, 2006), pp. 56-63.

Katz M, Reese E. T., (1968) Production of glucose by enzymatic hydrolysis of cellulose. *Applied Microbiology*, Vol. 16, No. 2, (Febraury 1968), pp.419–420.

Knutsen J. S., Davis R. H., (2004) Cellulase retention and sugar removal by membrane ultrafiltration during lignocellulosic biomass hydrolysis. *Applied Biochemistry and Biotechnology*, Vol. 113-116, pp.585–599.

Koon Ong, L. (2004).Conversion of Lignocellulosic Biomass to Fuel Ethanol- A brief review. Since cellulases are inhibited. *The Planter, Kuala Lumpur*, Vol. 80 (941), (Febraury 2009) pp. 517-524.

Kootstra, A., M., J.; Beeftink, H., H.; Scott, E., L. & Sanders, J., P., M. (2009). Comparison of dilute mineral and organic acid pretreatment for enzymatic hydrolysis of wheat straw. *Biochemical Engineering Journal*, Vol. 46, No. 2 (October, 2009), pp.126-131.

Krahe, M; Antranikian, G. & Märkl H (1996). Fermentation of extremophilic microorganisms. *FEMS Microbiology Reviews*, Vol.18, Nos.2-3 (May, 1996), pp.271-285.

Krenkova, J. & Foret, F. (2004) Immobilized microfluidic enzymatic reactors. *Electrophoresis* Vol. 25, Nos 21-22, (November 2004), pp. 3550–3563.

Kristensen J. B., Felby C., Jørgensen H., (2009) Determining yields in high solids enzymatic hydrolysis of biomass. *Applied Biochemistry and Biotechnology*, Vol. 156, N°. 1-3, (May 2009) pp..127-132.

Kumar, P.; Barrett, D.M. ; Delwiche, M.J., & Stroeve, P. (2009). Methods for Pretreatment of Lignocellulosic Biomass for Efficient Hydrolysis and Biofuel Production, *Industrial & engineering chemistry research*, Vol. 48, No.8, (January 2009), pp. 3713-3729.

Lalonde, J. & Margolin, A. (2002) *Immobilization of enzymes*. In: Drauz K, Waldmann H (eds), Enzyme catalysis in organic chemistry, 2nd edn. Wiley-VCH, Weinheim, pp.163–184, ISBN-10:3527299491, ISBN-13:978-352729949-2.

Lalonde, J. & Margolin, A. (2008) Immobilization of Enzymes, in Enzyme Catalysis in Organic Synthesis: A Comprehensive Handbook, Second Edition , eds.K. Drauz and H. Waldmann), Wiley-VCH Verlag GmbH, Weinheim, Germany. ISBN: 9783527299492; ISBN: 9783527618262

Larsson, S.; Palmqvist, E.; Hahn-Hägerdal, B.; Tengborg, C.; Stanberg, K., Zacchi, G. & Nilverbrant, N. (1999). The generation of fermentation inhibitors during dilute acid hydrolysis of softwood. *Enzyme and Microbiology Technology*, Vol.24,No.3-4 , (February, 1999), pp.151-159.

Lenihan, P.; Orozco, A.; O'Neil, E.; Ahmad,M.,N.,M.; Rooney, D.,W. & Walker G.,M. (2010). Dilute acid hydrolysis of lignocellulosic biomass. *Chemical Engineering Journal*, Vol. 156, No.2,(January, 2010), pp. 395-403.

Lenting, H.,B.,M. & Warmoeskerken M.,M.,C.,G.(2001). Mechanism of interaction between cellulase action and applied shear force, an hypothesis. *Journal of Biotechnology* Vol.89, Nos. 2-3 (August, 2001), pp.217-226

Li C., Yoshimoto M., Fukunaga K., Nakao K. (2007). Characterization and immobilization of liposome-bound cellulase for hydrolysis of insoluble cellulose. *Bioresouce technology*, Vol.98, No.7, (May 2007), pp. 1366-1372.

Lynd, L. R.; Weimer, P., J.; van Zyl, W., H. & Pretorius, I.,S. (2002). Microbial Cellulose Utilization: Fundamentals and Biotechnology. *Microbiology and Molecular Biology Reviews*, Vol. 66, No.3 (September, 2002), pp.506–577.

Lynd, L. R., Laser, M.S., Bransby, D., Dale, B.E., Davison, B., Hamilton, R., Himmel, M., Keller, M., McMillan, J.D., Sheehan, J., Wyman, C.E., (2008). How biotech can transform biofuels. *Nature Biotechnology*, Vol. 26, N°. 2, (February 2008), pp. 169–172.

Ma, K. & Adams, M., W. (1994). Sulfide dehydrogenase from the hyperthermophilic archaeon Pyrococcus furiosus: a new multifunctional enzyme involved in the reduction of elemental sulfur, *Journal of Bacteriology*, Vol.176, No.21, (November 1994), pp. 6509-17.

Mansfield, S.D., Saddler J.N., Gubitz G.M., 1998. Characterization of endoglucanases from the brown rot fungi Gloeophyllum sepiarium and Gloeophyllum trabeum. *Enzyme and Microbial Technology*, Vol. 23, Nos.1-2 (July, 1998), pp.133– 140.

Margolin, A. (1996). Novel crystalline catalysts. *Trends in Biotechnology*, Vol. 14, Issue 7, (July 1996), pp. 223-230.

Mateo, C.; Grazu, V.; Palomo, J., M.; Lopez-Gallego, F.; Fernandez-Lafuente, R. & Guisan, M., J., (2007). Immobilization of enzymes on heterofunctional epoxy supports. *Nature Protocols* Vol. 2, (April 2007), pp. 1022-1033.

Mateo, C.; Fernández-Lorente, G.; Cortés, E.; Garcia, J.L.; Fernández-Lafluente, R. & guisan, J., M.(2001).One step purification, covalent immobilization and additional stabilization of poly-His-tagged proteins using novel heterofunctional chelate-epoxy supports, *Biotechnology and Bioengineering*, Vol.27, No.3 (November, 2001), pp.269-276.

Mathew, G.,M.; Sukumaran, R.,K.; Singhania, R.,R. & Pamdey, A. (2008). Progress in research on fungal cellulases for lignocelluloses degradation. *Journal of Scientific & Industrial Research*, Vol.67, No.11, (November, 2008), pp. 898-907.

McCarter, J., D. & Whiters, S.,G. (1994). Mechanisms of enzymatic glycoside hydrolysis. *Current Opinion in Structural Biolology*, Vol.4, No.6 (February, 2004), pp.885-892

NIST, Standard Reference Materials, 12/2009, Avaible from : http://ts.nist.gov/measurementservices/referencematerials/index.cfm

Novozyme and NREL, Novozyme and reduce enzyme cost (2005), News on: *Sci Direct-Focus on Catalysts*, pp. 4-4

Olofsson K., Bertilsson M. & Lidén G., (2008). A short review on SSF - an interesting process option for ethanol production from lignocellulosic feedstocks. *Biotechnology for Biofuels*, Vol. 1, No.7 (May 2008), pp.1-14

Paljevac, M. ; Primožič, M. ; Habulin, M.; Novak, Z. & Knez, Ž. (2007). Hydrolysis of carboxymethyl cellulose catalyzed by cellulase immobilized on silica gels at low and high pressures. *Journal of Supercritical Fluids*, Vol.43, No.1 (November, 2007), pp.74-80.

Parry, N.; Beever, D.; Owen, E.; Nerinckx, W.; Claeyssens, M.; Van Beeumen, J.; Bhat, M. (2002). Biochemical characterization and mode of action of a thermostable endoglucanase purified from *Thermoascus aurantiacus*. *Archives of Biochemistry Biophysics*, Vol. 404, No. 2 (August, 2002), pp.243–253

Pessela, B.,C.; Mateo, C.;Carrascosa, A., V.; Vian, A.; García, J., L.; Rivas, G.;Alfonso, C.; Guisan, J.,M; Fernández-Lafluente, R. (2003). One step purification, covalent immobilization and additional stabilization of a thermophilic polyHis-tagged beta-galactosidase from thermos sp. Strain T2 by using novel heterofunctional chelate-epoxy Sepeabeds, *Biomacromolecules*, Vol.4,No.1 (January-February, 2003), pp.107-13.

Pinelo M, Jonsson G, Meyer A.S., (2009) Review: Membrane technology for purification of enzymatically produced oligosaccharides: molecular features affecting performance. *Separation and Purification Technology*, Vol. 70, No. 1, (November 2009), pp. 1-11. 2009;70:1-11.

Rahman, R., N., Z., A.; Fujiwara, S.; Takagi, M. & Imanaka, T. (1998) Sequence analysis of glutamate dehydrogenase (GDH) from the hyperthermophilic archaeon Pyrococcus sp. KOD1 and comparison of the enzymatic characteristics of native and recombinant GDHs, *Molecular and General Genetics MGG*, Vol. 257, No. 3, (May, 1997), pp. 338-347.

Ritter, J.,G. ; Seborg, M., R. & Mitchell, R.,L. (1932). Factors Affecting Quantitative Determination of Lignin by 72 Per Cent Sulfuric Acid Method. *Analytical Edition*, Vol. 4, No.2 (April 1932) pp. 202-204.

Romaniec, M.; Fauth, U.; Kobayashi, T.; Huskisson, N.; Barker, P. & Demain, A. (1992) Purification and characterization of a new endoglucanase from Clostridium thermocellum. *Biochemical Journal*, Vol.283, (Part1) (April, 1992), pp.69–73, ISSN

Rosgaard L, Andrić P, Dam-Johansen K, Pedersen S, Meyer AS. (2007a).Effects of substrate loading on enzymatic hydrolysis and viscosity of pretreated barley straw. *Applied Biochemistry and Biotechnology*, Vol. 143, No. 1 (April 2007). pp.27–40.

Rosgaard L, Pedersen S, Langston J, Akerhielm D, Cherry JR, Meyer AS. (2007b). Evaluation of minimal Trichoderma reesei cellulase mixtures on differently pretreated Barley straw substrates. *Biotechnology Progress*, Vol. 23, No. 6, (December 2007), pp. 1270–1276.

Rudolf A, Alkasrawi M, Zacchi G, Liden G.(2005) A comparison between batch and fed-batch simultaneous saccharification and fermentation of steam pretreated spruce. *Enzyme Microbial Technology*, Vol. 37, No. 2, (July 2005), pp.195–204.

Ruttersmith, L. & Daniel, R. (1991) Thermostable cellobiohydrolase from the thermophilic eubacterium Thermotoga sp. strain FjSS3-B.1: purification and properties. *Biochemical Journal*, Vol. 277, No. 3, pp.887–890

Saeman, J. F.; Moore, W. E.; Mitchell, R., L. & Millet, M., A. (1954). Techniques for the determination of pulp constituents by quantitative paper chromatography. *Tappi Journal*, Vol. 37, No. 8 (August, 1954), p.336

Sakon, J.; Adney, W.; Himmel, M.; Thomas, S.; Karplus, P. (1996) Crystal structure of thermostable family 5 endoglucanase EI from *Acidothermus cellulolyticus* in complex withcellotetraose. *Biochemistry*, Vol.35, No.33, (August, 1996), pp.10648–10660

Sassner, P.; Galbe, M. & Zacchi, G. (2006). Bioethanol production based on simultaneous saccharification and fermentation of steam-pretreated salix at high dry-matter content. *Enzyme and Microbial Technology*, Vol. 39, Issue 4, (August 2006) pp. 756–762.

Schiraldi, C. & De Rosa, M. (2002) The production of biocatalysts and biomolecules from extremophiles. *Trends in Biotechnology*, Vol.20,No. (12), pp.515-521.

Sheldon, R., A. (2007). Enzyme Immobilization: The quest for Optimumu Performance, *Advanced Synthesis & Catalysis*, Vol. 349, (February 2007), pp. 1289-1307.

Sheldon, R., A.; Schoevaart, R., van Langen, I.,M. (2005). Crosslinked enzyme aggregates (CLEAs): a novel and versatile method for enzyme immobilization (a review). *Biocatalysis and Biotransformation*, Vol. 23, No. 3 (August 2005), pp. 141-147.

Sluiter, J., B. ; Ruiz, O., R. ; Scarlata, C., J. ; Sluiter, A. (2010). Compositional Analysis of Lignocellulosic Feedstocks. 1.Review and Description of Methods, *Journal of Agricultural and food chemistry*, Vol.58 , No. 16, (July, 2010), pp.9043-9053.

Sun, Y.; Cheng, J. (2002). Hydrolysis of lignocellulosic materials for ethanol production: A review, *Bioresource technology Y.*, Vol. 83 , No. 1 (October 2001), pp. 1-11.

Taherzadeh, M.,J. & Karimi, K. (2007). Enzyme-based hydrolysis processes for ethanol from lignocellulosic materials: a review. *Bioresources*, Vol.2, No.4, pp. 707-738.

Tatsumi, H.; Katano H. & Ikeda, T. (2006). Kinetic analysis of enzymatic hydrolysis of crystalline cellulose by cellobiohydrolase using an amperometric biosensor. *Analytical Biochemistry*, Vol. 357,No.2 (October, 2006), pp. 257–261.

Te'o, V.; Saul, D. & Bergquist, P. (1995). CelA, another gene coding for a multidomain cellulases from the extreme thermophile Caldocellum saccharolyticum. *Applied Microbiology and Biotechnology.*, Vol. 43, No.2, (June, 1995) pp. 291–296.

Tengborg, C.; Galbe, M. & Zacchi, G. (2001). Influence of enzyme loading and physical parameters on the enzymatic hydrolysis of steam-pretreated softwood. *Biotechnology Progress*. Vol.17, No.1 (January-February, 2001), pp.110-117.

Ting, C., L.; Makarov, D.,E. & Wang, Z.,G. (2009). A Kinetic Model for the Enzymatic Action of Cellulase. *The Journal of Physical Chemistry*, Vol.113, No.14 (March,2009), pp. 4970–4977

Tolan, J. S., (2002). Iogen's process for producing ethanol from cellulosic biomass. *Clean Technologies and Environmental Policy*. Vol. 3, No 4, (February 2002), pp. 339-345.

Tu, M.; Zhang, X.; kurabi, A. et al. (2006). Immobilization of β-glucosidase on Eupergit C for lignocellulose hydrolysis. *Biotechnology Letters*,Vol.28, No.3 (February, 2006), pp.151-156.

Tuohy, M.; Walsh, J.; Murray, P.; Claeyssens, M.; Cuffe, M.; Savage, A. & Coughlan, M. (2002) Kinetic parameters and mode of action of the cellobiohydrolases produced by *Talaromyces emersonii*. *Biochemica et Biophysica Acta*, Vol.1596, No.2 (April, 2002), pp.366–380

Turner,P.; Mamo, G. & Karlsson, E., N. (2007). Potential and utilization of thermophiles and thermostable enzymes in biorefining, *Microbial Cell Factories*, Vol.6, No.9 (March, 2007), pp.1-23

Verardi, A.; De Bari, I.; Ricca, E.; Calabrò, V.; Hydrolysis of cellulose with immobilized cellulases: process analysis and control, *Proceedings of the 19th European Biomass Conference & Exhibition*, Berlin (Germany), 6-10 June 2011

Venardos, D.; Herbert, E.,K.&Donald, W.,S. (1980). Conversion of cellobiose to glucose using immobilized β-glucosidase reactors. *Enzyme and Microbial Technology*, Vol.2,No.2 (April, 1980),pp. 112-116.

Venturi, L.; Polizeli, M.; Terenzi, H.; Furriel, R. & Jorge, J. (2002) Extracellular β-d-glucosidase from Chaetomium thermophilum var. coprophilum: production, purification and some properties. *Journal of Basic Microbiology*, Vol. 42, No.1 (February, 2002) pp.55–66

Viikari, L.; Alapuranen, M.; Puranen, T.; Vehmaanperä, J. & Siika-aho, M. (2007). Thermostable Enzymes in lignocellulose Hydrolysis. *Advances in Biochemical Engineering/Biotechnology*, Vol.108, (June,2007), pp. 121-145

Watanabe, H.& Tokuda, G. (2010). Cellulolytic System in Insects. *Annual Review of entomology*,Vol.55, (January,2010), pp.609-632.

Werpy, T; Petersen, G. ; Aden, A. ; Bozell, J. ; Holladay, J. ; White, J. ; Manheim, Amy ; Eliot, D. ; Lasure, L. & Jones, S. (2004). *Top Value Added Chemicals from Biomass: Results of Screening for Potential Candidates from Sugars and Synthesis Gas.* Report (August, 2004), Department of Energy, Oak Ridge, TN

Wilson D.,B. (2008). Three microbial strategies for plant cell wall degradation. *Annals of the New York Academy of Science*, Vol.1125 (March, 2008), pp. 289–297.

Wyman, C.E. (1996). *Handbook on bioethanol: production and utilization.* Applied energy technologies series (Ed.), ISBN 1-56032553-4

Wyman, C.,E. (1999). Production of low cost sugars from biomass: progress, opportunites, and challenges. In Overerd RP, Cornet E. (eds.) Biomass- a growth opportunity in green energy and value-added products. *Proocedings of the 14th Biomass conference of the americas.*, Vol.1, pp. 867-872, Pergamon, Oxford.

Wyman C. E. (2007). What is (and is not) vital to advancing cellulosic ethanol. *Trends in Biotechnology*, Vol. 25, No. 4, (April 2007) pp.153–157.

Yang S, Ding W, Chen H. (2006). Enzymatic hydrolysis of rice straw in a tubular reactor coupled with UF membrane. *Process Biochemistry*, Vol. 41, No. 3, (March 2006), pp.721–725.

Yuan X., Shen N., Sheng J., Wei X., (1999). Immobilization of cellulase using acrylamide grafted acrylonitrile copolymer membranes. *Journal of Membrane Science*, Vol. 155,Issue 1, (March 1999), pp. 101-106.

Zhang, Y., H.; Berson, E.; Sarkanen, S. & Dale, B., E. (2009). Session 3 and 8: pretreatment and biomass recalcitrance: fundamentals and progress. *Applied Biochemistry and Biotechnology*,Vol. 153, Nos.1-3, (March, 2009), pp.80-83.

Zhang, Y.,P. & Lynd, L.,R. (2004). Toward an aggregated understanding of enzymatic hydrolysism of cellulose: noncomplexed cellulose systems. *Biotechnology and Bioengineering*, Vol.88, No.7 (December, 2004), pp.797-823

Zhao, X.,S.; Bao, X.,Y.; Guo, W. & Lee, F.,Y. (2006) Immobilizing catalysts on porous materials. Materialstoday, Vol. 9, No.3 (February, 2006),pp. 32-39

Zhu, J.Y; Pan, X.,J. (2010). Woody biomass pretreatment for cellulosic ethanol production: Technology and energy consumption evaluation. *Bioresource Technology*, Vol.101, No.13 (July, 2010), pp.4992-5002.

Zverlov, V.; Mahr, S.; Riedel, K. & Bronnenmeier, K. (1998). Properties and gene structure of a bifunctional cellulolytic enzyme (CelA) from the extreme thermophile Anaerocellum thermophilum with separate glycosyl hydrolase family 9 and 48 catalytic domains. *Microbiology*, Vol. 143, (Part2), pp. 3537–3542.

6

Second Generation Bioethanol from Lignocellulosics: Processing of Hardwood Sulphite Spent Liquor

Daniel L. A. Fernandes, Susana R. Pereira, Luísa S. Serafim, Dmitry V. Evtuguin and Ana M. R. B. Xavier
CICECO, Department of Chemistry, University of Aveiro
Portugal

1. Introduction

The world is facing a reduction of global fossil fuels resources, like petroleum, natural gas, or charcoal, while energy requirements are progressively growing up. Fossil fuels should be replaced, at least partially, by biofuels once the current fuel supply is suspected to be unsustainable in the foreseen future. In fact, the search for sustainable alternatives to produce fuel and chemicals from non-fossil feedstocks has attracted considerable interest around the world, to face the needs of energy supply and to response to climate change issues. Alternative resources of energy are being explored in order to reduce oil dependence and increase energy production by exploring of solar, wind, hydraulic and other natural phenomena. Besides these sources of energy, also biomass possesses a potential target for fuel and power production as well as for chemicals or materials feedstocks. Thus biomass can efficiently replace petroleum-based fuels for a long term. (Sanchez *et al.* 2008; Alvarado-Morales *et al.* 2009; Brehmer *et al.* 2009; Gonzalez-Garcia *et al.* 2009; Singhania *et al.* 2009; Mussatto *et al.* 2010; Sannigrahi *et al.* 2010).

Many countries in Europe, North and South America and Asia are replacing fossil fuels by biomass-based fuels according to international regulations. One of the directives of European Union (2009/28/CE) imposes a quota of 10% for biofuels on all traffic fuel until 2020 (Rutz *et al.* 2008; Xavier *et al.* 2010). Also economic incentives for research on biofuels are being implemented all over the world. Bioethanol can be produced from different raw materials containing simple sugars, starch or more complex substrates as lignocellulosics. New methodologies for biofuels (e.g. ethanol and biodiesel) production have been developed in the last years, to achieve new and non cost-intensive technologies for bioconversion of lignocellulosic renewable resources. The most common renewable fuel is ethanol, which is produced from direct fermentation of sugars (e.g. from sucrose of sugarcane or sugar beet) or polysaccharides (e.g. starch from corn and wheat grains) (Gonzalez-Garcia *et al.* 2009; Mussatto *et al.* 2010). The selection of the best raw material is strongly dependent on the local conditions where feedstock is obtained. Evidently, ethanol in Brazil is produced from sugarcane, whereas, in North America or Europe the ethanol industry is based on starchy materials. Besides, energy considerations should be attained: not only the energy input required for ethanol production and the content in fermentable sugars of the feedstock must be considered, but also the annual ethanol yield per cultivated hectare. As suggested, for beet molasses, the yield of ethanol per ton of feedstock is lower than that for corn, but on the other hand, when compared to starchy materials the beet productivity per cultivated hectare, expressed in L/(ha year), is considerably higher, (Sanchez *et al.* 2008).

The growth of the biofuels industry raised questions regarding the sustainability of these "first generation" biofuels. The feedstocks described play an essential role in human and animal food chains, therefore the rise of prices of food all over the world resulted in social disturbance (Gonzalez-Garcia *et al.* 2009; Mussatto *et al.* 2010; Xavier *et al.* 2010). These raw materials were also expected to be limited due to the reassign of arable lands from food to fuel production leading to competition for feedstocks (Gray *et al.* 2006; Bacovsky *et al.* 2010). Moreover, first generation biofuels were accused of not contributing to reduce gas emissions, therefore the use of this technology was highly criticized. For all these reasons additional research in this area is mandatory, in order to search for non-food crops, like wastes from agriculture and/or industry as sources of raw-material. European Union strongly incentives research focusing biotechnological solutions for energy and chemical demands from renewable resources, such as, forestry wastes, agricultural biomass residues and food industrial wastes for "second generation" biofuels production.

The great advantage for the choice of lignocellulosic biomass as feedstock is the non-interference with food chain, which allows the production of bioethanol without using arable lands (Sanchez *et al.* 2008; Zhang 2008). Lignocellulosic biomass is a complex raw material which can be processed in different ways to obtain other value-added compounds contributing to the possibility of establishing a biorefinery. Different value-added products such as lactic acid, acetic acid, furfural, methanol, hydrogen and many other products can be obtained from its sugars. Lignin, the non-carbohydrate component, can be used for the production of advanced materials, polymers and aromatic aldehydes (Sanchez *et al.* 2008; Zhang 2008; Sannigrahi *et al.* 2010; Santos *et al.* 2001). In this way, lignocellulosic biomass can be used as substrate for the production of second generation biofuels, contributing to the diversification of energy supply and gas mitigation, offering less competition for the food and feed industry (Rutz *et al.* 2008; Bacovsky *et al.* 2010). The use of these raw materials to produce fuel, power and value-added chemicals, fits well into the biorefinery concept invoked to decrease the dependence from fossil resources and to improve the economic sustainability (Alvarado-Morales *et al.* 2009; Xavier *et al.* 2010). However, for a world massive utilization of fuel ethanol, a cost-effective technology for ethanol production is also required. In other words, ethanol production costs should be lowered (Sanchez *et al.* 2008). In a biorefinery, different technologies, including fermentation, biocatalytic, thermal and chemical technologies, must be used simultaneously for biomass conversion for the production and the purification of different value-added products (Alvarado-Morales *et al.* 2009).

Bioethanol is one of the products that can be obtained via biorefinery using bio-based resources. It is one of the most attractive biofuels, since it can be easily produced in large amounts and blended with gasoline or used pure as a "green" fuel. Furthermore, due to the higher oxygen content, ethanol allows a better oxidation of the gasoline and reduces CO and particulate emissions. Other advantages of ethanol *versus* gasoline are the higher octane number, broader flammability limits, higher flame speeds, heat of vaporization and compression ratio and a shorter burn time (Balat *et al.* 2008; Mussatto *et al.* 2010). The use of bioethanol can also contribute for the reduction of CO_2 build-up, while the CO_2 content of fossil fuels will remain in storage (Sanchez *et al.* 2008; Gonzalez-Garcia *et al.* 2009; Chen *et al.* 2010a; Balat 2011). Moreover, combustion of ethanol results also in lower NOx emissions, being free of sulphur dioxide. However, as disadvantages, ethanol has an energy density lower than gasoline, it is fully miscible in water and its lower vapour pressure makes motor cold start more difficult (Balat *et al.* 2008; Gonzalez-Garcia *et al.* 2009; Chen *et al.* 2010a; Mussatto *et al.* 2010; Balat 2011). Simultaneously, bioethanol is a building block for the production of several other chemicals, usually petrochemical-based, like acetaldehyde, ethane, ethylene, propylene, butadiene, carbon monoxide or hydrogen (Idriss *et al.* 2000; Wang *et al.* 2008; Yu *et al.* 2009; Lippits *et al.* 2010; Oakley *et al.* 2010; Song *et al.* 2010). Today nearly 95% of hydrogen is produced from fossil-based materials such as methane and naphtha. Bioethanol as chemical reagent for hydrogen production could be a way to support hydrogen economy from a renewable and clean energy source (Yu *et al.* 2009; Lippits *et al.* 2010). Besides, the production of olefins from ethanol has attracted much attention since it valorises bioethanol production under a biorefinery context (Thygesen *et al.* 2010).

In this context Hardwood Sulphite Spent Liquor (HSSL) is a subproduct of pulp and paper industry that results from the acidic sulphite pulping process in high amounts per day. The main objective of acidic sulphite pulping process is to remove lignin and hemicelluloses from wood and to maintain cellulose integrity as much as possible. In this process, lignin and hemicelluloses are hydrolysed and released in the aqueous phase. HSSL can be a suitable substrate for 2nd generation bioethanol production as well as other biobased products since it is rich in monosaccharides obtained during the acidic sulphite pulping process.

2. Lignocellulosics: Variety and chemical composition

2.1 Lignocellulosic biomass as a renewable resource for energetic, chemicals and materials platform

Lignocellulosic biomass (LCB) is the most abundant renewable resource on Earth, comprising about 50% of world biomass. LCB is outside the human food chain and its energetic content exceeds many times world basic energy requirements. These features make it an important option as feedstock, as a relatively inexpensive raw-material, for bioethanol production, and for the development of other bioindustries, to face the international demand for biofuel market. In 2008 it was estimated that 200×10^9 tons of biomass were produced and only 3% were used in pulp and paper industries (Rutz *et al.* 2008; Sanchez *et al.* 2008; Zhang 2008).

The use of LCB as feedstock for bioethanol production results in significant reduction of gas emissions (Sanchez *et al.* 2008; Brehmer *et al.* 2009) and in economic profits increase due to low-cost raw-materials (Balat *et al.* 2008). LCB can be classified based on their origin: wood (softwoods and hardwoods) and shrubs, non-food agricultural crops (kenaf, reed, rapeseed, etc.) and residues (such as olive stones, wheat straw, corncobs, rise husk, sugarcane and winemaking residues, among others), and municipal solid wastes related to thinning, gardening, road maintenance, etc. (Demirbas 2005; Balat *et al.* 2008; Sanchez *et al.* 2008). Wastes from pulp and paper industries, as spent liquors, paper broke, fibres from primary sludge, waste newsprint and office paper or recycled paper sludge are another specific group of LCB to consider.

The conversion of LCB to fermentable monomeric sugars is much more difficult than the conversion of starch. Numerous studies on the development of large-scale production of

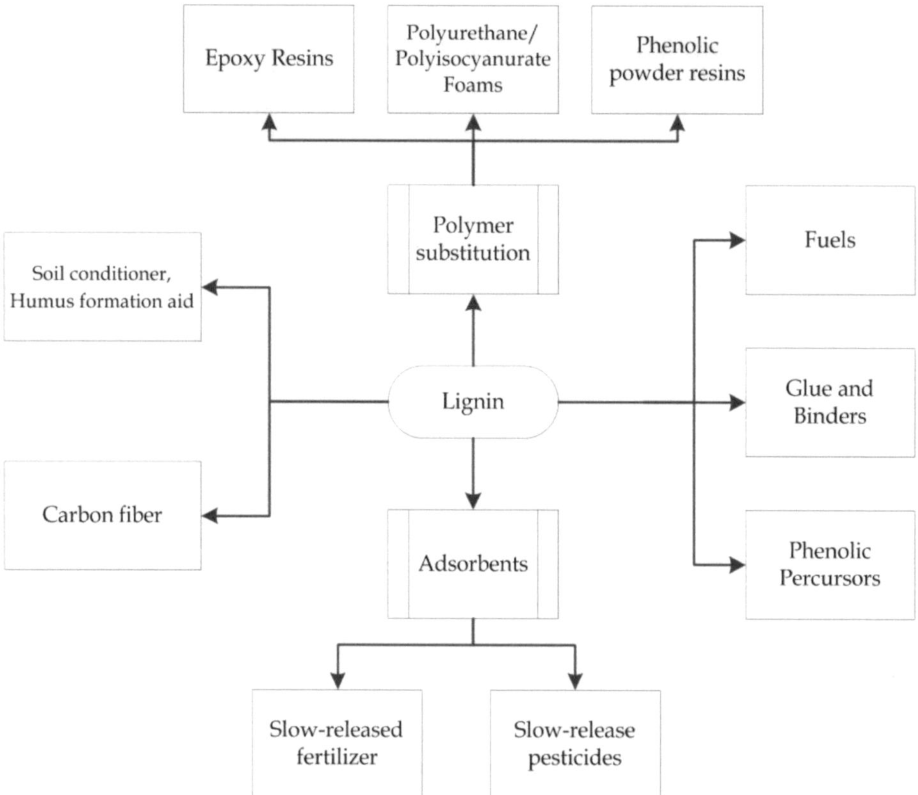

Fig. 1. Lignin potential utilization pathways, adapted from Zhang 2008

ethanol from LCB have been carried out around the world in the last years (Mussatto *et al.* 2004). The particular inherent structure of LCB is the main limiting factor of its conversion to ethanol. Besides cellulose, with a broad range of applications, lignin and hemicelluloses are also considered promising raw materials for the aforementioned purposes. The brief presentation of potential pathways of lignin and hemicelluloses is depicted in Fig. 1 and Fig. 2, respectively.

2.2 Major macromolecular components of lignocellulosic biomass

The composition of LCB depends on the plant species and consists primarily of cellulose, hemicelluloses and lignin, which are the integral part of cell wall in plant tissues (Fig. 3) (Fengel *et al.* 2003). Lignin is an amorphous aromatic biopolymer composed of phenyl propane structural units linked by ether and/or carbon-carbon bonds, supplying tissues stiffness, antiseptic, and hydrophobic properties amongst others (Fig. 4). The types of lignin structural units (*p*-hydroxyphenyl, guaiacyl and syringyl units), their abundance, types and frequency of inter-unit linkages vary significantly from plant to plant (Fengel *et al.* 2003). Lignin contributes to 15-30% of plant biomass and is the principal non-hydrolysable residue of LCB.

Cellulose and hemicelluloses are hydrolysable structural polymers of cell wall and the main sources of fermentable sugars (Lawford *et al.* 1993; Sanchez *et al.* 2008). Hemicelluloses contribute to 10-40% of plant material and are essentially heteropolysaccharides constituted

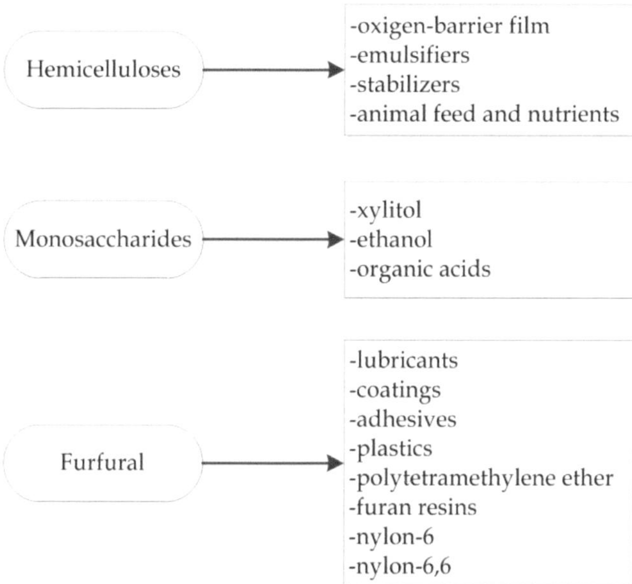

Fig. 2. Hemicelluloses potential utilization pathways, adapted from Zhang 2008

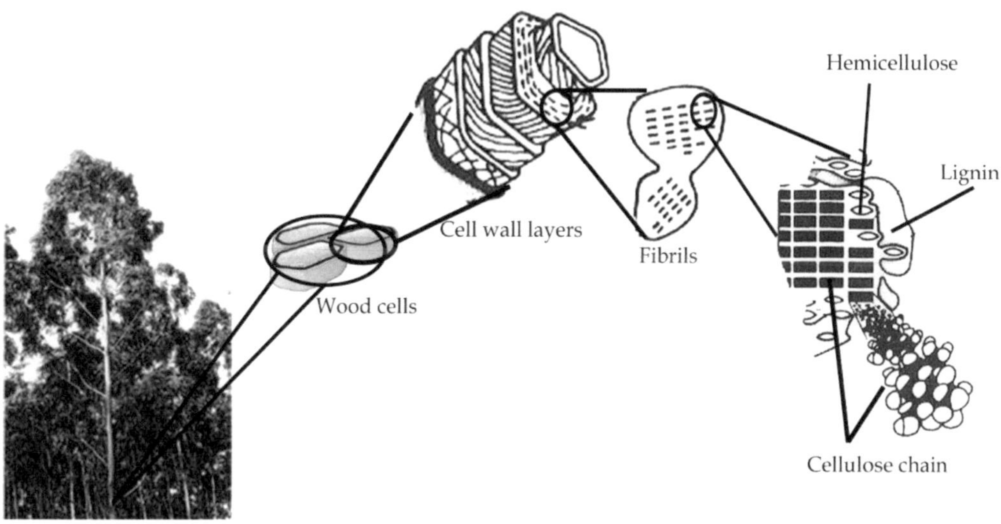

Fig. 3. Representation of wood plant cell wall and its macromolecular components

by pentoses, mainly D-xylose and L-arabinose, and hexoses, mainly D-mannose, D-galactose and D-glucose. These monosaccharides result from pentosans with a main backbone built by pentoses, and hexosans with a main backbone built by hexoses. Hemicelluloses possess an irregular structure and are chemically linked to lignins in the cell wall (Sjöström 1993). The structure and the composition of hemicelluloses vary significantly among plant species. The most abundant hemicelluloses are xylans followed by mannans and galactans (Fig. 5). Hemicelluloses play an important structural role in cell wall regulating the spatial distribution of principal macromolecular components (cellulose and lignin) and providing their compatibility.

Cellulose, the most abundant structural polysaccharide (30-50% abundance in the cell wall), is comprised by repeated β-D-glucopyranose units linked by β(1→4)-glycosidic bonds. In plant cell walls, cellulose chains aggregate into elementary fibrils (EF) which, in turn, are assembled into microfibrils (MF). MF are embedded into a matrix of lignin and hemicelluloses, thus becoming isolated of each other (Fig. 3). Plant cells assembled in different tissues are also separated by a layer enriched in lignin (middle lamella). This structural hierarchy hinders either chemical or enzymatic hydrolysis of cellulose, being the last one particularly difficult. Cellulose, the amorphous-crystalline polymer, is poorly accessible to hydrolysis due to the predominance of crystalline domains.

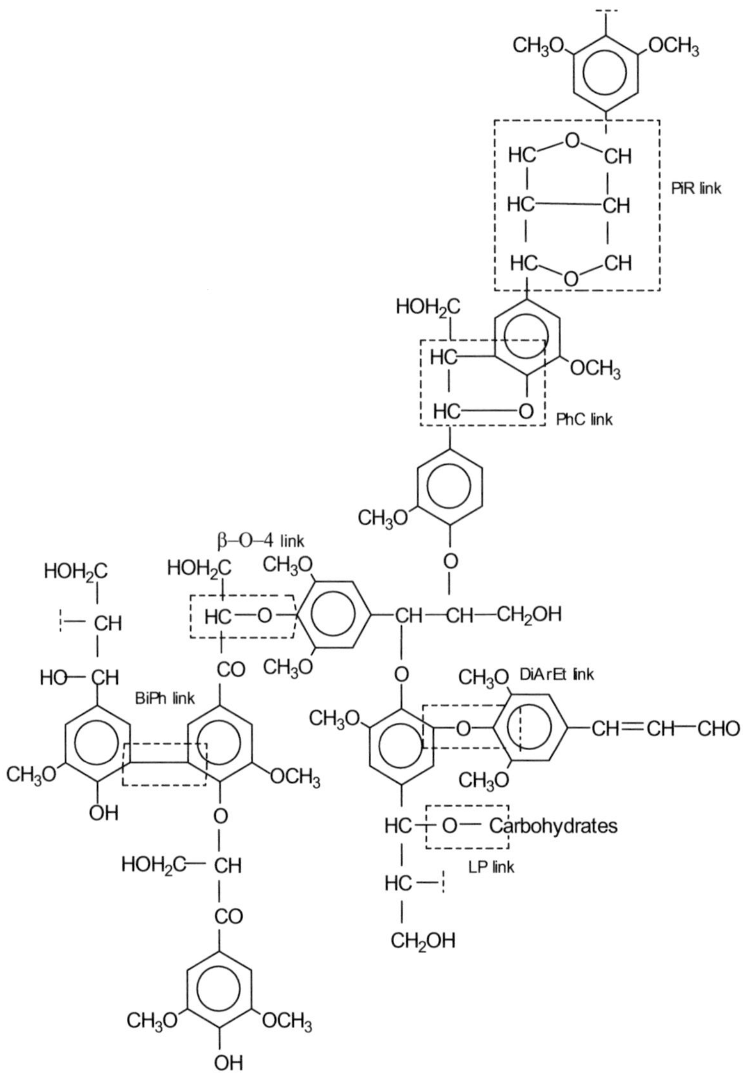

Fig. 4. Schematic representation of lignin (fragment of hardwood lignin)

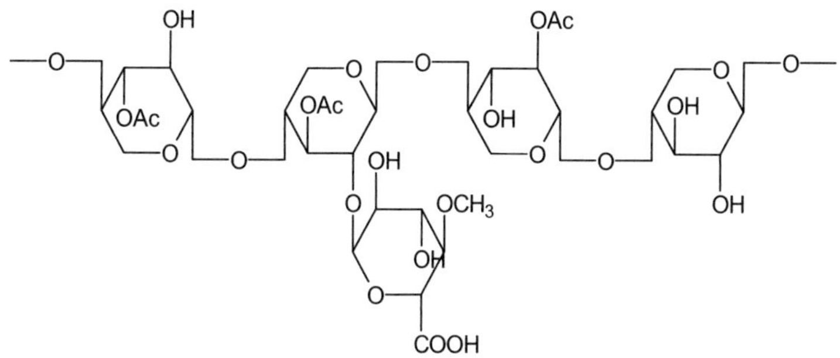

O-acetyl-4-*O*-methylglucuronoxylan

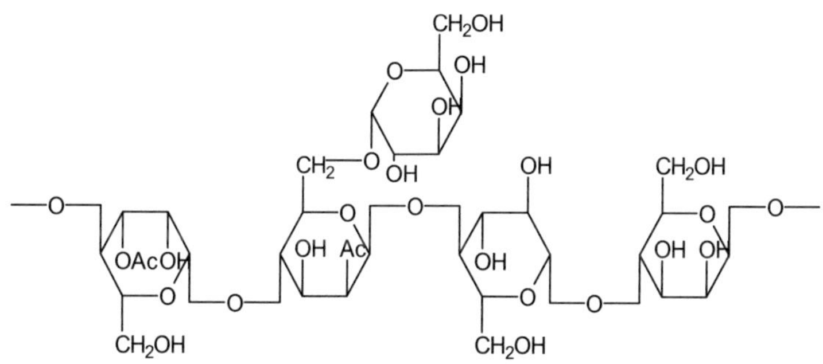

O-acetyl-galactoglucomannan

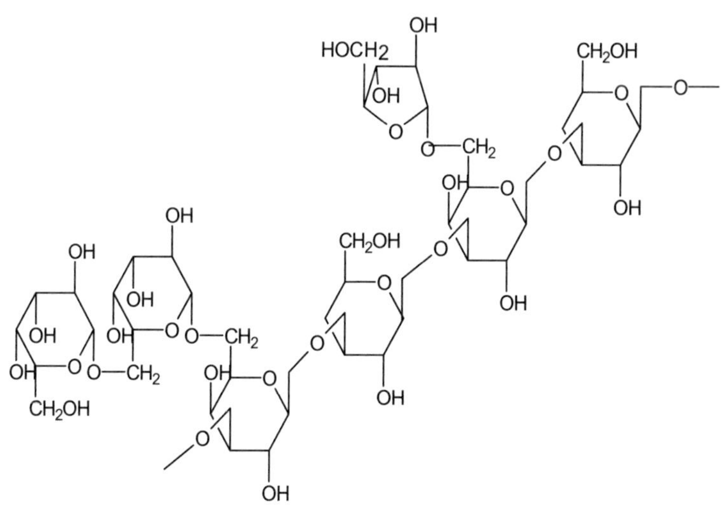

arabinogalactan

Fig. 5. Schematic representation of major hemicelluloses in lignocellulosics

2.3 Hydrolysis of LCB polysaccharides for the ethanol production

Bioethanol production from LCB includes basically the following steps: (1) hydrolysis of cellulose and hemicelluloses; (2) separation of released sugars from lignin residue (3) fermentation of sugars; (4) recovery and purification of ethanol to meet final specifications. The hydrolysis (saccharification) is one of the most important steps and is technically difficult to perform due to the poor accessibility of cellulose caused by many physical, chemical and structural factors mentioned above. It is an energy consuming task, contributing substantially to the economic costs of the process and is a subject of many research works (Mussatto *et al.* 2004; Sanchez *et al.* 2008; Alvira *et al.* 2010; Sannigrahi *et al.* 2010). Hydrolysis can be carried out using organic or strong inorganic acids or enzymes as cellulases and hemicellulases. Some characteristics of different conventional and prospective hydrolytic processes are summarised in Table 1.

Parameter	Hydrolytic processes		
	Dilute acid	Concentrated acid	Enzymatic
Yield of sugars, %	ca 50	80-90	ca 50
Acid consumption	Low	High	-
Reactivity of hydrolysis lignin	Low	Low	High
Technological status	Commercial in former USSR	Pilot scale	Pilot scale

Table 1. Process conditions and properties for different hydrolytic processes

Organic acids, mainly acetic and formic acids, are normally used in the autohydrolysis process and arisen upon hydrothermal treatment of LCB at high temperatures (170-220 °C) as the result of partial degradation of macromolecular components (acetylated xylan/mannan and lignin). These relatively weak organic acids at low concentration are more effective in the hydrolysis of hemicelluloses to a significant extent than of cellulose. Consequently a pre-hydrolysis step is widely used in the production of dissolving pulps by kraft cooking when wood chips are processed prior to pulping by hydrothermal treatment to eliminate significant part of hemicelluloses (Sjöström 1993). The pre-hydrolysis is also a part of pretreatment strategies aiming to hydrolyse selectively the hemicelluloses in LCB to obtain fermentable sugars and/or to improve cellulose accessibility towards hydrolytic enzymes. In this process, the monomeric sugars from hemicelluloses (xylose, galactose, glucose, mannose, and arabinose) and acetic acid are released in the medium (Lawford *et al.* 1993; Sanchez *et al.* 2008). Additionally, degradation of lignin/tannins and sugars originate biologically toxic compounds: gallic acid, syringic acid, pyrogallol, vanillic acid, furfural, 5-hydroxymethylfurfural, among others (Marques *et al.* 2009). Significant efforts were done to minimize the production of such highly toxic compounds, as well as acetic acid, for ethanol-producing microorganisms. The pretreatment should improve recovery of sugars from hemicelluloses, facilitate the cellulose hydrolysis step (when the main objective is the complete saccharafication of all polysccharides from LCB), and avoid the formation of inhibitors for subsequent fermentation processes (Mussatto *et al.* 2004; Alvira *et al.* 2010; Sannigrahi *et al.* 2010).

Inorganic acids (mainly H_2SO_4 and HCl) are effective hydrolysis catalysts and allow complete saccharification of LCB polysaccharides. There are some differences between the use of diluted (1-5%) and concentrated acids in the hydrolysis step. In the first case the complete saccharification takes place at high temperatures (160-180 °C) and leads to the formation of residual hydrolysis lignin (cellolignin) as a massive by-product (Sanchez *et al.* 2008). Due to drastic reaction conditions, sugars are readily degraded via intramolecular dehydration resulting in furfural from pentoses and 5-hydroxymethylfurfural from hexoses. All of these secondary products have a high inhibitory effect on the metabolism of microorganisms. In order to avoid sugars degradation, these compounds should be continuously removed from the reactor by continuous pumping of "fresh" acidic solution through the biomass bed (percolation hydrolysis). This process is used industrially since 1930th in former USSR and nowadays may be considered outdated due to its poor efficiency: low sugars recovery and production of high amounts of chemically inert hydrolysed lignin. The hydrolysis with concentrated acids (50-70% of H_2SO_4 or 30% HCl) allows for effective saccharification of LCB at moderate temperatures (30-80 °C) for short reaction time with high sugars yield. However, due to the technical difficulties and high consumption of acid, this hydrolysis method is not commercialized yet and is implemented only on pilot scale.

The hydrolysis of polysaccharides by hydrolytic enzymes (cellulases and hemicellulases) is one of the most promising tools for the saccharification of LCB. Hydrolytic enzymes permit highly selective hydrolysis of polysaccharides at relatively low temperatures (30-60 °C), practically without emission of products from sugars degradation. Endo-cellulases break internal bonds to disrupt both the amorphous and the crystalline structures of cellulose, exposing its polysaccharide chains. Exocellulase cleaves two to four units from the ends of the exposed chains produced by endocellulase, while β-glucosidase hydrolyses the exocellulase product into individual monosaccharides. Since no degradation of glucose occurs, more sugars could be available for a subsequent fermentation, which is the main advantage of this process. However, this process is slower when compared with acidic hydrolysis and hydrolytic enzymes have poor accessibility to polysaccharides of cell wall, especially cellulose. For these reasons this process is time consuming and results in low sugar yields. LCB enzymatic hydrolysis needs a preliminary treatment step to improve the accessibility of enzymatic attack. This preliminary step includes the application of physical methods (mechanical, hydrothermal, etc.) to disintegrate plant tissues and chemical/biochemical treatments to eliminate concomitant biopolymers, mainly lignin and hemicelluloses, hindering the cellulose accessibility. However, the enzymatic efficiency of cellulose conversion still needs to be improved.

The poor efficiency of mild acidic hydrolysis and, particularly, enzymatic biotreatment for direct saccharification of LCB, represents an obstacle for a successful production of second-generation biofuels. For this reason, the development of pretreatment techniques to improve cellulose accessibility and saccharification efficiency is a permanent challenge (Sanchez *et al.* 2008). A general perspective scheme for LCB conversion into ethanol is presented in Figure 6. The first step presumes LCB pretreatment invoked to degrade strong woody biomass matrix and thus blows away the integral tissues. Different lignocellulosic materials have different physic and chemical characteristics and consequently it is necessary to adopt a specific pretreatment suitable for each raw material. The selected pretreatment will have a determinant effect in the subsequent steps. The amount and type of simple sugars released, toxic compounds formed and their concentration, as well as the overall energy demand and wastewater required in the treatments, depend directly on the specific pretreatment applied (Mussatto *et al.* 2004; Alvira *et al.* 2010). Several methodologies for biomass pretreatments have been developed during the last decades. They can be classified into biological (using brown, white and soft-rot fungus or their lignolytic and cellulolytic enzymes to degrade lignin and hemicelluloses), physical (mechanical milling and extrusion), chemical (alkali or acid pretreatments, ozonolysis, organosolv and pretreatment with ionic liquids) and physicochemical (steam explosion, hydrothermal treatment, ammonia fibre explosion, wet oxidation, microwaves, ultrasound and CO_2 explosion) (Balat *et al.* 2008; Alvira *et al.* 2010; Sannigrahi *et al.* 2010). LCB pretreatment leads to partial or major removal of hemicelluloses in the form of mono- or oligosaccharides. Then, cellulose is prepared for the hydrolysis step, if the objective is fermentation of glucose from cellulose, or for further processing to obtain pulps for textile and paper products (Fig. 6). This extra step (dashed) can be catalysed by dilute or concentrated mineral acids or enzymes (cellulases).

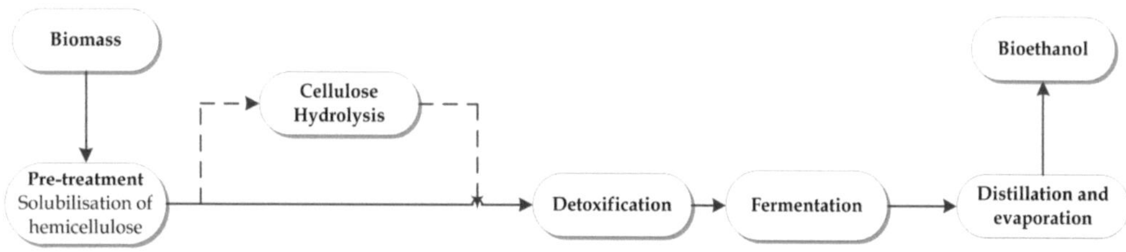

Fig. 6. Schematic steps for production of bioethanol from lignocellulosic biomass

Until now, fuel ethanol from LCB is not yet considered a viable alternative, mainly due to the high complexity involved on this process, compared with the cheaper oil derived fuels. However, in the last years, with the oil crisis, environmental concerns and the increased need for energy and fuels, bioethanol has become a realistic option in the energy market (Cardona *et al.* 2007; Sanchez *et al.* 2008). New research has been developed in order to overcome cellulosic to ethanol bioconversion problems and to make this process a cost-effective technology, with a process integration that combines different steps into one single unit (Lawford *et al.* 1993; Cardona *et al.* 2007). Furthermore, the process integration in other industrial plants, namely large scale industries, can be a good solution for reducing costs of bioethanol production, such as in pulp and paper mill industries, with the advantage of reduced release of subproducts.

3. HSSL as a source of fermentable sugars

3.1 Acidic sulphite wood pulping process and (H)SSL composition

In pulp-and-paper industry the removal of lignin (fibre consolidating material) from wood is carried out during the pulping process to obtain a fibre material (cellulose pulp) suitable for papermaking or as a chemical feedstock. About 10% of chemical pulps are produced worldwide employing sulphite methods. The acidic sulphite chemical pulping is carried out under acidic conditions (pH 1-2) at 135-145 °C for 6-12h in batch digesters using $SO_2/MeHSO_3$ (Me - pulping base) aqueous solution (Sjöström 1993). During sulphite pulping process, lignin and part of hemicelluloses (about 50% based on wood) are dissolved in sulphite spent liquors (SSLs) composed by monomeric sugars already in the fermentable form. Roughly 1 ton of solid waste is dissolved in the spent liquor (SSL 11-14% solids) per ton of pulp produced. SSLs are produced in large amounts, about 90 billion litres annually worldwide (Lawford 1993). SSL is usually burned, for chemical and energy recovery after its concentration by evaporation (Fig. 7). The utilization of SSL is considered for a long time to produce value-added products fitting well to the biorefinery concept (Lawford *et al.* 1993; Marques *et al.* 2009).

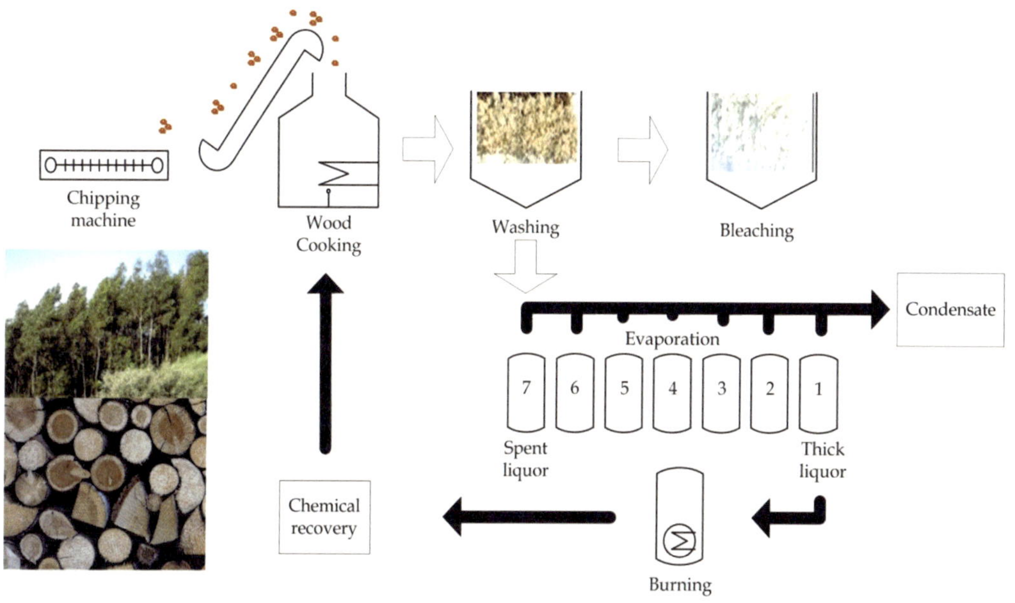

Fig. 7. Representation of acidic sulphite wood pulping process with Spent Sulphite Liquor release

For this reason, the use of raw materials like SSL is advantageous over other agro-forestry wastes, since the more complex lignocellulosic components were previously hydrolysed, releasing most of the sugars as monosaccharides. Consequently this process is already cost-effective for pulp production, improving the 2nd generation bioethanol process economy from SSL (Lawford *et al.* 1993; Helle *et al.* 2008; Marques *et al.* 2009; Xavier *et al.* 2010). However, besides monosacharides, SSL contains several fermentation inhibitors that require a preliminary detoxification step (Lawford *et al.* 1993; Xavier *et al.* 2010).

The major organic components of SSLs are, lignosulphonates, and sugars, and their composition varies notably among softwoods and hardwoods (Table 2). Softwood sulphite spent liquor (SSSL) from coniferous, yields a high proportion of hexose sugars content (about 76%), mainly mannose and glucose, while HSSL, from hardwood *Eucalyptus globulus*, produces a liquor with high content of pentose sugars (xylose about 70%). Hexoses bioprocessing is well studied and already implemented in different processes, while pentoses are difficult to use as feedstock for industrial bioprocesses, because pentoses are not fermented by the yeasts currently used on ethanol production, namely *Saccharomyces cerevisiae*. Therefore, while the use of SSSL has been studied since 1907, when SSSL was used in Sweden for bioethanol production and also during the World War II, for yeast production as a source of protein and vitamins, the HSSL bioprocessing only recently become investigated (Lawford *et al.* 1993; Helle *et al.* 2008; Marques *et al.* 2009; Xavier *et al.* 2010). *Pichia stipitis*, recently reclassified as *Scheffersomyces stipitis* (Kurtzman and Suzuki, 2010), is the most studied yeast capable to convert pentoses to ethanol. However, this yeast is highly sensitive to HSSL inhibitors, namely formic and acetic acids, furfural, levulinic acid and phenolics. For this reason, HSSL needs a special pretreatment for inhibitors removal, which is another technical issue to consider (Helle *et al.* 2008; Xavier *et al.* 2010).

Component	Spruce[1] 52%yield	Birch[1] 49% yield	Eucalyptus[2] 52% yield
Lignosulfonates	480	370	360
Carbohydrates	280	375	200
Xylose	60	340	135
Mannose	120	10	5
Arabinose	10	10	5
Galactose	50	10	30
Glucose	40	5	20
Acetic acid	40	40	50
Extractives	40	60	20

[1] (Sjöström 1993)
[2] (Marques *et al.* 2009)

Table 2. Chemical composition of Spent Sulphite Liquors of Spruce, Birch and Eucalyptus wood (approximate values given in kilograms per ton of pulp)

3.2 Fermentation inhibitors and their removal

As mentioned before, during the conversion of LCB into monomeric sugars, other type of products are formed and some of them can be strong inhibitors in fermentation bioprocesses. When compared to the fermentation of pure sugars, LCB hydrolysates present slower kinetics with a lower ethanol yield and productivity and in some cases a complete inhibition of growth and ethanol production can be observed. The variety and concentration of toxic compounds in feedstocks depend on both, the raw material and the pretreatment conditions applied for polysaccharides hydrolysis. The maximum concentration allowed for each inhibitor, without losing fermentation efficiency, depends on several factors: the origin of toxic compound, the inhibition mechanism, the microbial strain used and its physiological state, and also the fermentative process technology, the dissolved oxygen concentration in the medium and the pH (Mussatto *et al.* 2004).

Selection of a detoxification methodology for a specific feedstock is mandatory for attaining good results in 2nd generation bioethanol production. The identification of the main and relevant inhibitors present in the feedstocks is crucial in order to choose a specific, efficient and low-cost detoxification methodology. Besides, this knowledge can helps to establish the best conditions in hydrolysis pretreatment in order to minimize the inhibitors formation.

Fermentation inhibitors are conventionally classified in four groups according to their origin in lignocellulosics and hydrolysis processing: sugar degradation products, lignin degradation products, compounds derived from extractives and heavy metal ions (Parajó *et al.* 1998; Mussatto *et al.* 2004). Sugar degradation products are formed during hydrolysis and the main compounds produced are furfural from pentoses and 5-hydroxymethylfurfural (HMF) from hexoses as mentioned above. Furfural can inhibit cell growth, affecting the specific growth rate and cell-mass yield (Palmqvist *et al.* 2000b). However, it was noticed that some bioethanol-producing microorganisms like *Pichia stipitis* are not affected by furfural in low concentrations up to 0.5 $g.L^{-1}$ (Mussatto *et al.* 2004). Moreover it could have a positive effect on cell growth. Nigam (2001) referred that ethanol yield and productivity were not affected by 0.27 $g.L^{-1}$ of furfural. However concentrations above 1.5 $g.L^{-1}$ interfered in respiration and inhibited cell growth almost completely, decreasing ethanol yield in 90%

and productivity in 85% (Nigam 2001b). HMF has an inhibitory effect similar to that of furfural, but at a lower extension. Usually HMF is present in lower concentrations than furfural, due to its high reactivity and also due to the experimental conditions in the hydrolysis process that degrades lower amounts of hexoses. It was reported that HMF increases the lag phase extension and decreases cell growth (Delgenes *et al.* 1996; Palmqvist *et al.* 2000b). Mussatto *et al.* (2004) reported that a synergistic effect occurs when these compounds are combined with several other compounds formed during lignin degradation. Different compounds, aromatic, polyaromatic, phenolic, and aldehydic can be released from lignin during hydrolysis of LCB materials, and they are considered more toxic to microorganisms than furfural and HMF, even in low concentrations. Phenolic compounds are the most toxic products for microorganisms present in lignocellulosic hydrolyzates. They promote a loss of integrity in biological membranes, thus, affecting their ability as selective barriers and as enzyme matrices and decreasing cell growth and sugar assimilation (Parajó *et al.* 1998; Palmqvist *et al.* 2000b). Syringaldehyde and vanillic acid affect cell growth (Mussatto *et al.* 2004; Cortez *et al.* 2010) and the ethanolic fermentative metabolism of several microorganisms, like *P. stipitis* (Delgenes *et al.* 1996). In SSL, these compounds are normally present in the sulphonated form, due to the cooking process (Marques *et al.* 2009).

Extractives (acidic resins, taninic, and terpene acids) and also acetic acid derived from acetyl groups present in the hemicellulose are released during the hydrolytic processes. In terms of toxicity, the extractives are considered less toxic to microbial growth, than lignin derivatives or acetic acid (Mussatto *et al.* 2004). Gallic acid and pyrogallol are low molecular weight phenolic compounds normally formed from hydrolysable tannins (Marques *et al.* 2009) and some authors have shown anti-fungal properties of these phenolics (Dix 1979; Panizzi *et al.* 2002; Upadhyay *et al.* 2010). Acetic acid is also known as antimicrobial compound and the mechanism of inhibition is well-understood. At low pH, in the undissociated form, it can diffuse across the cell membrane, promoting the decrease of the cytoplasmatic cell activity and even causing cell death (Lawford *et al.* 1998; Mussatto *et al.* 2004). It has been reported that acetic acid inhibition degree depends not only on its concentration, but also on oxygen concentration and on pH of fermentation medium (Vanzyl *et al.* 1991). Another type of inhibitors are heavy metal ions, namely iron, chromium, nickel and copper, which result from reactors corrosion during the acidic hydrolysis pretreatment. Their toxicity acts at metabolic pathways level, by inhibiting enzyme activity (Mussatto *et al.* 2004).

As previously mentioned, a detoxification step is required before the hydrolysates undergo fermentation. Therefore, after identification of the toxic compounds, the choice of the best hydrolysate detoxification method is crucial for an effective and economical feasible detoxification methodology, in order to improve the fermentative process (Mussatto *et al.* 2004; Sanchez *et al.* 2008). Three different approaches have been described to decrease the concentration of inhibitors: (1) prevention of formation of inhibitors during the pretreatment step as mentioned before; (2) detoxification of the raw-material before fermentation; (3) development of microorganisms able to resist to inhibition.

Xavier and co-workers (2010) reported HSSL containing nearly 25 $g.L^{-1}$ of xylose to *P. stipitis* for bioethanol production. Four increasing concentrations of HSSL were accessed to evaluate its toxicity. The results showed that increasing HSSL content in the fermentation medium decreased dramatically the maximum cell growth rate (μ_{max}), ethanol yield ($Y_{p/s}$) and productivity (qp_m) attained. It was reported that HSSL content higher than 40% (v/v) was critical for bioethanol production (Table 3). Acetic acid has been appointed as the main inhibitor of *P. stipitis* and other microorganisms (Schneider 1996; Lawford *et al.* 1998; Nigam

HSSL content (%)	μ_{max} (h^{-1})	qp_m ($g.L^{-1}.h^{-1}$)	$Y_{p/s}$ ($g_e.g_s^{-1}$)	Acetic acid ($g.L^{-1}$)
0	0.37	0.77	0.37	0
20	0.32	0.40	0.30	1.6
40	0.12	0.10	0.23	3.3
60	0	0	0	4.9

Table 3. Results of bioethanol production by *P. stipitis* at different HSSL contents (Xavier *et al.* 2010)

2001a). After the removal of acetic acid, ethanol fermentations were still unsuccessful, meaning that other compounds present had a toxic effect (Xavier *et al.* 2010).

Several biological, physical and chemical detoxification methods were developed in order to reduce inhibitor concentrations. The efficiency of detoxification methodology depends on chemical composition of the hydrolysate, as well as on microorganism chosen for bioethanol production (Mussatto *et al.* 2004; Helle *et al.* 2008; Sanchez *et al.* 2008). For this reason, the detoxification methods cannot be directly compared since mechanisms of inhibition and degree of toxicity removal are completely different (Palmqvist *et al.* 2000a).

Evaporation with vapour and vacuum evaporation are physical detoxification methods, in order to reduce the concentration of volatile compounds present in the hydrolysates, such as acetic acid, furfural and formaldehyde, and at the same time, to increase sugars concentrations. However, these methods also increase the non-volatile toxic compounds content, such as extractives and lignin derivatives. A balance between these two effects should be achieved or, consequently, the degree of fermentation inhibition will increase. Furthermore, the energy required for these processes should be properly considered to attain a potential economical gain (Lawford *et al.* 1993; Mussatto *et al.* 2004). As mentioned above, in the particular case of HSSL, evaporation is already implemented in the pulp production process for liquor concentration, to prepare it to burn for energy and chemical recovery. This is an advantage for HSSL bioconversion, and it is possible to optimise the evaporation stage, in order to get a good balance between volatile and non-volatile toxic compounds and sugar concentration for the fermentation process. Additionally, the condensate obtained in this step is rich in furfural and acetic acid, that can be easily extracted and purified for selling purposes as added-value products (Evtuguin *et al.* 2010).

Alkali treatment, in particular overliming, is the most common detoxification method and is considered one of the best technologies. This method consists on the addition of lime ($Ca(OH)_2$), or other alkali compound such as sodium or potassium hydroxide, until pH 9-10 promoting the precipitation of toxic compounds. Acetic acid, furfural, HMF, soluble lignin and phenolic compounds are mostly removed with this methodology, increasing the fermentability of hydrolysates. Several authors obtained the best results with alkali treatment using calcium hydroxide (Lawford *et al.* 1993; Martinez *et al.* 2001; Helle *et al.* 2008; Sanchez *et al.* 2008). Martinez *et al.* (2001) reported for sugarcane bagasse hydrolysate at 60 °C that the addition of $Ca(OH)_2$ to adjust the pH to 9.0, promoted the precipitation of furanic and phenolic compounds. The obtained results showed a removal of nearly 51% and 41% respectively, of furans and phenolics with only 8.7% of sugars loss. Lawford *et al.* (1993) also used $Ca(OH)_2$ for HSSL treatment at pH 10, followed by neutralisation to pH 7 with 1N of H_2SO_4. This methodology resulted in the improvement of the volumetric productivity and conversion efficiency, 92%, of bioethanol production by a recombinant strain of *Escherichia coli*.

Toxic compounds can also be removed by adsorption. Several authors have studied the capacity of removal of toxic compounds using different materials as adsorbents such as, activated charcoal (Dominguez *et al.* 1996; Lee *et al.* 1999; Mussatto *et al.* 2001; Canilha *et al.* 2004) and ion-exchange resins (Vanzyl *et al.* 1991; Larsson *et al.* 1999; Lee *et al.* 1999; Nilvebrant *et al.* 2001; Xavier *et al.* 2010). In particular, a specific strategy of adsorption on ion-exchange resins was employed by Xavier *et al.* (2010) to toxic compounds removal from HSSL for subsequent sugar purification and then ethanol fermentation with *P. stipitis* (Fig. 8).

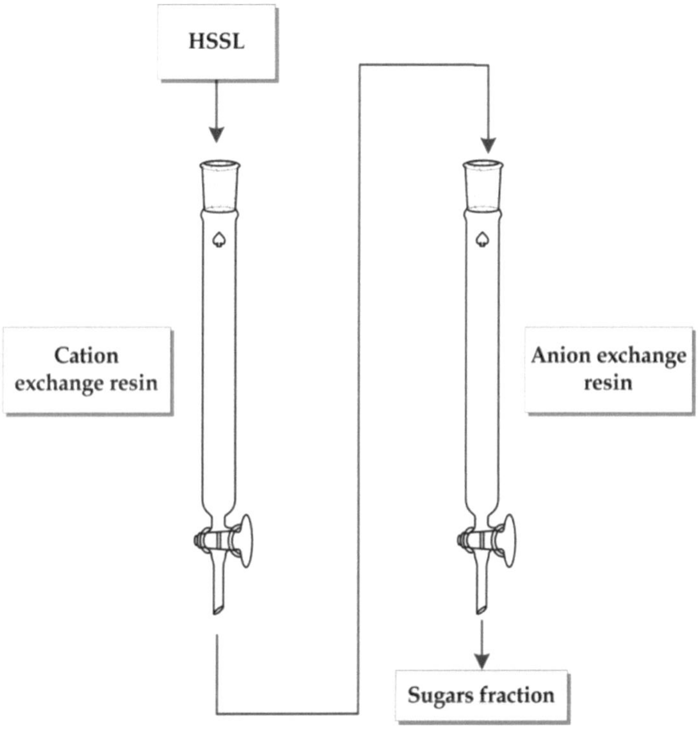

Fig. 8. Scheme of HSSL detoxification by adsorption of inhibitors using ion-exchange resins

In order to remove the cations added during pulping processing, namely Mg^{2+}, HSSL was initially treated with a cation-exchange resin column. Then free carboxylic acids and polyphenols, including lignosulphonates, were separated from sugars with an anion-exchange resin in the second column. This process provided a transparent solution (sugars faction) containing essentially neutral monomeric sugars with traces of neutral polyphenolics (Table 4). However, this separation process released the sugars with some dilution and a concentration step was required for fermentation. This procedure led to excellent results of ethanol production by *P. stipitis*: high fermentation efficiency, 96%, productivity, 1.22 $g.L^{-1}.h^{-1}$, and yield, 0.49 g of ethanol / g of sugar.
Biological methods for detoxification of hydrolysates involve the use of specific enzymes or microorganisms that can degrade or consume the toxic compounds present in the hydrolysates. Jönsson *et al.* (1998) reported an increasing glucose consumption and ethanol productivity when wood hydrolysates were detoxified with laccase and peroxidase enzymes from *Trametes versicolor*, a white-rot fungus. These oxidative enzymes have the capability to degrade acid and phenolic compounds (Jonsson *et al.* 1998). The use of

Compound	Concentration ($g.L^{-1}$)
Lignosulphonates	traces
Acetic acid	n.d.[a]
pH	5.4 ± 0.1
Xylose	5.7 ± 0.3
Glucose	0.5 ± 0.2

[a]not detected

Table 4. Chemical composition of sugars fraction after ion-exchange detoxification

microorganisms was also proposed to remove inhibitors from HSSL. Xavier and co-workers (2010) presented the first approach for HSSL biological detoxification, specifically for acetic acid removal. Four yeasts commonly used for acetic acid removal from wine were chosen, *Candida tropicalis, Candida utilis, S. cerevisiae and Pichia anomala,* and results are presented in Table 5.

Yeast	μ_0 (h^{-1})	Time of complete consumption of acetic acid (h)
Saccharomyces cerevisiae	0.15 ± 0.02	20
Candida tropicalis	0.14 ± 0.03	70
Candida utilis	0.16 ± 0.05	220
Pichia anomala	0.22 ± 0.03	72

Table 5. Results of biological deacidification of HSSL (Xavier *et al.* 2010)

According to these results, *S. cerevisiae* was selected for biological deacidification of HSSL. Sequential strategy of deacidification by *S. cerevisiae* and fermentation by *P. stipitis* on 60% of HSSL was carried out. Despite the acetic acid consumption by *S. cerevisiae,* xylose fermentation by *P. stipitis* produced only cell biomass, and no ethanol was detected in the medium. These results clearly showed the presence of other toxic compounds from HSSL, eventually phenolic compounds, probably inhibiting the sugars conversion to ethanol by *P. stipitis* (Xavier *et al.* 2010).

A different approach for performing biological detoxification of HSSL, with better results, was made in the same research group, using the *Paecilomyces variotti* filamentous fungus. This fungus can be found in air and soils of tropical countries, and has been studied for single cell protein (SCP) production, another important added-value product, normally used in animal feeding (Nigam 1999). Besides, *P. variotti* presents a good performance to grow in residues like HSSL and consumes substrates, including phenolic compounds, as carbon source. Pereira *et al.* (2011) showed for the first time the possibility of using this fungus to detoxify HSSL hydrolysates for subsequent ethanol fermentation. The biological treatment with *P. variotti* yielded HSSL with very low levels of acetic acid. Moreover, toxic compounds like gallic acid, pyrogalol and other low molecular phenolics were completely consumed and metabolized by *P. variotti,* indicating that this detoxification method can be suitable for treating HSSL into a proper feedstock for further bioprocessing. A successful fermentation of this detoxified HSSL by *P. stipitis* was performed, attaining an ethanol yield of 0.24 $g_{ethanol}{\cdot}g_{sugars}^{-1}$. However, more research is required in order to improve the ethanol fermentation yields and productivities (Pereira *et al.* 2011).

Comparing the four different detoxification methodologies described, ion-exchange resins provided the best results on subsequent bioethanol fermentation (Table 6). High percentages of different toxic compounds from the hydrolysate were removed and provided the highest ethanol yield (0.49 $g.g^{-1}$) and volumetric productivity (1.22 $g.L^{-1}.h^{-1}$). However, ion-exchange resins are expensive and difficult to implement and operate in large scale industries. *P. variotti* treatment, despite the fact of having promoted low ethanol fermentation yields in preliminary results (Table 6), appeared to be a very promising detoxification method. Furthermore the biomass of *P. variotti* can be used as SCP for animal feeding, increasing the economic potential of the process. More research work is being developed to combine this coupled strategy of biological detoxification of HSSL with simultaneous SCP production (Pereira *et al.* 2011). Other approaches for detoxification of hydrolysates were proposed and different methods can be used sequentially to improve their own capacity (Mussatto *et al.* 2004).

Treatment	Ethanol ($g.L^{-1}$)	$Y_{p/s}$ ($g_{et}.g_s^{-1}$)	Conversion Efficiency (%)	Strain and feedstock	Reference
Ion-exchanges Resins	8.10	0.49	96	*P. stipitis*/HSSL	(Xavier *et al.* 2010)
Evaporation + alkaline treatment	9.7	0.30	59	*P. stipitis*/HSSL	(Nigam 2001a)
P. variotti	2.36	0.24	47	*P. stipitis*/HSSL	(Pereira *et al.* 2011)
Ion-exchanges Resins	n.a.	0.45	88	*S. cerevisiae*/ Spruce hydrolysate	(Nilvebrant *et al.* 2001)
Alkaline treatment	10.0	0.40	78	*Escherichia coli*	(Lawford *et al.* 1993)
Alkaline treatment	12.2	0.25	49	*P. stipitis*	(Vanzyl *et al.* 1988)

Table 6. Results of bioethanol production for different detoxification methodologies

3.3 Microorganisms and their metabolism

The extension of substrate utilisation is critical to determine the economic viability of ethanol production from LCB. This presumes a complete conversion of sugars presented in feedstocks to ethanol under industrial conditions. Under an industrial context, the microorganism chosen should meet some requirements, which are discussed in relation to four benchmarks: (1) Process water economy; (2) Inhibitor tolerance; (3) Ethanol yield; (4) Specific ethanol productivity. Several species of bacteria, yeast and filamentous fungi naturally ferment sugars to ethanol. Each microorganism has its advantages and disadvantages, some can use only hexoses for producing ethanol and others can use both, hexoses and pentoses, but many times with low ethanol yields (Hahn-Hagerdal *et al.* 2007).
The mixture of sugars obtained after LCB hydrolysis, besides glucose, also contains other sugars e.g. xylose, mannose, galactose, arabinose and also some oligosaccharides. Therefore, in the fermentation process, microorganisms ferment these sugars into bioethanol according

to reactions presented below. The calculation of the theoretical maximum yield should follow equation 1 for pentoses or equation 2 for hexoses:

$$3C_5H_{10}O_5 \rightarrow 5C_2H_5OH + 5CO_2 \quad (1)$$

$$C_6H_{12}O_6 \rightarrow 2C_2H_5OH + 2CO_2 \quad (2)$$

According to these equations, the theoretical maximum yield is 0.51 g bioethanol and 0.49 g carbon dioxide per g of xylose and glucose.

In order to obtain an economically feasible conversion process of any biomass, it is imperative that the microorganisms chosen should be able to convert efficiently all the sugars present into the desired end product, in this case bioethanol (Chu *et al.* 2007; Hahn-Hagerdal *et al.* 2007; Matsushika *et al.* 2009). The ideal yeast for bioethanol production from LCB should consume the sugars present and provide high production yields as well as specific productivities. Moreover it should not suffer any inhibition from the other components of the raw material (Hahn-Hagerdal *et al.* 2007).

One of the most effective and well-known ethanol producing microorganisms from hexose sugars is the yeast *S. cerevisiae.* This yeast is successfully employed at industrial scale, allowing for high ethanol productivity, since it bears high tolerance to ethanol and to inhibitors normally present in lignocellulosic residues. However, this yeast is unable to ferment xylose to ethanol efficiently, though it can only ferment its isomer, xylulose (Jeppsson *et al.* 2006; Chu *et al.* 2007; Hahn-Hagerdal *et al.* 2007; Matsushika *et al.* 2009). Some yeasts were reported to be efficient in xylose conversion to ethanol, such as, *P. stipitis, Candida shehatae* and *Pachysolen tannophilus* (Huang *et al.* 2009). Among them, *P. stipitis* exhibits the best potential for industrial application due to the high ethanol yield obtained (Huang *et al.* 2009). Nevertheless, this yeast is sensitive to organic acids, including acetic acid, which are present in lignocellulosic residues. These compounds inhibit both cell growth and the bioethanol production (Bajwa *et al.* 2009; Huang *et al.* 2009). Although, wild type S. *cerevisiae* cannot ferment xylose to ethanol, several genetic engineered strains have been already developed (Hahn-Hagerdal *et al.* 2007; Mussatto *et al.* 2010). Other yeasts, like *P. stipitis,* can naturally utilize both types of sugars with high yields and its use for producing 2nd generation bioethanol from HSSL is being developed (Xavier *et al.* 2010). Hence, it is important to improve the yeast strain with the most promising characteristics in order to optimize ethanol production from LCB hydrolysates through genetic engineering and/or strain adaptation (Chu *et al.* 2007; Hahn-Hagerdal *et al.* 2007; Matsushika *et al.* 2009). Table 7 summarizes the fermentation performance of several yeasts in different media.

Among bacteria, the most promising for industrial implementation are *Escherichia coli, Klebsiella oxytoca* and *Zymomonas mobilis. Z. mobilis* is the bacteria which has the lowest energy efficiency resulting in a higher ethanol yield (up to 97% of theoretical maximum). However, this bacterium is only able to ferment glucose, fructose and sucrose to ethanol. Another problem appears when the medium has sucrose, due to the formation of the polysaccharide levan (made up of fructose), which increases the viscosity of fermentation broth, and of sorbitol, a product of fructose reduction that decreases the efficiency of the conversion of sucrose into ethanol (Lee *et al.* 2000). *K. oxytoca,* an enteric bacterium, found in paper, pulp streams and different sources of wood, is able to grow at low pH (minimum 5.0) and temperatures up to 35 °C. This bacterium is able to grow either on hexoses or pentoses, as well as on cellobiose and cellotriose (Lee *et al.* 2000; Cardona *et al.* 2007; Chen *et al.* 2010b).

Yeast	Strain	Description	Type of medium	Detoxification method	Fermentative process	$Xyl_{initial}$ ($g\ L^{-1}$)	Y_E ($g\ g^{-1}$)	Reference
Pichia stipitis	BCRC21777	Wild type	Rice straw hydrolysate	Overliming	Batch 30ºC;100rpm	21	0.37	(Huang *et al.* 2009)
	BCRC21777	Adapted	Rice straw hydrolysate	Overliming	Batch 30ºC;100rpm	21	0.45	(Huang *et al.* 2009)
	NRRL Y -7124	Adapted	HSSL	Overliming	Batch 30ºC	40	0.30	(Nigam 2001a)
	NRRL Y -7124	Wild type	HSSL	Ion-exchange resins	Batch 29ºC;180rpm	21	0.48	(Xavier *et al.* 2010)
Saccharomyces cerevisiae	ADAP8	XYLA[1],XKS1/ SUT1	Complex	None	Batch 30ºC;200rpm	20	0.35	(Madhavan *et al.* 2009)
	MA-N5	XYL1/XYL2/X KS1	Complex	None	Batch	45	0.36	(Matsushika *et al.* 2009)
	MA-R4	XYL1/XYL2/X KS1	Complex	None	Batch	45	0.35	(Matsushika *et al.* 2009)
	MA-R5	XYL1/XYL2/X KS1	Complex	None	Batch	45	0.37	(Matsushika *et al.* 2009)
	TMB 3400	n.a.[2]	Spruce hydrolysate	n.a[2].	Fed-batch	6	0.43	(Hahn-Hagerdal *et al.* 2004)
	TMB 3006	n.a[2].	Spruce hydrolysate	n.a[2].	Fed-batch	6	0.37	(Hahn-Hagerdal *et al.* 2004)
	MT8-1	XYL1/XYL2/X KS1	Lignocellulosic Hydrolysate	Biological with enzymes	Batch 30ºC;100rpm	9	0.41	(Katahira *et al.* 2006)
	F 12	XYL1/XYL2/X KS1	Vinasse residue	Biological with enzymes	Batch 30ºC;300rpm	6	0.27	(Olsson *et al.* 2006)

[1]xylose isomerase from P*iromyces sp.*
[2]not available

Table 7. Fermentation performance of several yeasts in different media

Several metabolic engineering and genetic modification strategies to enhance an efficient fermentation of xylose to ethanol were studied for *S. cerevisiae* (Chu *et al.* 2007; Hahn-Hagerdal *et al.* 2007; Matsushika *et al.* 2009). Although the genes that allow for xylose utilization are present in *S. cerevisiae,* they are expressed in low levels resulting in production rates of ethanol from xylose ten times lower than the verified for glucose as substrate (Chu *et al.* 2007; Hahn-Hagerdal *et al.* 2007). In pentose-fermenting yeasts, xylose catabolism begins with its reduction to xylitol by a NADH- or NADPH-dependent xylose reductase (XR), as seen in Fig. 9. Then, xylitol is oxidized to xylulose by NAD-dependent xylitol dehydrogenase (XDH) (Chu *et al.* 2007; Hahn-Hagerdal *et al.* 2007; Bengtsson *et al.* 2009). Xylulose is phosphorylated by the enzyme xylulokinase (XK) to produce xylulose-5-phosphate (X5P). This enters in glycolytic pathway and then in the pentose phosphate pathway (PPP). The formed intermediates are converted to pyruvate in the Embden–Meyerhof–Parnas pathway. Under anaerobic conditions, fermentation of pyruvate occurs by decarboxylation promoted by pyruvate decarboxylase to acetaldehyde which is then reduced to ethanol by alcohol dehydrogenase (Chu *et al.* 2007; Hahn-Hagerdal *et al.* 2007).

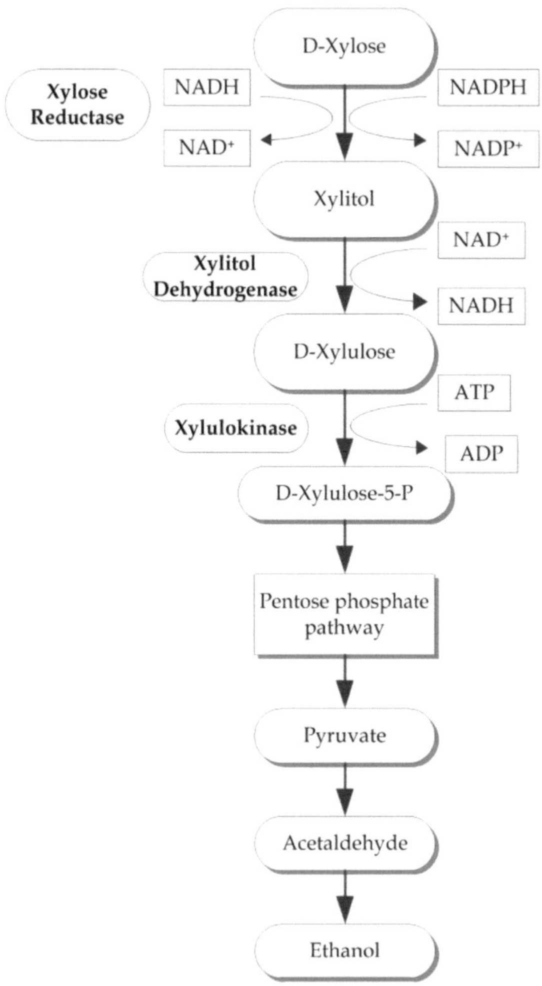

Fig. 9. Xylose metabolic pathway in yeasts (adapted from Matsushika *et al.* 2009)

The most straightforward metabolic engineering strategy was the expression of a bacterial xylose isomerase (XI) gene, so that xylose can directly be converted to xylulose (Jeppsson *et al.* 2006). The XI gene from the thermophilic bacterium *Thermus thermophilus* was successfully expressed in *S. cerevisiae,* generating xylose-fermenting recombinant strains (Karhumaa *et al.* 2005). Also, the genes of *Piromyces sp.* XI were also successfully expressed in *S. cereviasiae* (Kuyper *et al.* 2003). Another possible metabolic engineering strategy consisted in expressing fungal XR and XDH genes. Stable xylose-fermenting *S. cerevisiae* strains were obtained by integrating the *P. stipitis* XYL1 and XYL2 genes encoding XR and XDH, respectively, and over expressing the endogenous XKS1 gene encoding xylulokinase (XK) (Bengtsson *et al.* 2009; Matsushika *et al.* 2009). However, ethanol yield attained with these strains was far from the theoretical maximum of 0.51 $g.g^{-1}$, as can be seen in Table 7 because the metabolic pathway stopped in xylitol. This situation was attributed to the fact that since XR is NAD(P)H-dependent and XDH is strictly NAD+-dependent the relation between the two cofactors sometimes becomes unbalanced (Jeppsson *et al.* 2006; Chu *et al.* 2007; Bengtsson *et al.* 2009).

Wahlbom and Hahn-Hägerdal (2002) found that the addition of electron acceptors such as acetoin, furfural and acetaldehyde re-oxidized NAD^+ needed by XDH and decreased the amount of xylitol formed. Shifting the cofactor utilization in the XR step from NADPH to NADH was also a successful strategy for decreasing xylitol (Jeppsson *et al.* 2006). Since *S. cerevisiae* lacks the xylose-specific transporter, another common approach is to express in this microorganism the gene that encodes the transport of monosaccharides from *P. stipitis* (Van Vleet *et al.* 2009). Hence, xylose uptake occurs by facilitated diffusion mainly through non-specific hexose transporters, which have lower affinity for xylose (Matsushika *et al.* 2009). This approach enhanced xylose fermentation to ethanol by *S. cerevisiae* (Van Vleet *et al.* 2009).

In addition to metabolic engineering, natural selection of strains and random mutation are also alternatives to obtain improved xylose-fermentative yeasts. These evolutionary engineering approaches were successfully applied to several *S. cerevisiae* strains for effective xylose fermentation. These methods are particularly useful since they are non-invasive and can identify bottlenecks in the xylose metabolic pathway that can then be targeted to be overcome by genetic engineering (Chu *et al.* 2007; Matsushika *et al.* 2009). Chu and Lee (2007) suggested that an intense selection pressure will favour the presence of *S. cerevisiae* mutants able to grow slowly on xylose.

Recent studies have redirected their attention to the xylose-fermenting yeast, *P. stipitis.* In this case, the major issue is the inhibitors tolerance which can be critical when real raw materials are tested. Hence, an evolutionary strategy has been adopted. The strains adaptation was normally accomplished by sequential transfer of culture samples to different media composed by increasing concentrations of the residue in study (Mohandas *et al.* 1995; Bajwa *et al.* 2009; Huang *et al.* 2009). To accelerate the mutations, ultra violet radiation (UV) was also tested by Bajwa and co-workers (Bajwa *et al.* 2009).

Many challenges in ethanol production from xylose using metabolically engineered strains were being overcome. Several approaches were successfully employed to engineer xylose metabolism. Nevertheless, these approaches are insufficient for industrial bio-processes mainly due to the low fermentation rate of xylose when compared with glucose. Another bottleneck is the lack of tolerance to the major inhibitors present in lignocellulosic feedstocks. A successful fermentation of LCB hydrolysates requires not only a producing strain that consumes all the sugars present but with tolerance towards lignocellulose degradation products. Moreover, most of the methodologies tested were applied to defined synthetic media containing pure substrates and their applicability to real complex substrates should be validated. However, the composition of the inhibitors in raw materials as lignocellulosic wastes changes frequently and, consequently, the metabolic engineering method probably need some modifications to be applied (Hahn-Hagerdal *et al.* 2007; Matsushika *et al.* 2009). Metabolic engineering approaches to improve inhibitor tolerance were so far limited to the over expression of specific enzymes including laccase, phenylacrylic acid decarboxylase, glucose 6-phosphate dehydrogenase and alcohol dehydrogenase (Hahn-Hagerdal *et al.* 2007). These enzymes can transform some of the inhibitors (mainly the aromatic compounds) into products that microorganisms can assimilate.

In brief, the technical and economic issues related to the choice of fermenting microorganism are the conversion efficiency uniformity, the tolerance to inhibitors, the process requirements (aeration, temperature, pH, sterilization) and the bioprocess licensing (Lawford *et al.* 1993). Further intensive studies that combine functional genomics analysis with metabolic engineering are required for developing robust yeast strains, tolerant to several inhibitors and to the variability of the substrate and with the ability to ferment xylose from lignocellulosic feedstocks, in order to produce ethanol, at similar rates as those attained with glucose, to be applied at industrial level (Chu *et al.* 2007; Hahn-Hagerdal *et al.* 2007; Matsushika *et al.* 2009).

4. Biorefinery approach

With the depletion of petroleum resources and increasing demand on energy, lignocellulose derived ethanol seems to be the future of transportation fuels. Also, it is noticeable that the integrated biorefineries, which generate chemicals, materials, fuels and energy from the biomass, would replace the current petroleum refineries, moving the world toward a carbohydrate-based economy (Gnansounou 2009).

By-products like HSSL cannot be discharged into natural basins due to environmental concerns (211 g $COD.L^{-1}$) and must be processed (Evtuguin *et al.* 2010). The biochemical processing of HSSL is a well-known approach to produce value-added products such as SCP and ethanol, among others (Busch *et al.* 2006).

As seen previously, biological detoxification of HSSL by *P. variotti* was possible and the fungal biomass obtained (2.0 g biomass/g substrate consumed) can be sold as SCP, for animal nutrition. For process optimization a Sequential Batch Reactor (SBR) was chosen and the same inoculum was used during three batches to treat fresh HSSL. Each cycle was ended when the acetic acid reached a non-inhibitory concentration for *P. stipitis* and this operating strategy provided high volumes of detoxified HSSL, for subsequent bioethanol fermentation (Pereira *et al.* 2011). With this detoxification process as well as with the described ion-exchange process (Xavier *et al.* 2010) HSSL can be further bioprocessed by *P. stipitis*, as reviewed before. The maximum concentration of ethanol attained was 8.1 $g.L^{-1}$ with a yield of 0.49 g ethanol.g $sugars^{-1}$ (Xavier *et al.* 2010). The bioethanol produced from HSSL, regarding the aforementioned fermentation results, may be estimated as high as 100 litters per one ton of pulp (Evtuguin *et al.* 2010).

Biopolymers are also important value-added products that can be produced within a biorefinery concept, being capable to replace fossil-fuels based polymers. Microbial mixed cultures (MMC) under aerobic dynamic feeding conditions (ADF) in HSSL, can utilize acetic acid for polyhydroxyalkanoates (PHAs) production. PHAs are biodegradable plastics that can be stored intracellularly by bacteria from renewable resources. A MMC culture was selected in a SBR under ADF conditions using HSSL as substrate and was able to produce 37.7% of PHA per cell dry weight. The microorganisms were able to uptake the acetic acid and also xylose and store them as PHA. Another polymer that was possible to produce from HSSL is bacterial cellulose (BC). The majority of the cellulose available on earth is produced by plants but some microorganisms such as algae, fungi and bacteria are also able to produce an extra-cellular form of cellulose. Bacteria belonging to the genera *Gluconacetobacter, Sarcina or Agrobacterium* are able to produce BC. BC is highly pure, since it is not associated with hemicelluloses and lignin as in plants (Klemm *et al.* 2001). BC bears also unique physical and mechanical properties that arise from its tridimensional and branched nano and micro-fibrillar structure (Iguchi *et al.* 2000). Finally, BC shows biocompatibility, being an excellent material for biomedical applications (Carreira *et al.* 2011). Carreira *et al.* (2011) using HSSL and *Gluconacetobacter sacchari* was able to produce 0.29 $g.L^{-1}$ of BC with a conversion ratio of 28% and yield of 105%. Although the production of BC was low, when compared to the results obtained with pure compounds (2.70 $g.L^{-1}$ with glucose) these preliminary results showed that it was possible to produce BC from this kind of by-product (Carreira *et al.* 2011).

Even after all these bioprocesses, the remaining residues (biomass, sugars not consumed and other compounds) still represent a large amount of carbon oxygen demand (>100 $gCOD.L^{-1}$). In this way, anaerobic digestion (AD) has shown great potential in using renewable resources such as management residues. AD is a biological process by which organic matter is transformed into methane and carbon dioxide in the absence of oxygen (Mata-Alvarez *et al.* 2000). The digestion process begins with bacterial hydrolysis of the input materials in order to break down insoluble organic polymers, such as carbohydrates, and make them available for other bacteria. Then acidogenic bacteria convert sugars and amino acids into carbon dioxide, hydrogen, and organic acids. These are then converted into acetic acid, along with additional hydrogen, and carbon dioxide. Finally, methanogenic microorganisms convert these products to methane and carbon dioxide (Mata-Alvarez *et al.* 2000). Preliminary results showed that with a MMC, volatile fatty acids (VFAs) like acetic, proprionic and n-butiric acids, can be produced using HSSL after bioethanol production. A yield of 0.15 mg COD.mg COD^{-1} for the acetic acid was obtained using AD. Although AD in HSSL is a research area in progress, acidification of HSSL and wastes from bioethanol production is possible. Not only the remaining sugars but also other compounds present in the HSSL were converted into VFAs.

SCP, bioethanol, PHAs, BC and VFAs are some of the value-added products that so far, can be produced from HSSL the subproduct of acidic sulphite pulping process (Fig. 10). Although yields were low for industrial implementation, most of these data are preliminary results and would be useful to optimize the process and develop new strategies towards a comprehensive utilization of by-products from sulphite pulp production thus fulfilling the environmental concerns, improving the sustainability of pulp plant and contributing also for the pulp mill profits. Optimization of all these processes is a necessary step for improving productivity for the biorefinery implementation in the industrial process and the commercial application of the value-added products.

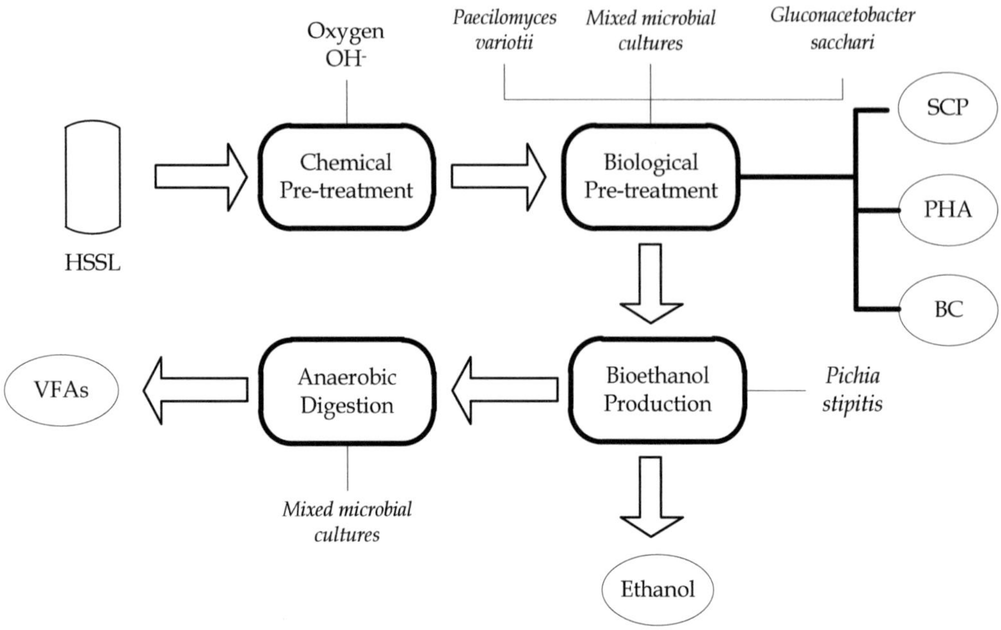

Fig. 10. Value-added products that can be produced using HSSL in the biorefinery concept

5. References

Alvarado-Morales, M., J. Terra, K. V. Gernaey, J. M. Woodley & R. Gani. (2009). Biorefining: Computer aided tools for sustainable design and analysis of bioethanol production. *Chemical Engineering Research & Design,* Vol. 87, No. 9A, pp. 1171-1183, ISSN 0263-8762.

Alvira, P., E. Tomas-Pejo, M. Ballesteros & M. J. Negro. (2010). Pretreatment technologies for an efficient bioethanol production process based on enzymatic hydrolysis: A review. *Bioresource Technology,* Vol. 101, No. 13, pp. 4851-4861, ISSN 0960-8524.

Bacovsky, D., W. Mabee & M. Worgetter. (2010). How close are second-generation biofuels? *Biofuels Bioproducts & Biorefining,* Vol. 4, No. 3, pp. 249-252, ISSN

Bajwa, P. K., T. Shireen, F. D'Aoust, D. Pinel, V. J. J. Martin, J. T. Trevors & H. Lee. (2009). Mutants of the Pentose-Fermenting Yeast *Pichia stipitis* With Improved Tolerance to Inhibitors in Hardwood Spent Sulfite Liquor. *Biotechnology and Bioengineering,* Vol. 104, No. 5, pp. 892-900, ISSN 0006-3592.

Balat, M. (2011). Production of bioethanol from lignocellulosic materials via the biochemical pathway: A review. *Energy Conversion and Management,* Vol. 52, No. 2, pp. 858-875, ISSN 0196-8904.

Balat, M., H. Balat & C. Oz. (2008). Progress in bioethanol processing. *Progress in Energy and Combustion Science,* Vol. 34, No. 5, pp. 551-573, ISSN 0360-1285.

Bengtsson, O., B. Hahn-Hagerdal & M. F. Gorwa-Grauslund. (2009). Xylose reductase from *Pichia stipitis* with altered coenzyme preference improves ethanolic xylose fermentation by recombinant Saccharomyces cerevisiae. *Biotechnology for Biofuels,* Vol. 2, No. 9, pp. 10, ISSN 1754-6834.

Brehmer, B., R. M. Boom & J. Sanders. (2009). Maximum fossil fuel feedstock replacement potential of petrochemicals via biorefineries. *Chemical Engineering Research & Design*, Vol. 87, No. 9A, pp. 1103-1119, ISSN 0263-8762.

Busch, R., T. Hirth, A. Liese, S. Nordhoff, J. Puls, O. Pulz, D. Sell, C. Syldatk & R. Ulber. (2006). The utilization of renewable resources in German industrial production. *Biotechnology Journal*, Vol. 1, No. 7-8, pp. 770-776, ISSN 1860-7314.

Canilha, L., J. B. de Almeida e Silva & A. I. N. Solenzal. (2004). Eucalyptus hydrolysate detoxification with activated charcoal adsorption or ion-exchange resins for xylitol production. *Process Biochemistry*, Vol. 39, No. 12, pp. 1909-1912, ISSN 1359-5113.

Cardona, C. A. & O. J. Sanchez. (2007). Fuel ethanol production: Process design trends and integration opportunities. *Bioresource Technology*, Vol. 98, No. 12, pp. 2415-2457, ISSN 0960-8524.

Carreira, P., J. A. S. Mendes, E. Trovatti, L. S. Serafim, C. S. R. Freire, A. J. D. Silvestre & C. P. Neto. (2011). Utilization of residues from agro-forest industries in the production of high value bacterial cellulose. *Bioresource Technology*, Vol. 102, No. 15, pp. 7354-7360, ISSN 0960-8524.

Chen, H. Z. & W. H. Qiu. (2010a). Key technologies for bioethanol production from lignocellulose. *Biotechnology Advances*, Vol. 28, No. 5, pp. 556-562, ISSN 0734-9750.

Chen, M. L. & F. S. Wang. (2010b). Optimization of a Fed-Batch Simultaneous Saccharification and Cofermentation Process from Lignocellulose to Ethanol. *Industrial & Engineering Chemistry Research*, Vol. 49, No. 12, pp. 5775-5785, ISSN 0888-5885.

Chu, B. C. H. & H. Lee. (2007). Genetic improvement of *Saccharomyces cerevisiae* for xylose fermentation. *Biotechnology Advances*, Vol. 25, No. 5, pp. 425-441, ISSN 0734-9750.

Cortez, D. V. & I. C. Roberto. (2010). Individual and interaction effects of vanillin and syringaldehyde on the xylitol formation by Candida guilliermondii. *Bioresource Technology*, Vol. 101, No. 6, pp. 1858-1865, ISSN 0960-8524.

Delgenes, J. P., R. Moletta & J. M. Navarro. (1996). Effects of lignocellulose degradation products on ethanol fermentations of glucose and xylose by Saccharomyces cerevisiae, Zymomonas mobilis, Pichia stipitis, and Candida shehatae. *Enzyme and Microbial Technology*, Vol. 19, No. 3, pp. 220-225, ISSN

Demirbas, A. (2005). Bioethanol from cellulosic materials: A renewable motor fuel from biomass. *Energy Sources*, Vol. 27, No. 4, pp. 327-337, ISSN 0090-8312.

Dix, N. J. (1979). Inhibition of fungi by gallic acid in relation to growth on leaves and litter. *Transactions of the British Mycological Society*, Vol. 73, No. 2, pp. 329-336, ISSN 0007-1536.

Dominguez, J., C. Gong & G. Tsao. (1996). Pretreatment of sugar cane bagasse hemicellulose hydrolysate for xylitol production by yeast. *Applied Biochemistry and Biotechnology*, Vol. 57-58, No. 1, pp. 49-56, ISSN 0273-2289.

Evtuguin, D. V., M. R. B. Xavier, C. M. Silva & A. Prates (2010). Towards comprehensive utilization of products from sulphite pulp production: A biorefinary approach. *XXI Encontro Nacional da TECNICELPA / VI CIADICYL*, Lisbon, Portugal 12-15 October, 2010.

Fengel, D. & G. Wegener (2003). Wood. Chemistry, Ultrastructure, Reactions, Kessel Verlag, ISBN 3-935638-39-6, Munich, Germany.

Gnansounou, E. (2009). Handbook of plant-based biofuels, CRC Press, ISBN 978-1-56022-175-3.

Gonzalez-Garcia, S., C. M. Gasol, X. Gabarrell, J. Rieradevall, M. T. Moreira & G. Feijoo. (2009). Environmental aspects of ethanol-based fuels from Brassica carinata: A case study of second generation ethanol. *Renewable & Sustainable Energy Reviews,* Vol. 13, No. 9, pp. 2613-2620, ISSN 1364-0321.

Gray, K. A., L. S. Zhao & M. Emptage. (2006). Bioethanol. *Current Opinion in Chemical Biology,* Vol. 10, No. 2, pp. 141-146, ISSN 1367-5931.

Hahn-Hagerdal, B., K. Karhumaa, C. Fonseca, I. Spencer-Martins & M. F. Gorwa-Grauslund. (2007). Towards industrial pentose-fermenting yeast strains. *Applied Microbiology and Biotechnology,* Vol. 74, No. 5, pp. 937-953, ISSN 0175-7598.

Hahn-Hagerdal, B. & N. Pamment. (2004). Microbial pentose metabolism. *Applied Biochemistry and Biotechnology,* Vol. 113, No., pp. 1207-1209, ISSN 0273-2289.

Helle, S. S., T. Lin & S. J. B. Duff. (2008). Optimization of spent sulfite liquor fermentation. *Enzyme and Microbial Technology,* Vol. 42, No. 3, pp. 259-264, ISSN 0141-0229.

Huang, C.-F., T.-H. Lin, G.-L. Guo & W.-S. Hwang. (2009). Enhanced ethanol production by fermentation of rice straw hydrolysate without detoxification using a newly adapted strain of *Pichia stipitis*. *Bioresource Technology,* Vol. 100, No. 17, pp. 3914-3920, ISSN 0960-8524.

Idriss, H. & E. G. Seebauer. (2000). Reactions of ethanol over metal oxides. *Journal of Molecular Catalysis a-Chemical,* Vol. 152, No. 1-2, pp. 201-212, ISSN 1381-1169.

Iguchi, M., S. Yamanaka & A. Budhiono. (2000). Bacterial cellulose—a masterpiece of nature's arts. *Journal of Materials Science,* Vol. 35, No. 2, pp. 261-270, ISSN 0022-2461.

Jeppsson, M., O. Bengtsson, K. Franke, H. Lee, B. Hahn-Hägerdal & M. F. Gorwa-Grauslund. (2006). The expression of a Pichia stipitis xylose reductase mutant with higher KM for NADPH increases ethanol production from xylose in recombinant *Saccharomyces cerevisiae*. *Biotechnology and Bioengineering,* Vol. 93, No. 4, pp. 665-673, ISSN 1097-0290.

Jonsson, L. J., E. Palmqvist, N. O. Nilvebrant & B. Hahn-Hagerdal. (1998). Detoxification of wood hydrolysates with laccase and peroxidase from the white-rot fungus Trametes versicolor. *Applied Microbiology and Biotechnology,* Vol. 49, No. 6, pp. 691-697, ISSN 0175-7598.

Karhumaa, K., B. Hahn-Hägerdal & M.-F. Gorwa-Grauslund. (2005). Investigation of limiting metabolic steps in the utilization of xylose by recombinant *Saccharomyces cerevisiae* using metabolic engineering. *Yeast,* Vol. 22, No. 5, pp. 359-368, ISSN 1097-0061.

Katahira, S., A. Mizuike, H. Fukuda & A. Kondo. (2006). Ethanol fermentation from lignocellulosic hydrolysate by a recombinant xylose- and cellooligosaccharide-assimilating yeast strain. *Applied Microbiology and Biotechnology,* Vol. 72, No. 6, pp. 1136-1143, ISSN 0175-7598.

Klemm, D., D. Schumann, U. Udhardt & S. Marsch. (2001). Bacterial synthesized cellulose -- artificial blood vessels for microsurgery. *Progress in Polymer Science,* Vol. 26, No. 9, pp. 1561-1603, ISSN 0079-6700.

Kurtzman, C. P. & M. Suzuki. (2010). Phylogenetic analysis of ascomycete yeasts that form coenzyme Q-9 and the proposal of the new genera *Babjeviella, Meyerozyma,*

Millerozyma, Priceomyces, and Scheffersomyces. Mycoscience, Vol. 51, No. 1, pp. 2-14, ISSN 1340-3540.

Kuyper, M., H. R. Harhangi, A. K. Stave, A. A. Winkler, M. S. M. Jetten, W. T. A. M. de Laat, J. J. J. den Ridder, H. J. M. Op den Camp, J. P. van Dijken & J. T. Pronk. (2003). High-level functional expression of a fungal xylose isomerase: the key to efficient ethanolic fermentation of xylose by *Saccharomyces cerevisiae*? *FEMS Yeast Research,* Vol. 4, No. 1, pp. 69-78, ISSN 1567-1364.

Larsson, S., A. Reimann, N.-O. Nilvebrant & L. Jönsson. (1999). Comparison of different methods for the detoxification of lignocellulose hydrolyzates of spruce. *Applied Biochemistry and Biotechnology,* Vol. 77, No. 1, pp. 91-103, ISSN 0273-2289.

Lawford, H. & J. Rousseau. (1998). Improving fermentation performance of recombinant zymomonas in acetic acid-containing media. *Applied Biochemistry and Biotechnology,* Vol. 70-72, No. 1, pp. 161-172, ISSN 0273-2289.

Lawford, H. G. & J. D. Rousseau. (1993). Production of ethanol from pulp-mill hardwood and softwood spent sulfite liquors by genetically-engineered *Escherichia-coli. Applied Biochemistry and Biotechnology,* Vol. 39, No., pp. 667-685, ISSN 0273-2289.

Lee, W.-C. & C.-T. Huang. (2000). Modeling of ethanol fermentation using Zymomonas mobilis ATCC 10988 grown on the media containing glucose and fructose. *Biochemical Engineering Journal,* Vol. 4, No. 3, pp. 217-227, ISSN 1369-703X.

Lee, W., J. Lee, C. Shin, S. Park, H. Chang & Y. Chang. (1999). Ethanol production using concentrated oak wood hydrolysates and methods to detoxify. *Applied Biochemistry and Biotechnology,* Vol. 78, No. 1, pp. 547-559, ISSN 0273-2289.

Lippits, M. J. & B. E. Nieuwenhuys. (2010). Direct conversion of ethanol into ethylene oxide on copper and silver nanoparticles Effect of addition of CeOx and Li2O. *Catalysis Today,* Vol. 154, No. 1-2, pp. 127-132, ISSN 0920-5861.

Madhavan, A., S. Tamalampudi, A. Srivastava, H. Fukuda, V. Bisaria & A. Kondo. (2009). Alcoholic fermentation of xylose and mixed sugars using recombinant <i>Saccharomyces cerevisiae</i> engineered for xylose utilization. *Applied Microbiology and Biotechnology,* Vol. 82, No. 6, pp. 1037-1047, ISSN 0175-7598.

Marques, A. P., D. V. Evtuguin, S. Magina, F. M. L. Amado & A. Prates. (2009). Chemical Composition of Spent Liquors from Acidic Magnesium-Based Sulphite Pulping of Eucalyptus globulus. *Journal of Wood Chemistry and Technology,* Vol. 29, No. 4, pp. 322-336, ISSN 0277-3813.

Martinez, A., M. E. Rodriguez, M. L. Wells, S. W. York, J. F. Preston & L. O. Ingram. (2001). Detoxification of dilute acid hydrolysates of lignocellulose with lime. *Biotechnology Progress,* Vol. 17, No. 2, pp. 287-293, ISSN 8756-7938.

Mata-Alvarez, J., S. Macé & P. Llabrés. (2000). Anaerobic digestion of organic solid wastes. An overview of research achievements and perspectives. *Bioresource Technology,* Vol. 74, No. 1, pp. 3-16, ISSN 0960-8524.

Matsushika, A., H. Inoue, T. Kodaki & S. Sawayama. (2009). Ethanol production from xylose in engineered *Saccharomyces cerevisiae* strains: current state and perspectives. *Applied Microbiology and Biotechnology,* Vol. 84, No. 1, pp. 37-53, ISSN 0175-7598.

Mohandas, D. V., D. R. Whelan & C. J. Panchal. (1995). Development of xylose-fermenting yeasts for ethanol production at high acetic acid concentrations. *Applied Biochemistry and Biotechnology,* Vol. 51, No. 2, pp. 307-318, ISSN 0273-2289.

Mussatto, S., es, I. Roberto & esConceição. (2001). Hydrolysate detoxification with activated charcoal for xylitol production by Candida guilliermondii. *Biotechnology Letters,* Vol. 23, No. 20, pp. 1681-1684, ISSN 0141-5492.

Mussatto, S. I., G. Dragone, P. M. R. Guimaraes, J. P. A. Silva, L. M. Carneiro, I. C. Roberto, A. Vicente, L. Domingues & J. A. Teixeira. (2010). Technological trends, global market, and challenges of bio-ethanol production. *Biotechnology Advances,* Vol. 28, No. 6, pp., ISSN 0734-9750|1873-1899.

Mussatto, S. I. & I. C. Roberto. (2004). Alternatives for detoxification of diluted-acid lignocellulosic hydrolyzates for use in fermentative processes: A review. *Bioresource Technology,* Vol. 93, No. 1, pp. 1-10, ISSN

Nigam, J. N. (2001a). Ethanol production from hardwood spent sulfite liquor using an adapted strain of Pichia stipitis. *Journal of Industrial Microbiology & Biotechnology,* Vol. 26, No. 3, pp. 145-150, ISSN 1367-5435.

Nigam, J. N. (2001b). Ethanol production from wheat straw hemicellulose hydrolysate by Pichia stipitis. *Journal of Biotechnology,* Vol. 87, No. 1, pp. 17-27, ISSN 0168-1656.

Nigam, P. (1999). SINGLE-CELL PROTEIN | Mycelial Fungi. Encyclopedia of Food Microbiology. K. R. Richard. Oxford, Elsevier: 2034-2044.

Nilvebrant, N.-O., A. Reimann, S. Larsson & L. Jönsson. (2001). Detoxification of lignocellulose hydrolysates with ion-exchange resins. *Applied Biochemistry and Biotechnology,* Vol. 91-93, No. 1, pp. 35-49, ISSN 0273-2289.

Oakley, J. H. & A. F. A. Hoadley. (2010). Industrial scale steam reforming of bioethanol: A conceptual study. *International Journal of Hydrogen Energy,* Vol. 35, No. 16, pp. 8472-8485, ISSN 0360-3199.

Olsson, L., H. R. Soerensen, B. P. Dam, H. Christensen, K. M. Krogh & A. S. Meyer. (2006). Separate and simultaneous enzymatic hydrolysis and fermentation of wheat hemicellulose with recombinant xylose utilizing Saccharomyces cerevisiae. *Applied Biochemistry and Biotechnology,* Vol. 129, No. 1-3, pp. 117-129, ISSN 0273-2289.

Palmqvist, E. & B. Hahn-Hagerdal. (2000a). Fermentation of lignocellulosic hydrolysates. I: inhibition and detoxification. *Bioresource Technology,* Vol. 74, No. 1, pp. 17-24, ISSN 0960-8524.

Palmqvist, E. & B. Hahn-Hagerdal. (2000b). Fermentation of lignocellulosic hydrolysates. II: inhibitors and mechanisms of inhibition. *Bioresource Technology,* Vol. 74, No. 1, pp. 25-33, ISSN 0960-8524.

Panizzi, L., C. Caponi, S. Catalano, P. L. Cioni & I. Morelli. (2002). In vitro antimicrobial activity of extracts and isolated constituents of Rubus ulmifolius. *Journal of Ethnopharmacology,* Vol. 79, No. 2, pp. 165-168, ISSN 0378-8741.

Parajó, J. C., H. Domínguez & J. Domínguez. (1998). Biotechnological production of xylitol. Part 3: Operation in culture media made from lignocellulose hydrolysates. *Bioresource Technology,* Vol. 66, No. 1, pp. 25-40, ISSN 0960-8524.

Pereira, S. R., S. Ivanusa, D. V. Evtuguin, L. S. Serafim & A. M. R. B. Xavier. (2011). Detoxification of eucalypt spent sulphite liquors: a process to enhance bioethanol production. *Bioresource Technology,* Vol. No., pp., ISSN In press DOI: 10.1016/j.biortech.2011.09.095

Rutz, D. & R. Janssen (2008). Biofuel Technology Handbook, WIP Renewable Energies, ISBN Contract No. EIE/05/022/SI2.420009, München, Germany.

Sanchez, O. J. & C. A. Cardona. (2008). Trends in biotechnological production of fuel ethanol from different feedstocks. *Bioresource Technology,* Vol. 99, No. 13, pp. 5270-5295, ISSN 0960-8524.

Sannigrahi, P., Y. Pu & A. Ragauskas. (2010). Cellulosic biorefineries--unleashing lignin opportunities. *Current Opinion in Environmental Sustainability,* Vol. 2, No. 5-6, pp. 383-393, ISSN 1877-3435.

Santos, S. G., A. P. Marques, D. L. D. Lima, D. V. Evtuguin & V. I. Esteves. (2011). Kinetics of Eucalypt Lignosulfonate Oxidation to Aromatic Aldehydes by Oxygen in Alkaline Medium. *Industrial & Engineering Chemistry Research,* Vol. 50, No. 1, pp. 291-298, ISSN 0888-5885.

Schneider, H. (1996). Selective removal of acetic acid from hardwood sulphite spent liquor using mutant yeast. *Enzyme and Microbial Technology,* Vol. 19, No., pp. 94-98, ISSN

Singhania, R. R., B. Parameswaran & A. Pandey (2009). Handbook of Plant-Based Biofuels, CRC Press, ISBN 978-1-56022-175-3, Boca Raton, United States of America.

Sjöström, E. (1993). Wood Chemistry. Fundamentals and Applications, Academic Press, ISBN 0126474818, New York, USA.

Song, Z. X., A. Takahashi, I. Nakamura & T. Fujitani. (2010). Phosphorus-modified ZSM-5 for conversion of ethanol to propylene. *Applied Catalysis a-General,* Vol. 384, No. 1-2, pp. 201-205, ISSN 0926-860X.

Thygesen, A., A. Thomsen, S. Possemiers & W. Verstraete. (2010). Integration of Microbial Electrolysis Cells (MECs) in the Biorefinery for Production of Ethanol, H₂ and Phenolics. *Waste and Biomass Valorization,* Vol. 1, No. 1, pp. 9-20, ISSN 1877-2641.

Upadhyay, G., S. P. Gupta, O. Prakash & M. P. Singh. (2010). Pyrogallol-mediated toxicity and natural antioxidants: Triumphs and pitfalls of preclinical findings and their translational limitations. *Chemico-Biological Interactions,* Vol. 183, No. 3, pp. 333-340, ISSN 0009-2797.

Van Vleet, J. H. & T. W. Jeffries. (2009). Yeast metabolic engineering for hemicellulosic ethanol production. *Current Opinion in Biotechnology,* Vol. 20, No. 3, pp. 300-306, ISSN 0958-1669.

Vanzyl, C., B. A. Prior & J. C. Dupreez. (1988). Production of Ethanol from Sugar-Cane Bagasse Hemicellulose Hydrolyzate by *Pichia-Stipitis*. *Applied Biochemistry and Biotechnology,* Vol. 17, No., pp. 357-369, ISSN 0273-2289.

Vanzyl, C., B. A. Prior & J. C. Dupreez. (1991). Acetic-Acid Inhibition of D-Xylose Fermentation by *Pichia-Stipitis*. *Enzyme and Microbial Technology,* Vol. 13, No. 1, pp. 82-86, ISSN 0141-0229.

Wang, W. & Y. Wang. (2008). Thermodynamic analysis of hydrogen production via partial oxidation of ethanol. *International Journal of Hydrogen Energy,* Vol. 33, No. 19, pp. 5035-5044, ISSN 0360-3199.

Xavier, A. M. R. B., M. F. Correia, S. R. Pereira & D. V. Evtuguin. (2010). Second-generation bioethanol from eucalypt sulphite spent liquor. *Bioresource Technology,* Vol. 101, No. 8, pp., ISSN 0960-8524(print) | 1873-2976(electronic).

Yu, C.-Y., D.-W. Lee, S.-J. Park, K.-Y. Lee & K.-H. Lee. (2009). Study on a catalytic membrane reactor for hydrogen production from ethanol steam reforming. *International Journal of Hydrogen Energy,* Vol. 34, No. 7, pp. 2947-2954, ISSN 0360-3199.

Zhang, Y. H. P. (2008). Reviving the carbohydrate economy via multi-product lignocellulose biorefineries. *Journal of Industrial Microbiology & Biotechnology,* Vol. 35, No. 5, pp. 367-375, ISSN 0169-4146.

7

Bioethanol Production from Steam Explosion Pretreated Straw

Heike Kahr, Alexander Jäger and Christof Lanzerstorfer
University of Applied Sciences Upper Austria
Austria

1. Introduction

1.1 Motivation and environmental aspects

The combustion of fossil fuels is responsible for 73% of carbon dioxide emissions into the atmosphere and therefore contributes significantly to global warming. Interest in the development of methods to reduce greenhouse gases has increased enormously. In order to control such emissions, many advanced technologies have been developed, which help in reducing energy consumption, increasing the efficiency of energy conversion or utilization, switching to lower carbon-content fuels, enhancing natural sinks for carbon dioxide, capture and storage of carbon dioxide, reducing the use of fossil fuels in order to decrease the amount of carbon dioxide and minimizing the levels of pollutants. In the last few years, research on renewable energy sources that reduce carbon dioxide emissions has become very important. Since the 1980s, bioethanol has been recognized as a potential alternative to petroleum-derived transport fuels in many countries. Today, bioethanol accounts for more than 94% of global biofuel production, with North America (mainly the US) and Brazil as the overall leading producers in the world (about 88% of the world bioethanol production in 2009).

Generally, biofuel production can be classified into three main types, depending on the converted feedstocks used: biofuel production of first, second and third generation. Bioethanol production of the first generation is either from starchy feedstocks, e.g. seeds or grains such as wheat, barley and corn (North America, Europe) or from sucrose-containing feedstocks (mainly Brazil). The feedstocks used for bioethanol production of the second generation are lignocellulose-containing raw materials like straw or wood as a carbon source. Biofuel production of the third generation is understood as the production of lipolytic compounds mainly from algae.

The feedstocks of bioethanol production of the first generation could also enter the animal or human food chain. Therefore, bioethanol production of the first generation is regarded critically by the global population, worrying about food shortages and price rises. Other reasons which lead to research and developments in bioethanol production of the second generation are: a shortage of world oil reserves, increasing fuel prices and reduction of the greenhouse effect. In addition to this, the renewable energy directive (EC 2009/28 RED) demands a reduction for Europe of 6% in the greenhouse gases for the production and use of fuels. This reduction is only possible if biofuels are added to diesel fuel or gasoline by the year 2020. It also seems that the target for greenhouse gas reduction for Europe can only be achieved if the biofuels are mainly from biothanol of the second generation. Outside Europe (Brasil, USA) the targets can be achieved using first generation biofuels. Hence, research and development on the production of bioethanol of the second generation needs to be intensively promoted, particularly in the European countries.

1.2 State of science and technology

Bioethanol production of the first generation from sugar cane and from wheat or corn is well established in Brazil as well as in the US and Europe. The world´s ethanol production in more than 75 countries amounted in 2008 to more than 77 billion litres of ethanol (Sucrogen bioethanol, 2011).
Bioethanol production of the second generation can use lignocelluloses from non-food crops (not counted in the animal or the human food chain), including waste and remnant biomass e.g. wheat straw, corn stover, wood, and grass. These feedstocks are composed mainly of lignocellulose (cellulose, hemicelluloses and lignin).
The process of bioethanol production of the first generation is well established and shown in Fig. 1.

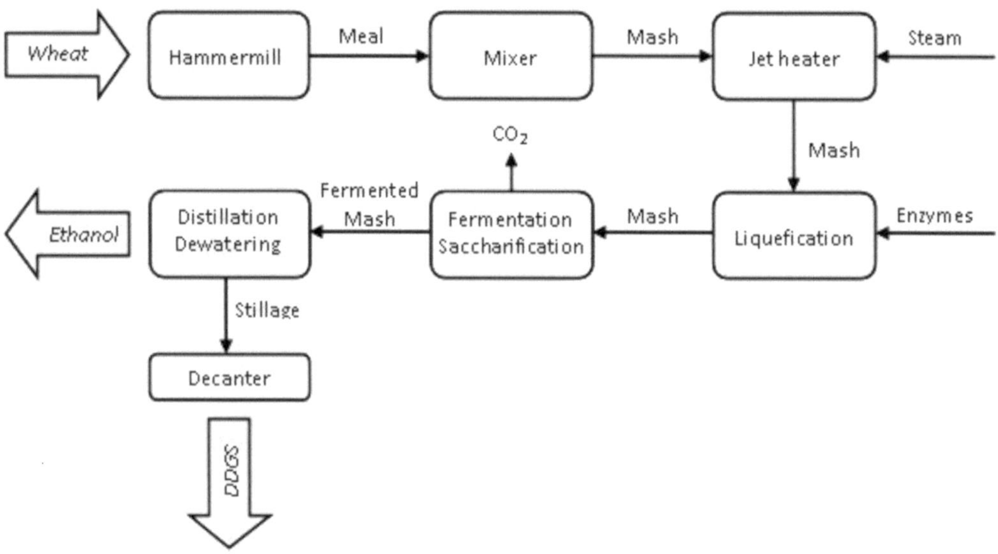

Fig. 1. Flow chart showing bioethanol production from starchy raw materials

The process of bioethanol production from wheat normally consists of five major process steps:
1. Milling of the grain
2. Liquefication at high temperatures
3. Saccharification (enzymatic degradation of starch)
4. Fermentation with yeast
5. Distillation (rectification) of ethanol

The production of bioethanol from lignocelluloses follows more or less the same principle and is composed of the following sub-steps: milling, thermophysical pretreatment hydrolysis, fermentation, distillation and product separation/processing (Fig. 2).

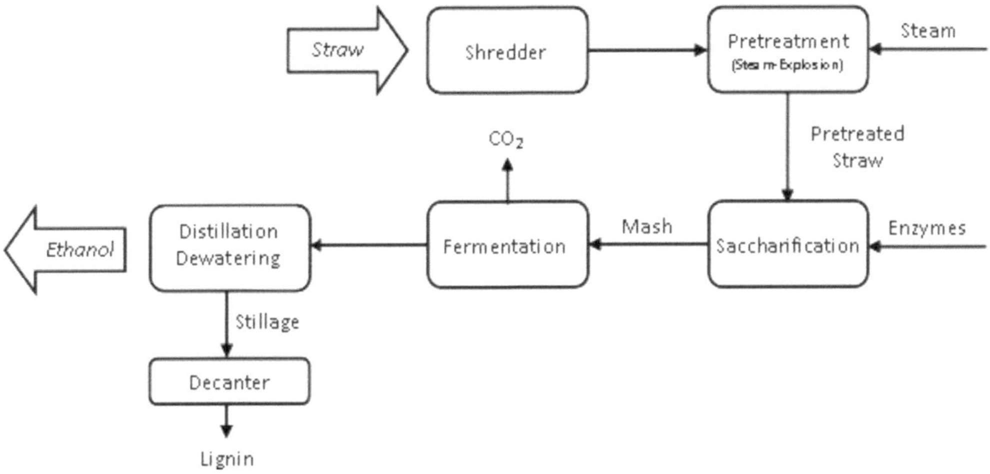

Fig. 2. Flow chart showing bioethanol production from lignocelluloses

The cellulose in the lignocellulose is not accessible to enzymes. Therefore, lignin and/or hemicelluloses have to be removed in order to make the enzymatic degradation of the cellulose possible. Ideal pretreatment should lead to better performance during bioethanol production from lignocelluloses.

The pretreament should cause the hydrolysis of hemicelluloses, high recovery of all carbohydrates, and high digestibility of the cellulose in enzymatic hydrolysis. No sugars should either be degraded or converted into inhibitory compounds. A high solid matter content and high concentration of sugars should be possible. The process should have low energy demands and require low capital and operational cost.

The pretreament methods can be classified roughly into three types: thermophysical methods, acid-based methods and alkaline methods. Thermophysical methods like steam pretreament, steam explosion or hydrothermolysis solubilise most of the cellulose and hemicelluloses. There is only a low level of sugar conversion. Cellulose and hemicelluloses have to be converted enzymatically into C6 sugars (mainly glucose) and to C5 sugars (mainly xylose). Acid-based methods use mineral acids like sulphuric acid and phosphoric acid. Hemicelluloses are degraded to sugar monomers, cellulose has to be converted to glucose enzymatically. Alkaline methods like ammonia fibre explosion leave some of the hydrocarbons in the solid fraction. Hemicellulases acting both on solid and dissolved hemicelluloses are required as well as the celluloytic enzymes.

Lignocellulose containing substrates are mainly composed of cellulose (40-50%), hemicellulose (25-35%) and lignin (15-20%). Cellulose is a glucose polymer, hemicellulose is a heteropolymer of mainly xylose and arabinose, and lignin is a complex poly-aromatic compound. The different pretreatment methods are necessary to loosen the close bonding between cellulose, hemicellulose and lignin. Wheat (*Triticum aestivum* L.) straw is composed of 45% cellulose, 26% hemicellulose and 19% lignin. Maize (*Zea mays)* straw is composed of 39% cellulose, 30% hemicellulose and 17% lignin.

The high percentage of hemicelluloses and the resulting pentoses, e.g. xylose from the hydrolysis of the polymer, are a further challenge to a cost-competitive bioethanol process with lignocelluloses as carbon source.

Yeasts used for the conversion of sugars into ethanol (mostly *Saccharomyces spec.)* usually only convert glucose into ethanol. C5 sugars like xylose are only converted into ethanol at low rates by very few yeast (*Pichia spec.)* strains. Research programs are underway either to adapt yeasts for the use of both C5 and C6 sugars or to modify *Saccharomyces* genetically to obtain yeast that produces ethanol simultaneously from C5 and C6 sugars.

Nevertheless, because of its ready availability and low costs, lignocellulosic biomass is the most promising feedstock for the production of fuel bioethanol. Large-scale commercial production of bioethanol from lignocellulose containing materials has still not been implemented.

2. Potential of second generation bioethanol

The world-wide availibility of feedstock has to be taken into account if bioethanol from lignocelluloses is to contribute significantly to the world fuel market. A report by Bentsen & Felby (2010) shows existing agricultural residue of 1.6 Gt/year cellulose and 0.8 Gt/year hemicelluloses (figures do not include Africa and Australia). This gives a theoretical quantity of 1.24 Gm^3 bioethanol from cellulose (690 l/t using *Saccharomyces spec.*) and 0.480 Gm^3 (600 l/t using *Zymomonas spec.*) from hemicelluloses. For comparison: the worldwide production of crude oil is estimated to reach not more than 4.8 Gm^3 pa (83 million bbl/day) and is supposed/predicted to decline to under 2.4 Gm^3 pa (41.5 million bbl/day) by 2040 (Zittel, 2010). The potential of bioethanol from agricultural residues seems to be high. But not all residues will be available and the conversion rates will not be 100%. Therefore, it is thought that lignocellulose-containing materials have to be produced on agricultural land possibly in combination with the production of feedstocks like wheat, corn or sugar cane. These crops would serve as feedstocks for bioethanol production of the first generation.

The yield per hectare is conservatively estimated at 3000 l and 1500 l per hectare of agricultural land for bioethanol of the first and second generation, respectively. A replacement of 41.5 million bbl/day of crude oil would require an area of land of around 5 million km^2.

Using DDGS (distillers dried grain solubles) as protein-rich animal feed, taking into account an increase in productivity in agriculture and using intermediate crops as feedstock, the required area could be reduced to under 2.5 million km^2. This represents approximately 3% of the world's land (Bentsen & Felby, 2010).

3. The production of bioethanol from lignocelluloses

3.1 Pretreatment

Lignocellulose containing biomass has to be pretreated prior to hydrolysis to improve the accessibility of the biomass. For this pretreatment, several processes are available: mechanical treatment for size reduction (e.g. chopping, milling, grinding), hydrothermal treatment (e.g. uncatalysed steam treatment with or without steam explosion, acid catalysed steam treatment, liquid hot water treatment) and chemical treatment (e.g. dilute acid, concentrated acid, lime, NH_3, H_2O_2). Diverse advantages and drawbacks are associated with each pretreatment method (Mosier et al., 2005; Hendriks & Zeeman, 2009; Chen & Qui, 2010; Talebnia et al., 2010).

Steam explosion is a widely-employed process for this pretreatment. This process combines chemical effects due to hydrolysis (autohydrolysis) in high temperature water and acetic acid formed from acetyl groups, and mechanical forces of the sudden pressure discharge (explosion). The steam explosion process offers several attractive features when compared to other technologies. These include less hazardous process chemicals and significantly lower environmental impact (Alvira et al., 2010). Typical operation conditions for steam explosion treatment of straw - temperature and duration of treatment - are summarised in Table 1.

Biomass	Temperature in °C	Duration of pretreatment in minutes	Catalyst	Reference
Wheat straw	220	2.5	none	Tomás-Pejó et al., 2009
Wheat straw	190	8	none	Ballesteros et al., 2004
Wheat straw	190	10	H_2SO_4	Jurado et al., 2009
Wheat straw	200	10	none	Sun et al., 2005
Wheat straw	200	4.5	none	Chen et al., 2007
Barley straw	210	5	none	Garcia-Aparicio et al., 2006
Barley straw	210	5	H_2SO_4	Linde et al., 2007
Corn stover	200	10	none	Yang et al., 2010
Corn stover	200	5	H_2SO_4	Varga et al., 2004
Rice straw	220	4	none	Ibrahim et al., 2011

Table 1. Typical operation data for steam explosion of straw

According to Overend and Chornet (1987), the severity of the pretreatment can be quantified by the severity factor R_0. The severity factor combines the temperature of the pretreatment (T in degree Celsius) and the duration of the pretreatment (t in minutes) thus:

$$R_0 = \int_0^t \exp\frac{T(t)-100}{14.75}.dt \tag{1}$$

The severity factor is based on the observation that it is possible to trade duration of treatment and the temperature of treatment so that equivalent final effects are obtained. However, it is not intended to give mechanistic insight into the process.

3.2 Hydrolysis

Clearly, the hydrolysis step is affected by the type of pretreatment and the quality of this process - particularly by the accessibility of the lignocellulose.

Lignoculluloses can be solubilised by enzymatic or chemical hydrolysis (mainly with acids). Both the pretreatment and hydrolysis are performed in a single step during acid hydrolysis. Two types of acid hydrololysis are usually applied: concentrated and dilute acid hydrolysis (Wyman et al., 2004, Gray et al., 2006, Hendriks & Zeeman, 2009).

Cellulase enzymes from diverse fungi (e.g. like *Trichoderma, Aspergillus*) (Dashtban et al., Sanchez, 2009) and bacteria (e.g *Clostridium, Bacillus*) (Sun & Cheng, 2002) can release sugar from lignocellulose at moderate temperatures (45-50°C) with long reaction times (one to several days) (reviewed in Brethauer & Wyman, 2010; Balat, 2011).

Three different enzymes work synergistically - the endo-β-1,4-glucanases (EC 3.1.2.4), exo-β-1,4-glucanases (EC 3.2.1.91) and β-glucosidase (EC 3.2.1.21) – to generate glucose molecules from cellulose (Lynd et al., 2002). In addition, enzymes like hemicellulases and ligninases improve the hydrolysis rate and raise the content of the fermentable sugar (Palonen & Viikari, 2004; Berlin et al., 2005).

Diverse factors inhibit the activity of the cellulase and thereby decrease the rate of hydrolysis and the effectiveness of the hydrolysis step: end-product inhibition, easily degradable ends of molecules are depleted, deactivation of the enzymes, binding of enzymes in small pores of the cellulose and to lignin (Brethauer & Wyman, 2010; Balat, 2011).

Hemicellulose is a highly complex molecule and multi-enzyme systems are needed like endoxylanase, exoxylanase, β-xylanase, α-arabinofuranosidase, α-glucoronidase, acetyl xylan esterase and ferulic acid esterase (all produced by diverse fungi e.g. *Aspergillus* and bacteria e.g. *Bacillus*) for the enzymatic hydrolysis (reviewed in Balat, 2011).

3.3 Fermentation

The microorganisms for the ethanolic fermentation process for lignocellulose-containing hydrolysates should ferment both hexoses and pentoses (if both cellulose and hemicellulose are solubilised) to achieve efficient bioethanol production. Unfortunately, no known natural microorganisms can efficiently ferment both pentoses and hexoses, which are generated during hydrolysis from lignocelluloses (Ragauskas et al., 2006). The perfect microorganism for fermentation should exhibit several properties: sugar tolerance, ethanol and thermotolerance, resistance against diverse inhibitors, fermentation of hexoses and pentoses and stability during industrial application.

Diverse microorganisms like *Saccheromyces cervisiae, Pichia stipitis, Escherichia coli* and *Zymomonas mobilis* are typically applied in the bioethanol process from lignocellulose. Both generally used microorganisms, the yeast *Saccheromyces cervisiae* and the bacterium *Zymominas mobilis,* can convert hexoses into bioethanol offering high ethanol tolerance and ethanol yields. Genetically modified yeast strains from *Saccheromyces cervisiae* converting both pentoses and hexoses into bioethanol have been generated (reviewed in Vleet & Jeffries, Bettiga et al., Matsushika et al., 2009). *Zymomonas mobilis* was also genetically altered converting xylose into ethanol (reviewed in Girio et al., 2010, Balat, 2011).

Pentoses (xylose, the main sugar from hemicellulose) can be utilized from the yeast strains *Pichia stipitis, Pachysolen tannophilus* and *Candidae shetatae.* The main disadvantage of these yeast strains is their low ethanol tolerance and ethanol yield. Bacteria like *Escherichia coli* and *Klebsiella oxytoca* take up hexoses and pentoses but lead to very low ethanol yields. Successful genetic modifications have been performed in these bacteria leading to higher ethanol yields (reviewed in Girio et al., 2010; Balat, 2011).

Enzymatic hydrolysis and fermentation can be carried out simultaneously (SSF). This process has several advantages: lower enzyme concentrations, higher sugar yields (no end product inhibition of cellulase), higher product yields, shorter process times and lower risk of contamination. The main disadvantage is the different optimal conditions for the hydrolysis and fermentation reactions (reviewed in Balat, 2011). Performing enzymatic hydrolysis and fermentation separately is known as SHF. Each step has to be carried out under optimal reaction conditions but the end product inhibition of the cellulase reduces the rate of hydrolysis and this type of process is costlier (reviewed in Balat, 2011).

3.4 Distillation

With conventional distillation at atmospheric pressure, the maximum achievable ethanol concentration is 90-95%, because in the system ethanol-water there is an azeotrope at 95.6% (w/w) ethanol, boiling at a temperature of 78.2°C. For the production of anhydrous ethanol

further dehydration of the concentrated ethanol is required. This can be achieved by employing azeotropic distillation, extractive distillation, liquid-liquid extraction, adsorption, membrane separation or molecular sieves (Hatti-Kaul, 2010; Huang et al., 2008).
Separation of ethanol from water is an energy-intensive process. The energy required for production of concentrated ethanol by distillation also depends very much on the feed concentration (Zacchi & Axelsson, 1989). The search for solutions for the reduction of the energy required is a field of intensive research. Membrane separation processes need much less energy for ethanol separation but are not in operation on an industrial scale. First results from a pilot plant using the Siftek™ membrane technology show a reduction of the energy required for dehydration of about 50% (Côté et al., 2010). Process and heat integration techniques also play an important role in energy saving in the bioethanol process (Alzate & Toro, 2006; Wingren et al., 2008). Maximum energy saving in the distillation of about 40% is possible by applying mechanical vapour recompression (Xiao-Ping et al., 2008). Solar distillation of ethanol is under investigation for distillation of bioethanol in smaller plants (Vorayos et al., 2006). The production of solid biofuel or biogas for thermal energy supply also reduces the net energy requirement of bioethanol production (Eriksson & Kjellström, 2010; Šantek et al., 2010).

3.5 Use of residues for energy supply

The stillage from distillation can be separated in a liquid-solid separation step into two fractions. The solid fraction is usually used for solid fuel production. The liquid fraction is either fed to an anaerobic digestion process, generating biogas with a methane concentration of about 60% (Prakash et al., 1998) or is used for solid fuel production together with the solid fraction after evaporation of most of the water. In this case the concentrated liquid fraction is mixed with the solid fraction before drying and pelletizing.
Biogas is used for heat generation or combined heat and power generation for the bioethanol process, whereas solid biofuels can also be sold on the market.

4. Our results for bioethanol production from steam explosion pretreated straw

4.1 Steam explosion pretreatment

4.1.1 Operation of steam explosion reactor

The bulk density of the straw in the steam explosion reactor depends very much on the condition of the straw and the feeding method. When filling the pilot reactor with chopped straw manually, a bulk density of about 60 kg m^{-3} was achieved. Loading baled straw would lead to a bulk density of approximately 150 kg m^{-3} (bulk density of straw bales according Jenkins (1989): 100 - 200 kg m^{-3}). The bulk density of straw pellets is 500 kg m^{-3} and higher (Theerarattananoon et al., 2011). For reliable discharge of the treated straw from the reactor in the explosion step, addition of water to the dry straw is usually required. The thermal energy requirement of the steam explosion treatment is met by steam directly fed into the reactor. In small steam explosion units, steam is also optionally used for jacket heating of the reactor. In adiabatic operation, the thermal energy is required for heating up the biomass and the added water. The steam in the vapour phase of the reactor is lost through a vent during the sudden pressure discharge of the reactor. The steam required for heating up the biomass and the added water $m_{st,1}$ (in kg) can be calculated thus:

$$m_{st,1} = m_S \cdot \left(c_{p,S} \cdot \Delta T_S + \frac{m_W}{m_S} \cdot c_{p,W} \cdot \Delta T_W + \Delta h_R \right) \cdot \frac{1}{\Delta h_V} \quad (2)$$

The mass of straw m_S and the mass of the added water m_W are in kg. The specific heat capacity of straw $c_{p,S}$ and the specific heat capacity of water $c_{p,W}$ are in kJ kg^{-1} K^{-1}. The temperature difference between pretreatment temperature and feed temperature for straw ΔT_S and water ΔT_W are in K. The enthalpy of vaporization for water Δh_V at pretreatment temperature and the net reaction enthalpy Δh_R of the pretreatment process are in kJ kg^{-1}.
The venting loss of steam $m_{st,2}$ (in kg) can be calculated thus:

$$m_{st,2} = m_S \cdot \left[\frac{1}{\rho_{S,b} \cdot \eta_V} - \frac{1}{\rho_S} - \left(\frac{m_W}{m_S} + \frac{m_{st,1}}{m_S} \right) \cdot \frac{1}{\rho_W} \right] \cdot \rho_{st} \cdot \quad (3)$$

The bulk density of the straw in the reactor $\rho_{S,b}$ as well as the density of straw ρ_S, the density of water ρ_W and the density of steam ρ_{st}, all at operation temperature and pressure, are in kg m^{-3}. The factor for the volumetric use of reactor volume is η_V.
An increase in steam consumption of 10% can be estimated because of non-adiabatic operation of the steam explosion system and steam leakages (Sassner et al., 2008). The total steam consumption is therefore calculated thus:

$$m_{st} = 1.1 \cdot \left(m_{st,1} + m_{st,2} \right) \quad (4)$$

A reduction in the cost of pretreatment can be achieved by minimisation of the specific steam demand. Ahn et al. (2009) determined the specific heat capacity of wheat straw with a water content of 4.3 g water/g dry sample to be 1.63±0.07 kJ kg^{-1} K^{-1}. The specific heat capacities of other types of straw were in the same range. The specific heat capacity of water is about 2.5 times higher than the specific heat capacity of straw. Therefore, the total water content of the input material is a main influencing factor on the thermal energy consumption of steam explosion pretreatment. Minimizing the rate of water addition to the straw is a way to reduce the steam consumption. Preheating of the added water using waste heat e.g. from the condenser of the distillation or increasing the bulk density of the straw in the reactor are also ways to reduce the steam consumption (Fig. 3).
A reduction in steam temperature would reduce the steam demand too, but at the same time reduce the effect of steam explosion treatment.
For the discharge of the treated straw from the reactor in the explosion step a certain fraction of the reactor volume has to remain filled with uncondensed steam. The remaining steam- filled fraction of the reactor volume under various operation conditions is shown in Fig. 4.
The steam explosion pretreatment of straw pellets is restricted by the pore volume available for the addition of water and condensing steam. From this point of view, a type of compacted straw with a density between 150 kg m^{-3} and 500 kg m^{-3} would be preferable.

4.1.2 Steam explosion experiments

The pretreatment of the straw was carried out in a steam explosion pilot unit using a reactor with a reaction volume of 0,015 m^3. Explosion was carried out into a cyclonic separator to separate the treated straw from the vapour phase. The vapour was then condensed in a regenerative cooler. The maximum steam temperature of the steam generator was 200°C

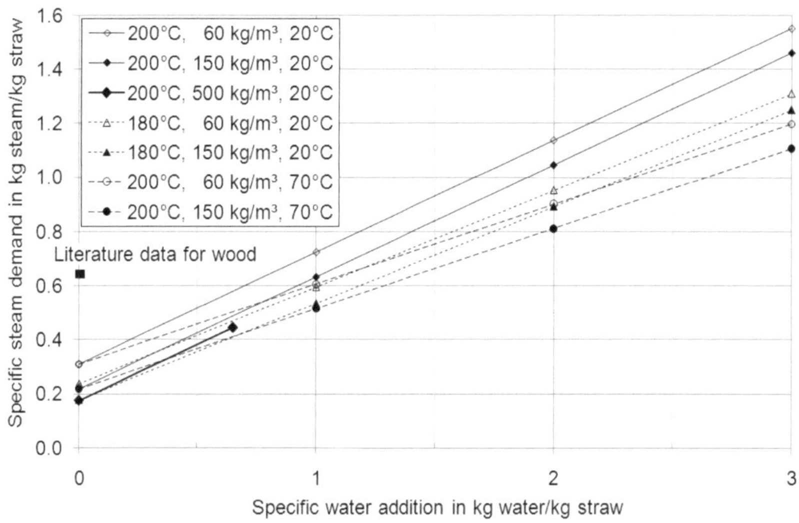

Fig. 3. Specific steam demand in steam explosion pretreatment of straw; general operating data: assumed volumetric use of reactor volume: 0.95; density of straw: 1290 kg m^{-3} (Shaw & Tabil, 2005); net reaction enthalpy neglected; individual operating data (as shown in the legend): temperature of treatment, bulk density of straw, temperature of added water; literature data: thermal energy demand without indication of water content (Zhu & Pan, 2010).

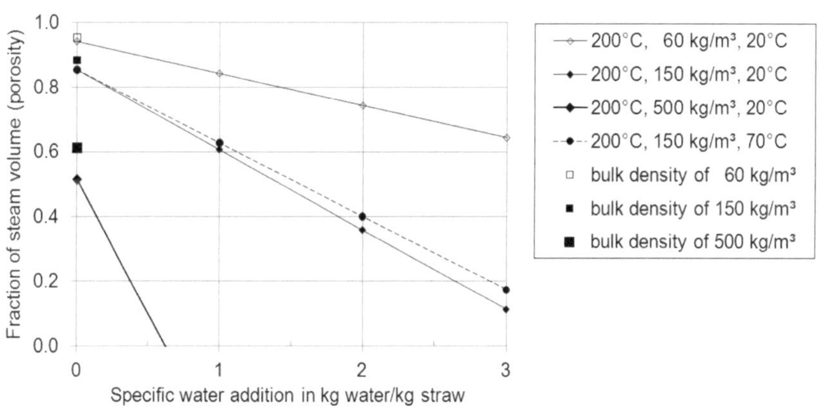

Fig. 4. Remaining steam-filled fraction of the reactor volume under various operating conditions; individual operating data (as shown in the legend): temperature of treatment, bulk density of straw, temperature of added water

(equivalent to a steam vapour pressure of 1.55 MPa). The operation temperature in the reactor is reached via a temperature ramp. In most experiments the mass of added water was 1.0 kg per kg of straw. The operation temperature was generally 200°C and the duration of the treatment was usually 10 minutes. This results in a severity factor of 9500 (log (R_0) = 3.98).

The bulk density of the straw in the reactor was 60 kg m^{-3} for chopped straw. When straw pellets (mixture of Triticale (*Triticosecale Wittmack*) and wheat straw) were pretreated, the

bulk density increased to 520 kg m^{-3}. However, in these cases the volumetric use of reactor volume had to be reduced. Also the ratio of added water was lower.
The steam consumption in the pilot tests was more than two times the calculated value due to only partial thermal insulation of the reactor. In the case of a cold start of the system, steam consumption was even higher.
Figs. 5a and 5b show an example of wheat straw before and after pretreatment. The scanning electron microscope (SEM) images show wheat straw with intact bundles of fibres before preatrement (Fig. 5a) and the same material after pretreatment (Fig. 5b), where the morphological structure has been broken down. This material is now accessible to the cellulytic enzyme complex.

Fig. 5a. Wheat straw untreated (SEM)

Fig. 5b. Wheat straw treated (SEM)

4.1.3 Recycling of low ethanol concentration solutions into the steam explosion reactor

The outcome of an economic study shows that the most important factor for economic bioethanol production is maximum ethanol output (von Sivers & Zacchi, 1996). A possibility to increase the ethanol output would be the recycling of effluents with low ethanol concentration, e.g. the stillage from the distillation, which contains about 1% ethanol (Cortella & Da Porto, 2003) or low concentration effluents from membrane separation steps via the steam explosion reactor. In this case, the added water would be replaced by the effluent to be recycled. During the steam treatment, vapour-liquid equilibrium of the ethanol-water system will be reached. Due to the fact that ethanol is more volatile than water, the concentration of ethanol in the vapour phase will be much higher than in the liquid phase. The vapour-liquid equilibrium of the ethanol-water system at 1.5 MPa is shown in Fig. 6.

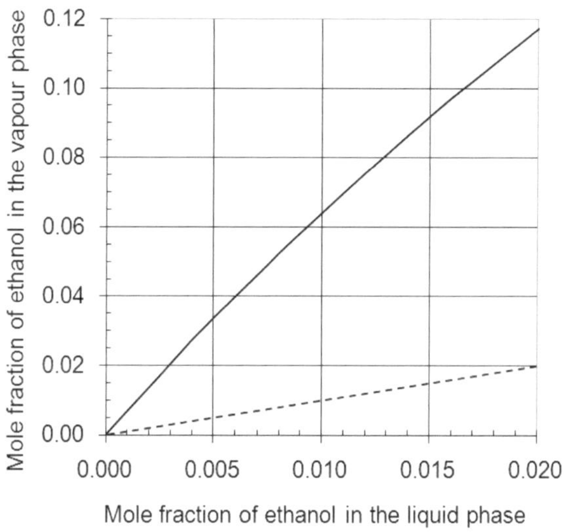

Fig. 6. Vapour-liquid equilibrium of the ethanol-water system at 1.5 MPa, calculated with the Wilson equation (Gmehling & Brehm, 1996)

When the reactor is vented, the exploded biomass is separated from the vapour phase in a cyclonic separator. In the separator secondary vapour is also produced by evaporation cooling of the wet biomass. The vapour phase has to be condensed by cooling at the separator outlet to recapture the ethanol. The collected condensate can be added to the feed of the distillation column.

In a first series of experiments on the recycling of ethanol-containing effluent, the added water in the feed to the steam explosion reactor was replaced by a solution containing 10% (w/w) ethanol. Analyses of the pretreated wet straw are shown in Table 2. The samples were taken from the treated straw heap in the separator immediately after the explosion step and transferred into a gastight bottle. With the exception of ethanol no significant differences were found when 10% ethanol (w/w) solution was used. The ethanol content of 31.6 g/kg feed straw (d.b.) in the treated straw from the experiment with the addition of 10% ethanol solution (w/w) is equivalent to 31.6% of the added ethanol; the remaining 68.4% is expected to be in the condensate. It was not possible to verify this due to limitations in the drainage of such small amounts of condensate from the installed regenerative cooler.

Treated straw samples taken from the separator about five minutes after the explosion step showed a significantly lower ethanol content. The average ethanol content in these samples was 13.5 g/kg of feed straw (d.b.), whereas the concentrations of the other components were

Added water 1 kg/kg wheat straw	Ethanol	Formic acid	Acetic acid	HMF	Furfural
Water	3.7	3.8	16.8	0.3	1.9
10% ethanol (w/w)	31.6	6.1	20.1	0.2	1.0

Table 2. Analyses of steam-exploded wheat straw (pretreatment conditions: 200°C, 10 min); all values in g/kg feed straw (d.b.); averages of two pretreatment experiments; wet straw samples were leached with deionised water, analysis of the filtrate by HPLC

very much the same. This can be explained by the evaporation of ethanol during the cooling of the treated straw. For example, the recycling of a 1% ethanol (w/w) solution would result in a condensate with about 5% ethanol (w/w) considering also the dilution of the liquid phase in the reactor by condensation of steam.

However, recycling of low ethanol concentration effluents could be limited by inhibitors contained in the effluent. Further tests with real effluents are therefore required.

4.2 Hydrolysis and fermentation

4.2.1 Description of the experiments

Bioethanol production from wheat straw was investigated. Several improvements, particularly one washing step and the recirculation strategy, were made. The washed wheat straw was named inhibitor-controlled wheat straw. These improvements increase both the sugar concentration and the bioethanol yield by up to 7%(vol). Also, the lignocellulose-containing substrate corn stover was tested for its potential in bioethanol production. Furthermore, recirculation of bioethanol was performed to ultimately raise the end concentration of bioethanol. Therefore, ethanol was added during the pretreatment process and a possible effect on the hydrolysis and fermentation steps was examined.

The enzyme mixture Accellerase TM1000 from Genencor® was used with enzyme activities of 775 IU cellulase (CMC)/g solids and 138 IU beta-glucosidase/g solids. Suspensions with various dry substances (10-20%) were produced with the pretreated substrate in citrate buffer (50 mM, pH 5.0) and incubated at 50°C for 96 hours in a shaking incubator (100 rpm). The hydrolysis of pretreated substrate was repeated three times in a recirculation process. Sample analysis was performed with HPLC. Diverse salts were added to the straw hydrolysate for fermentation. A wild-type strain of *Saccharomyces cerevisiae* was used exclusively for all experiments. The fermentation process was conducted at 30°C in a shaking incubator for one week (110 rpm).

4.2.2 Results

The glucose concentration obtained after hydrolysis from wheat straw pretreated with different levels/degrees of severity (conditions ranging from 160°C, 10 minutes to 200°C, 20 minutes) is demonstrated in Fig. 7. The pretreatment at 200°C over 20 minutes (severity factor 18000; $\log(R_0)=4.26$) achieved the highest sugar concentration, converting about 100% cellulose during the hydrolysis. Recirculation strategies with wheat straw were developed, where the sugar solution of a first hydrolysis reaction was recycled twice to fresh straw and the subsequent hydrolysis reaction. The glucose concentration was further increased by a recirculation process to fresh washed solids and subsequent hydrolysis from 30 g/l to 143 g/l

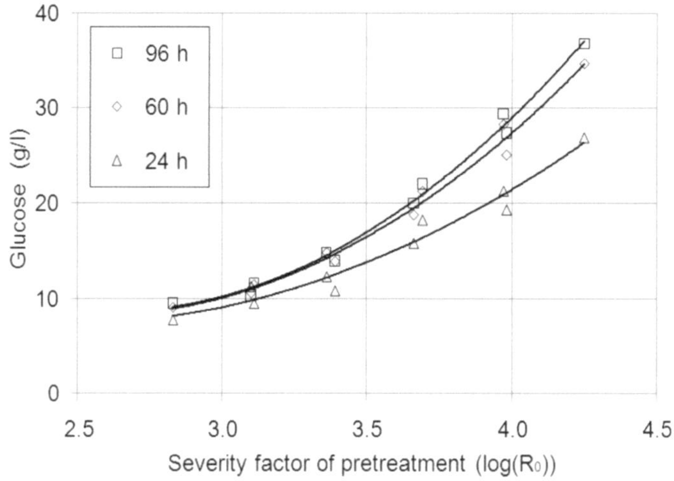

Fig. 7. Glucose concentration after hydrolysis of pretreated wheat straw

(20% solids, third hydrolysis). After fermentation with *Saccharomyces cerevisiae*, an ethanol concentration of 7.5%(vol) was obtained. In Fig. 8, the final glucose concentrations after recirculation processes with inhibitor-controlled wheat straw as well as bioethanol yields after fermentation are shown.

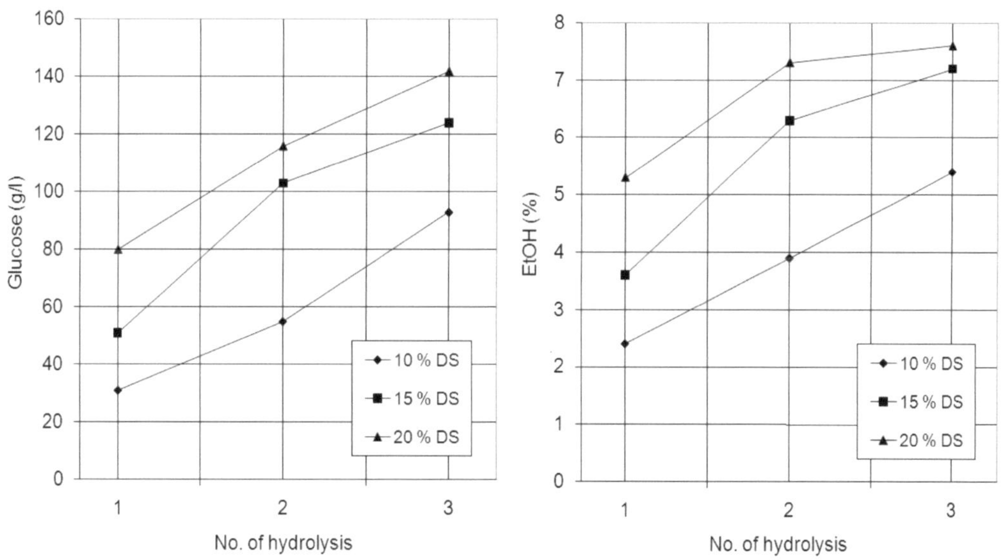

Fig. 8. Produced glucose concentration and bioethanol yields after fermentation of inhibitor controlled wheat straw

Corn stover was pretreated at 190°C for 10 minutes. Initially, 10% of the dry substance corn stover was hydrolyzed and fermented. Here, the sugar concentration was 32 g/l glucose and 10 g/l xylose yielding 1.9% bioethanol (Table 3). The dry substance was increased to 15% and 20%, yielding considerably higher sugar and bioethanol concentrations (Table 3).

	10 % dry substance	15 % dry substance	20 % dry substance
Glucose (g/l)	32	47	58
Xylose (g/l)	10	16	20
EtOH (%(vol))	1.9	2.8	3.9

Table 3. Sugar concentration and ethanol content from corn stover (10, 15 and 20 % solids)

Wheat straw was moistened with water before steam explosion pretreatment. Ethanol was added during pretreatment (10 minutes at 200°C) to test for a possible effect on the hydrolysis and fermentation step. The wet straw was hydrolyzed with the enzymes and fermented with yeast. Additional ethanol during the pretreatment process did not influence the sugar and bioethanol content (Table 4).

	Standard pretreatment	Pretreatment with 10% ethanol (w/w)
Glucose (g/l)	41	40
Xylose (g/l)	20	19
EtOH (%(vol))	2.1	2.2

Table 4. Sugar concentration and ethanol yields after fermentation of standard pretreatment and pretreatment with ethanol (from 10 % dry substance)

Alternatively, pellets from mixed straw were used to increase the dry substance already during the pretreatment step. It was possible to increase the glucose concentration from wet straw pellets to 60 g/l resulting in 2.5%(vol) bioethanol (from 10 % dry substance).

5. Other concepts for the use of lignocellulosic feedstocks

Diverse concepts for the use of lignocellulose-containing plants for bioethanol production are available. In the simplest concept, only the glucose is fermented to bioethanol, with the by-products xylose solution and lignin pellets. The xylose sugars can be used as barrier films, hydrogels, paper additives (Söderqvist et al., 2001; Lima et al., 2003; Grönholm et al., 2004) or in xylitol production (reviewed in Chen et al., 2010). At the moment, the utilization of lignin is unsatisfactory; therefore, the lignin pellets are used as solid biofuel.

The economy of bioethanol production from lignocellulose-containing materials can be improved in a cost-effective concept by simultaneous fermentation of both sugars (glucose and xylose) to bioethanol by diverse microorganisms. In the last twenty years, diverse microorganisms were genetically modified to ferment both glucose and xylose, with good results (reviewed in Hahn-Hägerdal et al., 2007; Matsushika et al., 2009; Jojima et al., Kim et al., Mussatto et al., Weber et al, Young et al., 2010). Furthermore, diverse adaptation programs, mutagenesis and breeding were performed to produce yeasts and other microorganisms with improved xylose fermentation (reviewed in Hahn-Hägerdal et al., 2007; Matsushika et al., 2009; Mussatto et al., 2010). However, in several countries production with GMO is only possible under strict standards and acceptance of GMO in these countries is poor.

In a biorefinery concept, co-production of biofuels, bioenergy and marketable chemicals from renewable biomass sources take place simultaneously. Diverse biorefinery concepts for wheat straw were developed such as: bioethanol from glucose, biohydrogen from xylose and the residual effluents from bioethanol and biohydrogen processes being used for biogas production (Kaparaju et al., 2009). The biorefinery concept including higher-value chemical by-products and autonomous power supplies will enhance economic competitiveness of second generation plants and, therefore, will make this type of plant economical in the near future.

6. Outlook

Research on bioethanol production from lignocellulose-containing substrates has made great progress over the last decades. As shown by other authors and our own results, the theoretical yield of bioethanol from cellulose (690 l/t cellulose, 283 l/t straw) is almost achievable. The yield of bioethanol from hemicelluloses still has to be increased. Compared to bioethanol production of the first generation, cost-effectiveness also has to be improved. No commercial bioethanol plant using lignocellulose-containing residues as feedstock is in operation in 2011. However, diverse pilot plants are in operation and the first demonstration plants have been completed and running succesfully.

The production of biofuels such as bioethanol is often criticized because of the negative impact of the feedstock on biodiversity. The competition of the raw materials for use either as biofuel or for food production is also a major obstacle to increasing bioethanol production capacity. Therefore, lignocellulose-containing residues offer a possibility to satisfy part of the increasing demand for fuel by means of biofuel.

Diverse scenarios are possible - only using first generation fuel, resulting in dramatic increases in world prices for feedstock crops. The stimulation of the second generation results in reduced pressure on world prices for feedstock crop. It is the authors' opinion that the higher demand for biofuels will necessarily lead to the use of lignocelluloses as feedstock to produce biofuels. In order to replace fossil fuels to a larger extent, not only agricultural residue must be used as feedstock. Agriculture has to be geared towards food as well as towards energy production. This will only be possible in the context of a coordinated international effort.

7. Acknowledgment

This work was supported by the following projects: FH Plus in Coin SteamExplo 818383; Bioethanolproduktion aus Lignocellulosen mit Steamexplosion (Fabrik der Zukunft, Projekt 814953); REGIO 13/ EFRE regional production of energy and by the country of Upper Austria and FH OOE basic financing - bioenergy.

8. References

Ahn, H.K.; Sauer, T.J.; Richard, T.L. & Glanville, T.D. (2009). Determination of thermal properties of composting bulking materials. *Bioresource Technology,* Vol.100, No.17, (September 2009), pp. 3974-3981, ISSN 0960-8524

Alzate, C.A.C. & Toro, O.J.S. (2006). Energy consumption analysis of integrated flowsheets for production of fuel ethanol from lignocellulosic biomass. *Energy,* Vol.31, No.13, (October 2006), pp. 2447-2459, ISSN 0360-5442

Alvira, P.; Tomás-Pejó, E.; Ballesteros, M. & Negro, M.J. (2010). Pretreatment technologies for an efficient bioethanol production process based on enzymatic hydrolysis: A review. *Bioresource Technology,* Vol.101, No.13, (July 2010), pp. 4851-4861, ISSN 0960-8524

Balat, M. (2011). Production of bioethanol from lignocellulosic materials via the biochemical pathway: A review. *Energy Conversion and Management,* Vol. 52, No. 2, (Februar 2011), pp. 858-875, ISSN 0196-8904

Ballesteros, M.; Oliva, J.M.; Negro, M.J.; Manzanares, P. & Ballesteros, I. (2004). Ethanol from lignocellulisic materials by a simultaneous saccharification and fermentation process (SFS) with Kluyveromyces marxianus CECT 10875. *Process Biochemistry,* Vol.39, No.12, (October 2004), pp. 1843-1848, ISSN 1359-5113

Bentsen, N. S. & Felby, C. (2010). Technical potentials of biomass for energy services from current agriculture and forest in selected countries in Europe, *The Americas and Asia. Forest&Landscape Working Papers,* No. 55, (2010), pp. 31, Forest & Landscpae Denmark, Frederiksberg, ISBN 9878779035072

Berlin, A.N.; Bengtsson, O.; Hahn-Hägerdal, B. & Gorwa-Grauslund, M.F. (2005). Evaluation of novel fungal cellulase preparations for ability to hydrolyze softwood substrates-evidence for the role of accessory enzymes. *Enzyme and Microbiology Technology,* Vol. 37, No.2, (July 2005), pp. 175-184, ISSN 0141-0229

Bettiga, M.; Bengtsson, O.; Hahn-Hägerdal, B. & Garwa-Grauslund, M.F. (2009) Arabinose and xylose fermatation by recombianant *Saccheroymces cerevisiae* expressing a fungal utilization pathway. *Microbial Cell Factories* Vol. 8, 40, (July 2009), ISSN: 1475-2859

Brethauer, S. & Wyman, C.E. (2010). Review: Continuous hydrolysis and fermentation for cellulosic ethanol production. *Bioresource Technology,* Vol. 101, No. 13, (July 2010), pp. 4862-4874, ISSN 0960-8524

Chen, H. & Lui, L. (2007). Unpolluted fractionation of wheat straw by steam explosion and ethanol extraction. *Bioresource Technology,* Vol.98, No.3, (February 2007), pp. 666-676, ISSN 0960-8524

Chen, H. & Qui, W. (2010). Key technologies for bioethanol production from lignocellulose. *Biotechnology Advances,* Nol. 28, No. 5, (September/October 2010), pp. 556-562, ISSN 0734-9750

Chen, X.; Jiang X.-H.; Chen, S. & Qin W. (2010). Microbial and Bioconversion Production of D-Xylitol and Its Detection and Application. *International Journal of Biological Sciences,* Vol. 6, No. 7, (December 2010), pp. 834-844, ISSN 1449-2288

Cortella, G. & Da Porto, C. (2003) Design of a continuous distillation plant for the production of spirits originated from fermented grape. *Journal of Food Engineering,* Vol.58, No.4, (August 2003), pp. 379-385, ISSN 0260-8774

Côté, P.; Noël, G. & Moore, S. (2010). The Chatham demonstration: From design to operation of a 20 m^3/d membrane-based ethanol dewatering system. *Desalination,* Vol.250, No.3, (January 2010), pp. 1060-1066, ISSN 0011-9164

Dashtban, M.; Schraft H. & Qin, W. (2009). Fungal bioconversion of lignocellulosic residues: Opportunities and perspectives. *International Journal Biological Science,* Vol. 5, No.4, (September 2009), pp. 578-595, ISSN 1449-2288

EC OF THE EUROPEAN PARLIAMENT AND OF THE COUNCIL (2009) DIRECTIVE 2009/28/of 23 April 2009 on the promotion of the use of energy from renewable sources. Official Journal of the European Union L 140/61, 5.6.2009

Eriksson, G. & Kjellström, B. (2010). Assessment of combined heat and power (CHP) integrated with wood-based ethanol production. *Applied Energy*, Vol.87, No.12, (December 2010), pp. 3632-3642, ISSN 0306-2619

Garcia-Aparicio, M.P.; Ballesteros, I.; Gonzáles, A.; Oliva; J.M.; Ballesteros, M. & Negro, M.J. (2006). Effect of Inhibitors Released During Steam-Explosion Pretreatment of Barley Straw on Enzymatic Hydrolysis. *Applied Biochemistry and Biotechnology*, Vol.129, No.1-3, (March 2006), pp. 278-288, ISSN 0273-2289

Girio, F.M.; Fonseca, C.; Calvalheiro, F.; Duarte, L.C.; Marques, S. & Bogel-Lukasik, R. (2010). Hemicelluloses for fuel ethanol: A review. *Bioresource Technology*, Vol. 101, No. 13, (July 2010) pp. 4775-4800, ISSN 0960-8524

Gmehling, J. & Brehm, A. (1996). Grundoperationen, Georg Thieme Verlag Stuttgart

Gray, K.A.; Zhao, L. & Emptage, M. (2006). Bioethanol. *Current Opinion Chemical Biology*, Vol. 10, No. 2 (April 2006), pp. 141-146, ISSN 1367-5931

Grönholm, M.; Eriksson, L. & Gatenholm P. (2004). Material properties of plasticized hardwood xylans for potential application as oxygen barrier films, *Biomacromolecules*, Vol. 5, No. 4, (July-August 2004), pp. 1528-1535, ISSN 1525-7797

Hahn-Hägerdal, B.; Karhumaa, K.; Fonseca, C.; Spencer-Martins, I. & Gorwa- Grauslund, M.F. (2007). Towards industrial pentose-fermenting yeast strains. *Applied Microbiology and Biotechnology*, Vol. 74, No. 5, (April 2007), pp. 937-953, ISSN 0175-7598

Hatti-Kaul R. (2010). Biorefineries - a path to sustainablility?. *Crop Science*, Vol. 50, No.2, pp.152-156, (January 2010), ISSN 0011-183X

Hendriks, A.T.W.M. & Zeeman, G. (2009). Pretreatments to enhance the digestability of lignocellulosic biomass. *Bioresource Technology*, Vol.100, No.1, (January 2009), pp. 10-18, ISSN 0960-8524

Huang, H.-J.; Ramaswamy, S., Tschirner, U.W. & Ramraro, B.V. (2008). A review of separation technologies in current and future biorefineries. *Separation and Purification Technology*, Vol.62, No.1, (August 2008), pp. 1-21, ISSN 1383-5866

Ibrahim, M.M.; El-Zawawy W.K.; Abdel-Fattah, Y.R.; Soliman, N.A. & Agblevor, F.A. (2011). Comparison of alkaline pulping with steam explosion for glucose production from rice straw. *Carbohydrate Polymers*, Vol.83, No.2, (January 2011), pp. 720-726, ISSN 0144-8617

Jenkins B.M. (1989) Physical properties of biomass. In: Biomass handbook, O. Kitani & C.W. Hall, (Eds.), 860-891, Gordon and Breach, ISBN 2881242693, New York, USA

Jojima, T.; Omumasaba, C.A.; Inui, M. & Yukawa, H. (2010). Sugar transporters in efficient utilization of mixed sugar substrates: current knowledge and outlook. *Applied Microbiology and Biotechnology*, Vol. 85, No. 3, (January 2010), 471-480, ISSN 0175-7598

Jurado, M.; Prieto, A.; Martínez-Alcalá, Á.; Martínez, Á.T. & Martínez, M.J. (2009). Laccase detoxification of steam-exploded wheat straw for second generation bioethanol. *Bioresource Technology*, Vol. 100, No. 12, (December 2009), pp. 6378-6384, ISSN 0960-8524

Kaparaju, P.; Serrano, M.; Thomsen, A.B.; Kongjan, P. & Angelidaki, I. (2009). Bioethanol, biohydrogen and biogas production from wheat straw in a biorefinery concept. *Bioresource*, Vol.100, No.9, (May 2009), pp. 2562-2568, ISSN 0960-8524

Kim, HJ.; Block, D.E. & Mills, D.A (2010). Simultaneous consumption of pentose and hexose sugars: an optimal microbial phenotype for efficient fermentation of lignocellulosic biomass. *Applied Microbiology and Biotechnology*, Vol. 88, No.5, (November 2010), pp. 1077-1085, ISSN 0175-7598

Lima, D.U.; Olimeira, R. C. & Buckeridge, M.S. (2003). Seed storage hemicelluloses as wet-end additives in papermaking. *Carbohydrate Polymers*, Vol.52, No.4, (June 2003), pp. 367-373, ISSN 0144-8617

Linde, M.; Galbe, M. & Zacchi, G. (2007). Simultaneous saccharification and fermentation of steam-pretreated barley straw at low enzyme loadings and low yeast concentration. *Enzyme and Microbial Technology*, Vol.340, No.5, (April 2007), pp. 1100-1107, ISSN 0141-0229

Lynd, L.R.; Weimer, P.J.; van Zyl W.H. & Pretorius, I.S. (2002). Microbial cellulose utilization: Fundamentals and biotechnology. *Microbiology and Molecular Biology Reviews*, Vol. 66, No. 3, (September 2002), pp. 506-577, NO. 3, ISSN 1098-5557

Matsushika, A.; Inoue, H.; Kodaki, T. & Sawayama, S. (2009) Ethanol production from xylose in engineered *Saccharomyces cerevisiae* strains: current state and perspectives. *Applied Microbiology and Biotechnology*, Vol. 84, No. 1, (Nov 2009), pp. 37-53, ISSN 0175-7598

Mosier, N.; Wyman, C.; Dale, B.; Elander, R.; Lee, Y.Y.; Holtzapple, M. & Ladisch, M. (2005), Features of promising technologies for pretreatment of lignocellulosic biomass. *Bioresource Technology*, Vol.96, No.6, (April 2005), pp. 673-686, ISSN 0960-8524

Mussatto, S.I.; Dragone, G.; Guimarães, P.M.; Silva, J.P.; Carneiro, L.M.; Roberto, I.C.; Vicente, A.; Domingues, L. & Teixeira, J.A. (2010) Technological trends, global market, and challenges of bio-ethanol production. *Biotechnology Advances, Vol.* 28, No. 6, (November-December 2010), pp. 817-830, ISSN 0734-9750

Overend, R.P. & Chornet, E. (1987). Fractionation of Lignocellulosics by Steam-Aqueous Pretreatments. *Philosophical Transactions for the Royal Society of London*, Series A, Vol.321, No.1561, (April 1987) pp. 523-536

Palonen, H. & Viikari, L. (2004). Role of oxidative enzymatic treatments on enzymatic hydrolysis of softwood. *Biotechnology and bioengineering*, Vol. 86, No. 5 (June 2004), pp. 550-557, ISSN 0006-3592

Prakash, R.; Henham, A. & Bhat, I.K. (1998). Net energy and gross pollution from bioethanol production in India. *Fuel*, Vol.77, No.14, (November 1998), pp. 1629-1633, ISSN 0016-2361

Ragauskas, A.J. (2006). The Path Forward for Biofuels and Biomaterials. *Science*, Vol.311, No.5 (June 2006), pp. 484-489, ISSN 0036-8075

Sanchez, C. (2009). Lignocellulosic residues: Biodegradation and bioconversion by fungi. Biotechnology Advances Vol. 27, No. 2, (March-April 2009), pp. 185-194, ISSN 0734-9750

Šantek, B.; Gwehenberger, G.; Šantek, M.I.; Narodoslawsky, M. & Horvat, P. (2010). Evaluation of energy demand and the sustainability of different bioethanol production processes from sugar beet. *Resources, Conservation and Recycling*, Vol.54, No.11, (September 2010), pp. 872-877, ISSN 0921-3449

Sassner, P.; Galbe, M. & Zacchi, G. (2008). Techno-economic evaluation of bioethanol production from three different lignocellulosic materials. *Biomass and Bioenergy*, Vol.32, No.5, (May 2008), pp. 422-430, ISSN 0961-9534

Shaw, M.D. & Tabil, L.G. (2005). Compression Studies of Peat Moss, Wheat Straw, Oat Hulls and Flax Shives. *Powder Handling and Processing,* Vol.17, No.6, (November/December 2005), pp. 344-350, ISSN 0934-7348

Söderqvist Lindblad, M.; Ranucci, E. & Albertsson, A.-C. (2001). Biodegradable polymers from renewable sources. New hemicellulose-based hydrogels. *Micromolecular Rapid Communications,* Vol. 22, No. 12, (March-April 2001), pp. 962-976, ISSN 1022-1336

Sucrogen bioethanol. (2011). http://ethanolfacts.com.au/globaloverview [06.09.2011]

Sun, Y. & Cheng, J. (2002). Hydrolysis of lignocellulosic materials for ethanol production: a review, *Bioresource Technology,* Vol. 83, No. 1 (May 2002), pp. 1-11, ISSN 0960-8524

Sun, X.F.; Xu, F.; Sun, R.C.; Geng, Z.C.; Fowler, P. & Baird, M.S. (2005). Characteristics of degraded hemicellulosic polymers obtained from steam exploded wheat straw. *Carbohydrate Polymers,* Vol. 60, No. 1, (April 2005), pp. 15-26, ISSN 0144-8617

Talebnia, F.; Karakashev, D. & Angelidaki, I. (2010). Production of bioethanol from wheat straw: An overview on pretreatment, hydrolysis and fermentation. *Bioresource Technology,* Vol. 101, No. 13, (July 2010), pp. 4744-4753, ISSN 0960-8524

Theerarattananoon, K.; Xu, F.; Wilson, J.; Ballard, R.; Mckinney, L.; Straggenborg, S.; Vadlani, P.; Rei, Z.J. & Wang, D. (2011). Physical properties of pellets made from sorghum stalk, corn stover, wheat straw and big blustem. *Industrial Crops and Products,* Vol.33, No.2, (March 2011), pp. 325-332, ISSN 0926-6690

Tomás-Pejó, E.; Oliva, J.M.; Gonzáles, A.; Ballesteros, I. & Ballesteros, M. (2009). Bioethanol production from wheat straw by the thermotolerant yeast Kluyveromyces marxianus CECT 10875 in a simultaneous saccharification and fermentation fed-batch process. *Fuel,* Vol.88, No.11, (November 2009), pp. 2142-2147, ISSN 0016-2361

Varga, E.; Réczey, K. & Zacchi, G. (2004). Optimization od Steam Pretreatment of Corn Stover to Enhance Enzymatic Digestibility. *Applied Biochemistry and Biotechnology,* Vol. 114, No. 1-3, (March 2004), pp. 509-523, ISSN 0273-2289

Vleet, J. & Jeffries, T. (2009). Yeast metabolic engineering for hemicellulosic ethanol production. *Current Opinion in Biotechnology,* Vol. 20, No. 3, (June 2009), pp. 300-306, ISSN 0958-1669

von Sievers, M. & Zacchi, G. (1996). Ethanol from lignocellulosics: a review of the economy. *Bioresource Technology,* Vol.56, No.2-3, (May-June 1996), pp. 131-140, ISSN 0960-8524

Vorayos, N.; Kiatsiriroat, T. & Vorayos, N. (2006). Performance analysis of solar ethanol distillation. *Renewable Energy,* Vol.31, No.15, (December 2006), pp. 2543-2554, ISSN 0960-1481

Weber, C.; Farwick, A.; Benisch, F.; Brat, D.; Dietz, H.; Subtil, T. & Boles, E. (2010). Trends and challenges in the microbial production of lignocellulosic bioalcohol fuels. *Applied Microbiology and Biotechnology,* Vol. 87, (July 2010), pp. 1303-1315, ISSN 0175-7598

Windgren, A.; Galbe, M. & Zacchi, G. (2008). Energy considerations for a SSF-based softwood ethanol plant. *Bioresource Technology,* Vol.99, No.7, (May 2008), pp. 2121-2131, ISSN 0960-8524

Wyman, C.E. (2004). Ethanol Fuel, In: *Encyclopedia of Energy,* Cutler, J. Cleveland, (Eds), pp. 541–555, Elsevier Inc., New York, ISBN 978-0-12-176480-7

Xiao-Ping, J.; Fang, W.; Shu-Guang, X.; Xin-Sun, T. & Fang-Yu, H. (2008). Minimum energy consumption process synthesis for energy saving. *Resources, Conservation and Recycling,* Vol.52, No.7, (May 2008), pp. 1000-1005, ISSN 0921-3449

Yang, M.; Li, W.; Liu, B.; Li, Q. & Xing, J. (2010). High-concentration sugars production from corn stover based on combined pretreatments and fed-batch process. *Bioresource Technology*, Vol.101, No.13, (July 2010), pp. 4884-4888, ISSN 0960-8524

Young, E.; Lee, S.-M. & Alper, H. (2010). Optimizing pentose utilization in yeast: the need for novel tools and approaches. *Biotechnology for Biofuels*, Vol. 3, (November 2010), pp. 24, ISSN 1754-6834

Zacchi, G. & Axelsson, A. (1989). Economic evaluation of preconcentration in production of ethanol from dilute sugar solutions. Biotechnology and Bioengineering, Vol. 34, No. 2, (June 1989), pp. 223-233, ISSN 1432-0614

Zhu, J.Y. & Pan, X.J. (2010). Woody biomass pretreatment for cellulosic ethanol production: Technology and energy consumption evaluation. *Bioresource Technology*, Vol.101, No.13, (July 2010), pp. 4992-5002, ISSN 0960-8524

Zittel, W. (2010) Studie „Save our Surface" im Auftrag des Österreichischen Klima- und Energiefonds, p 15, Klagenfurt, Austria. Available from: http://www.umweltbuero-klagenfurt.at/sos/wp-content/uploads/Argumentarium_SOS_08062011_Endversion.pdf

8

Towards Increasing the Productivity of Lignocellulosic Bioethanol: Rational Strategies Fueled by Modeling

Hyun-Seob Song, John A. Morgan and Doraiswami Ramkrishna
School of Chemical Engineering, Purdue University, West Lafayette, IN USA

1. Introduction

Bioethanol is not only currently the most widely used biofuel, but also potentially the most promising alternative to fossil fuels. The majority of bioethanol in today's use is made from sucrose-containing (e.g., sugarcane, sugar beet, and sweet sorghum) or starch-based feedstocks (e.g., corn, wheat, rice, barley, and potatoes). The excessive production of such crop-based (first generation) bioethanol, however, imposes an adverse effect on global food supply. A sustainable alternative feedstock which can be used for non-crop (second generation) bioethanol is lignocellulosic biomass such as rice straw (Binod et al., 2010), wheat straw (Talebnia et al., 2010), corn stover (Kadam & McMillan, 2003), switchgrass (Keshwani & Cheng, 2009), sugarcane bagasse (Cardona et al., 2010), and various other agriculture and forest residues.

Lignocellulose primarily consists of cellulose, hemicellulose and lignin. Cellulose is a homopolymer of glucose, while hemicellulose is a heteropolymer of pentoses (i.e., xylose and arabinose) and hexoses (i.e., glucose, mannose, and galactose) sugars. Lignin is a rich source of aromatic carbon compounds but extremely recalcitrant. Lignocellulose is decomposed via pretreatment and hydrolysis into a spectrum of sugars in which glucose and xylose are the first and second most dominant. These cellulosic sugars are finally converted to bioethanol by fermentation. The lignocellulosic bioethanol has not yet been produced on a commercial scale due to lack of cost-effectiveness. For ensuring its economical viability, comprehensive efforts are required to reduce cost (and maximize the profit) throughout the entire process from biomass to bioethanol.

In the current discussion, we limit ourselves to the fermentation step only and examine various issues with increasing bioethanol productivity. Cost-benefit analysis of the fermentation process shows that the processing cost is more dominant (two-thirds of the total cost) than the feed cost (Lange, 2007; Wingren et al., 2003). It is thus important to improve the processing efficiency, not just the sugar conversion alone. In this regard, increasing the *productivity* should be a preferred target over increasing the *yield*, not only in the reactor optimization, but also in strain improvement.

The yeast *Saccharomyces cerevisiae* has typically been used for the production of crop-based bioethanol. This wild-type strain is, however, not suitable for converting cellulosic sugars as it can efficiently ferment glucose but hardly xylose. Considerable effort has been made to endow *S. cerevisiae* with the ability to utilize xylose (Hahn-Hagerdal et al., 2007). Basic approaches to this end are to "push" and "pull" xylose into the central metabolism of *S. cerevisiae*. Push strategies introduce the transport and initial metabolic routes of xylose by expressing exogenous (i.e., foreign) genes. In pull strategies, reactions in the central metabolism are selectively overexpressed. Introduction of foreign plasmids imposes a "metabolic burden" or "metabolic load" on the host cell by consuming a significant amount of internal resources, hurting the normal metabolic functioning of the host cell (Glick, 1995). The most common observation is the decrease of cell growth rate (Bentley et al., 1990; Ricci & Hernandez, 2000). It is often (while not always) that as the product yield is increased, the production rate is reciprocally low (Chu & Lee, 2007).

Most of the recombinant yeast strains currently available show a sequential pattern in their consumption of mixed sugars (i.e., glucose and xylose). They preferably consume glucose with xylose on standby as denoted by the vertical line in Fig. 1.1(a). Then, simultaneous consumption take places along the tilted line only when the preferred substrate is depleted to a very low level (say, one tenth or one fifth of xylose level). Obviously, the productivity can be increased if simultaneous consumption occurs earlier (i.e. at higher concentrations of glucose).

To achieve this, two different strategies can be considered. First, we may develop a more efficient fermenting organism through further pathway modifications of existing recombinant yeast. The goal of this attempt at the genetic level corresponds to making the slope of the tilted line steeper (Fig. 1.1(b)). Alternatively, we may design a more efficient fermentation process through optimization of operating conditions or reconfiguration of reactors. For example, if we change initial sugar composition in batch culture by increasing relative portion of xylose in the culture medium, this also leads to earlier start of the simultaneous consumption (Fig. 1.1(c)).

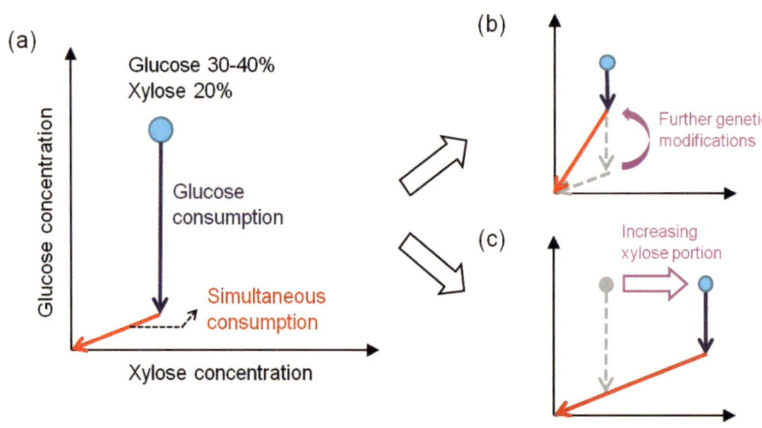

Fig. 1.1. (a) Sequential consumption of mixed sugars by existing recombinant yeast. Two possible ways to promote the simultaneous consumption: (b) metabolic pathway modification of fermenting organisms and (c) adjustment of sugar composition in the culture medium. Adapted from Song and Ramkrishna (2010) with minor modification.

In this chapter, we present model-based strategies for increasing the bioethanol productivity both at the genetic and reactor levels. Metabolic models help not only reduce trial and error, but also discover fresh strategies (Bailey, 1998). In view of the issues discussed above, there are two essential aspects of metabolic models required for the application to reactor and metabolic engineering. First, the mathematical models should be able to address productivity as well as yield. Second, it should be possible to account for metabolic burden. While diverse modeling approaches have been suggested as a tool, the cybernetic framework (Ramkrishna, 1983) is unique in this regard (Maertens & Vanrolleghem, 2010).

The cybernetic modeling approach describes cellular metabolism from the viewpoint that a microorganism is an optimal strategist making frugal use of limited internal resources to maximize its survival (Ramkrishna, 1983). Metabolic regulation of enzyme synthesis and their activities is made as the outcome of such optimal allocation of resources. This unique feature of accounting for metabolic regulation endows cybernetic models with the capability to accurately predict peculiar metabolic behaviors such as sequential or simultaneous consumption of multiple substrates. Further, in view of the constraint placed on resources, the cybernetic model provides a mechanism to account for metabolic burden imposed on the organism as a result of genetic changes.

After a brief sketch of the model structure (Section 2), we will see how metabolic models are used to establish rational strategies for increasing the productivity. In Section 3, basic guidelines for genetic modification of fermenting organisms are provided by identifying the potential target pathway and reactions. Diverse reactor-level strategies are also discussed in Section 4.

2. Metabolic model

The *hybrid cybernetic* approach (Kim et al., 2008; Song et al., 2009; Song & Ramkrishna, 2009) is used for modeling of recombinant yeast consuming glucose and xylose. The hybrid cybernetic model (HCM) incorporates the concept of elementary modes (EMs) (Schuster et al., 2000) into the cybernetic framework. EM is a metabolic pathway (or subnetwork) composed of a minimal set of reactions supporting a steady state operation of metabolism. Any feasible metabolic state can be represented by nonnegative combinations of EMs. HCM views EMs as cell's metabolic options, the choice of which is optimally modulated under dynamic environmental conditions such that a prescribed metabolic objective (such as the total carbon uptake flux) is maximized.

2.1 Basic structure

A hybrid cybernetic model can be given in a general form as follows:

$$\begin{aligned}\frac{d\mathbf{x}}{dt} &= \mathbf{S_x Z r_M} c + \frac{F_{IN}}{V}(\mathbf{x_{IN}} - \mathbf{x}) \\ \frac{dV}{dt} &= F_{IN} - F_{OUT}\end{aligned} \tag{1}$$

where $\mathbf{x}$ is the vector of n_x concentrations of extracellular components in the reactor (such as substrates, products and biomass), $\mathbf{S_x}$ is the $(n_x \times n_r)$ stoichiometric matrix, and $\mathbf{Z}$ is the $(n_r \times n_z)$ EM matrix, $\mathbf{r_M}$ is the vector of n_z fluxes through EMs, F_{IN} and F_{OUT} are volumetric feed rates at the inlet and outlet, V is the culture volume, $\mathbf{x_{IN}}$ is the vector of n_x concentrations of extracellular components in the feed. Eq. (1) can also represent batch operation by setting $F_{IN} = F_{OUT} = 0$ (i.e., V is constant), and fed-batch systems by setting $F_{OUT} = 0$. In chemostat operations, $F_{IN} = F_{OUT} = F$, and F/V is often given as dilution rate D. With $\mathbf{Z}$ normalized with respect to a reference substrate, $\mathbf{r_M}$ implies *uptake* fluxes through EMs. Fluxes through EMs are given as below:

$$r_{M,j} = v_{M,j}\left(e_{M,j} / e_{M,j}^{\max}\right) r_{M,j}^{kin} \tag{2}$$

where the subscript j denotes the index of EM, $v_{M,j}$ is the cybernetic variable controlling enzyme activity, $e_{M,j}$ and $e_{M,j}^{\max}$ are the enzyme level and its maximum value, respectively, and $r_{M,j}^{kin}$ is the kinetic term. Enzyme level $e_{M,j}$ is obtained from the following dynamic equation, i.e.,

$$\frac{de_{M,j}}{dt} = \alpha_{M,j} + u_{M,j} b r_{ME,j}^{kin} - \beta_{M,j} e_{M,j} - \mu e_{M,j} \tag{3}$$

where the first and second terms of the right-hand side denote constitutive and inducible rates of enzyme synthesis, and the last two terms represent the decrease of enzyme levels by degradation and dilution, respectively. In the second term of the right-hand side, $u_{M,j}$ is the cybernetic variable regulating the induction of enzyme synthesis, b is the fraction of internal resources (such as DNA, RNA, protein, lipid and other components) involved in the enzyme synthesis process, and $r_{ME,j}^{kin}$ is the kinetic part of inducible enzyme synthesis rate. In the third and fourth terms, $\beta_{M,j}$ and μ are the degradation and specific growth rates, respectively.

The cybernetic control variables, $u_{M,j}$ and $v_{M,j}$ are computed from the following the "Matching Law" and the "Proportional Law"(Kompala et al., 1986; Young & Ramkrishna, 2007), respectively:

$$u_{M,j} = \frac{p_j}{\sum_k p_k}; \quad v_{M,j} = \frac{p_j}{\max_k (p_k)} \tag{4}$$

where the return-on-investment p_j denotes the carbon uptake flux through the jth EM.

The structure of HCMs is illustrated using Fig. 2.1. In this tutorial example, we get three EMs from the network. The uptake flux is split into three individual fluxes thorough EMs, which are catalyzed by enzymes E_1, E_2 and E_3, respectively. HCMs view that the uptake fluxes are optimally distributed (by the cybernetic variables **u** and **v**) among three EMs for maximizing a metabolic objective function (such as the carbon uptake flux or growth rate). The uptake and excretion rates are represented by nonnegative combinations of individual fluxes through EMs.

2.2 Recombinant yeast strain 1400 (pLNH33)

Among many recombinant yeast strains currently available, we specifically choose *S. cerevisiae* 1400 (pLNH33) developed by Ho and coworkers (Krishnan et al., 1997). The strain was constructed by transforming the recombinant plasmids with two exogenous genes XYL1 and XYL2 (introduced from xylose-metabolizing *Pichia stipitis*), and one endogenous gene XKS1 (introduced from *S. cerevisiae*) into the host strain *Saccharomyces* yeast 1400 with high ethanol tolerance (Krishnan et al., 1997). The first two genes encode xylose reductase (XR) and xylitol dehydrogenase (XDH), which convert xylose to xylitol, and xylitol to xylulose, respectively, and the last one encodes xylulokinase (XK), which converts xylulose to xylulose-5-phophaste.

The HCM for the recombinant yeast 1400 (pLNH33) is presented below. The model has been previously developed by the authors (Song et al., 2009). The formulation of HCM is

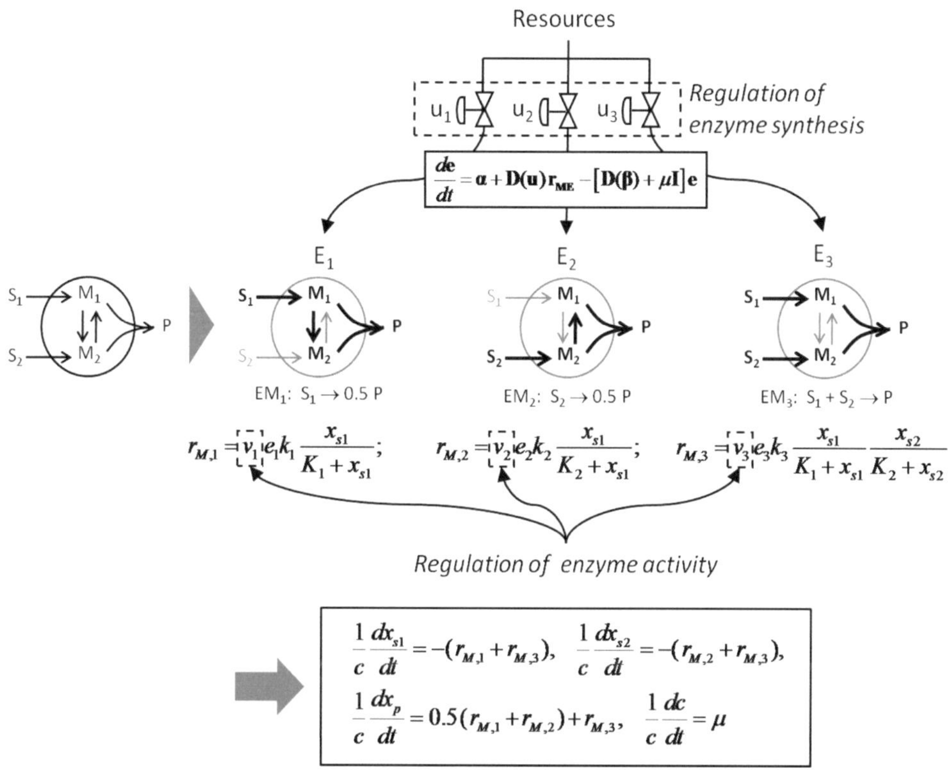

Fig. 2.1. Schematic illustration of the HCM concept. Adapted from from Song et al. (2009).

composed of (i) construction of metabolic network, (ii) computation and selection of EMs, and (iii) parameter identification by model fitting.

2.3 Construction of network model

The metabolic network encompasses all the primary reaction routes involved in the anaerobic growth of recombinant yeast such as glycolytic and pentose phosphate pathways, citric acid cycle, and reactions for pyruvate metabolism. In addition, two oxidoreductase reactions from xylose to xylulose catalyzed by the heterologous expression of XR and XDH enzymes are incorporated. Biochemical reactions participating in the metabolism of recombinant yeast are listed up in Table 2.1.

2.4 EM decomposition and reduction

Using METATOOL v5.0 (von Kamp & Schuster, 2006), the network is decomposed into 201 EMs, which are too many to be incorporated in the model. In general, as the network size increases, the number of EMs undergoes combinatorial explosion (Klamt & Stelling, 2002), leading to overparameterization (which implies an excessive number of parameters relative to the measurements available to determine them). This problem can be avoided using the Metabolic Yield Analysis (MYA) developed by Song and Ramkrishna (2009) by which an original set of EMs is condensed to a much smaller subset. As a result, 201 EMs are reduced to 12 EMs which can be classified into three groups depending on the substrate associated with them (Table 2.2).

	GLYCOLYSIS		
1	GLC + ATP → G6P + ADP	6	GOL → GOL$_x$
2	G6P ↔ F6P	7	GAP + NAD + ADP ↔ PG3 + NADH + ATP
3	F6P + ATP ↔ DHAP + GAP + ADP	8	PG3 ↔ PEP
4	DHAP ↔ GAP	9	PEP + ADP ↔ PYR + ATP
5	DHAP + NADH → GOL + NAD		
	PYRUVATE METABOLISM		
10	PYR → ACD + CO_2	14	ACT → ACT$_x$
11	ACD + NADH → ETH + NAD	15	ACT + CoA + 2ATP → AcCoA + 2ADP
12	ACD + NADH$_m$ → ETH + NAD$_m$	16	PYR + ATP + CO_2 → OAA + ADP
13	ACD + NADP → ACT + NADPH		
	PENTOSE PHOSPHATE PATHWAY		
17	G6P + 2NADP → Ru5P + CO_2 + 2NADPH	20	R5P + X5P ↔ S7P + GAP
18	Ru5P ↔ X5P	21	X5P + E4P ↔ F6P + GAP
19	Ru5P ↔ R5P	22	S7P + GAP ↔ F6P + E4P
	CITRIC ACID CYCLE		
23	PYR+NAD$_m$+CoA$_m$ → AcCoA$_m$+CO_2+NADH$_m$	27	ICT + NADP$_m$ → AKG + CO_2 + NADPH$_m$
24	OAA+NAD$_m$+NADH ↔ OAA$_m$+NADH$_m$+NAD	28	AKG+NAD$_m$+ADP → SUC+ATP+CO_2+NADH$_m$
25	OAA$_m$ + AcCoA$_m$ → ICT + CoA$_m$	29	SUC + 0.5NAD$_m$ ↔ MAL + 0.5NADH$_m$
26	ICT + NAD$_m$ → AKG + CO_2 + NADH$_m$	30	MAL + NAD$_m$ ↔ OAA$_m$ + NADH$_m$
	XYLOSE METABOLISM		
31	XYL + NADH → XOL + NAD	34	XOL + NAD → XUL + NADH
32	XYL + NADPH → XOL + NADP	35	XUL + ATP → X5P + ADP
33	XOL → XOL$_x$		
	BIOMASS FORMATION		
36	1.04AKG + 0.57E4P + 0.11GOL + 2.39G6P + 1.07OAA + 0.99PEP + 0.57PG3 + 1.15PYR + 0.74R5P + 2.36AcCoA + 0.31AcCoA$_m$ + 2.68NAD + 0.53NAD$_m$ + 11.55NADPH + 1.51NADPH$_m$ + 30.48 ATP + 0.43CO_2 → "1 g BIOM" + 2.36CoA + 0.31CoA$_m$ + 2.68NADH + 0.53NADH$_m$ + 11.55NADP + 1.51NADP$_m$ + 30.48ADP		
	OTHERS		
37	ATP → ADP + MAINT	38	NADH → NAD

Table 2.1. List of biochemical reactions included in the metabolic network model of recombinant yeast 1400 (pLNH33). Adapted from Song and Ramkrishna (2009).

Substrate	EM	Net reaction
Glucose	1	GLC → 2 CO2 + 2 ETH + 2 MAINT
	2	25.31 GLC → BIOM + 41.43 CO2 + 33.21 ETH
	3	40.41 GLC → BIOM + 56.52 CO2 + 48.31 ETH + 15.10 GOLx
Xylose	4	XYL → 1.833 CO2 + 1.583 ETH + 1.583 MAINT
	5	2 XYL → 2 CO2 + 1.5 ETH + 1.5 MAINT + XOLx
	6	31.97 XYL → BIOM + 49.42 CO2 + 33.21 ETH
	7	138.5 XYL → BIOM + 160.4 CO2 + 117.6 ETH + 84.37 GOLx
Mixture	8	GLC + 4 XYL → 2 ACTx + 2 CO2 + 2 MAINT + 4 XOLx
	9	GLC + 4 XYL → 9.333 CO2 + 8.333 ETH + 8.333 MAINT
	10	2.39 GLC + 25.99 XYL → 22.19 ACTx + BIOM + 37.82 CO2 + 9.037 ETH
	11	5.333 GLC + 2 XYL → ACTx + 8.5 CO2 + 4.5 ETH + 7.5 GOLx
	12	81.62 GLC + XYL → 12.03 ACTx + 1.754 BIOM + 105.9 CO2 + 85.25 ETH + 39.01 GOLx

Table 2.2. EMs represented in terms of extracellular metabolites. Acronyms for metabolites: ACTx = acetate, BIOM = biomass, CO2 = carbon dioxide, ETH = ethanol, GLC = glucose, GOL = glycerol, MAINT = Dissipated ATP for maintenance, XOLx = xylitol, XYL = xylose.

2.5 Model fit to experimental data

The model was compared with four different sets of anaerobic growth data on single and mixed sugars (Fig. 2.2). As a measure for the quality of model fit, coefficient of determination (also referred to as R^2) is presented for each component of Figs. 2.2(a) to (d) (Table 2.3). R^2 is defined as follows:

$$R^2 \equiv 1 - \frac{SS_{err}}{SS_{tot}}; \quad SS_{err} = \sum_i \left(y_{i,\mathrm{exp}} - y_{i,\mathrm{model}}\right)^2, \quad SS_{tot} = \sum_i \left(y_{i,\mathrm{exp}} - \overline{y}_{\mathrm{exp}}\right)^2, \qquad (5)$$

where $y_{i,\mathrm{exp}}$, $y_{i,\mathrm{model}}$, and $\overline{y}_{\mathrm{exp}}$ denote experimental data, their associated modeled value, and the mean of the observed data, respectively. R^2 values are very high (i.e., over 0.9) for major components (such as glucose, xylose, biomass and ethanol). R^2 values of minor components (such as glycerol and xylitol) are relatively low which is possibly due to the error introduced in data reading from literature graphs. Average R^2 values are over 0.8 in all cases.

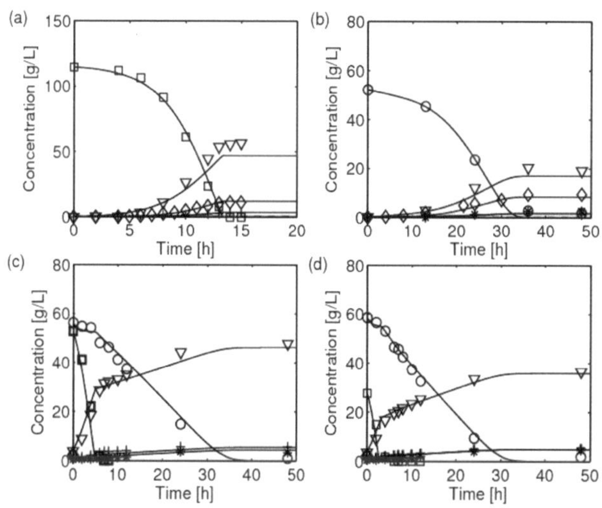

Fig. 2.2. Comparison of model simulations with experimental data. Substrates: (a) glucose only, (b) xylose only, (c) and (d) mixed sugars. Symbols: □ glucose, ○ xylose, ▽ ethanol, ◇ cell dry weight, + glycerol, ✱ xylitol, — simulations.

	Fig. 2.2			
	(a)	(b)	(c)	(d)
Glucose	0.997	–	0.972	0.982
Xylose	–	0.974	0.985	0.990
Cell dry weight	0.988	0.929	–	–
Ethanol	0.936	0.926	0.956	0.957
Glycerol	0.788	0.857	0.501	0.480
Xylitol	–	0.828	0.838	0.729
Average	0.927	0.903	0.850	0.828

Table 2.3. Coefficient of determination (or R^2) for individual components of Figs. 2.2(a) to (d).

3. Strategies for metabolic pathway modification

Comprehensive *in silico* analysis is carried out to establish rational guidelines for further genetic modification of recombinant yeast. The basic strategy is to identify the effective target mode for the genetic change. To this end, we examine the effect of overexpressing enzymes (catalyzing the throughput flux of EMs) on the ethanol productivity (P_{ETH}) which is computed as follows:

$$P_{ETH} = [x_{ETH}(t_f) - x_{ETH}(0)] / t_f \tag{6}$$

where x_{ETH} is the (molar or mass) concentration of ethanol, and t_f is the batch fermentation time.

For realistic simulations, incorporation of metabolic burden is critical. Metabolic burden is ascribed to the lower availability of internal resources for host cells because the same resources are competitively used by plasmids for their replication and more importantly, the synthesis of exogenous proteins. While several empirical correlations are available to consider the change of growth rate with the plasmid content (e.g., Lee et al., 1985; Satyagal & Agrawal, 1989), cybernetic models are able to directly take into account of the reduction of internal resources (b), for example, as follows:

$$b = \frac{b_0}{1+\phi} \tag{7}$$

where ϕ is the parameter depending on the overexpressed level of heterologous proteins as well as the plasmid copy number, and b_0 denotes the fraction of internal resources when no genetic modification is made (i.e., $\phi = 0$). We simulate enzyme overexpression by increasing the constitutive enzyme synthesis rate ($\alpha_{M,j}$'s) in Eq. (3) and relate ϕ to the ratio of "the total incremental of $\alpha_{M,j}$'s due to plasmids" to "the summation of inducible enzyme synthesis rates."

3.1 Identification of target pathway

Sensitivity analysis reveals the dependence of the ethanol productivity on the overexpression of enzymes catalyzing EM fluxes. The sensitivity of the ethanol productivity is calculated as follows:

$$\text{Sensitivity of } P_{ETH} = \frac{\alpha_{M,j}}{P_{ETH}} \frac{dP_{ETH}}{d\alpha_{M,j}}, \quad j \in \{1,2,\ldots,12\} \tag{8}$$

The sensitivity plot (Fig. 3.1(a)) shows that all xylose-consuming EMs (EM4 to EM7) are effective in increasing the ethanol productivity but the highest sensitivity is found among glucose-consuming modes (i.e., EM2). Both can contribute to increasing the productivity but in different ways. The former (i.e., amplifying fluxes of EM4 to EM7) promotes the simultaneous consumption of mixed sugars as illustrated in Fig. 1.1(b). On the other hand, the latter (i.e., amplifying EM2 flux) effectively increases the biomass formation as the growth rate of EM2 is the highest among others.

It should be noted that information provided from the sensitivity analysis is local because it shows only the change of productivity with respect to the "infinitesimal" change of enzyme expression level. It is more important to know how the productivity will change with respect to the "appreciable" change of enzyme levels. This information on *nonlinear* cellular behaviors can be acquired from dynamic simulations. The results are shown in Fig. 3.1(b) where mixed-sugar-consuming modes (EM8 to EM12) are excluded due to their negligible level of activation (Song et al., 2009). From this investigation, EM6 (red line) is chosen as the "best" mode, while EM2 is the second.

Non-monotonic profiles are observed in Fig. 3.1(b). For example, as the overexpression level of mode 5 increases, the productivity goes up initially but comes down afterwards. This may be seen as the outcome of competition between amplification of throughput flux of EM4 (i.e., benefit) and metabolic burden (i.e., cost).

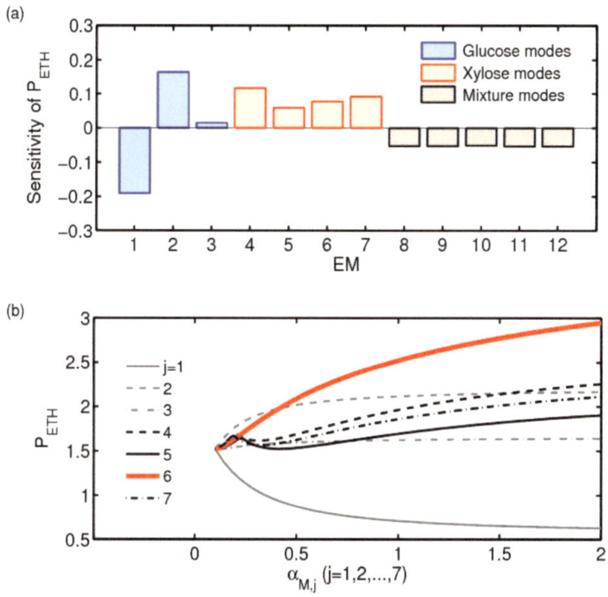

Fig. 3.1. The effect of enzyme overexpression on the ethanol productivity: (a) sensitivity of the productivity, (b) change of productivity subject to appreciable change of enzyme level.

3.2 Effect of amplifying the flux of the target pathway

The effect of amplifying EM6 flux on ethanol productivity is more clearly presented in batch fermentation profiles (Fig. 3.2).

Obviously, overexpression of enzymes has a limit due to the finite internal resources and other reasons. Although it is difficult to estimate the upper limit to overexpression level, we constrain the constitutive synthesis rates of enzymes to be less than a certain threshold, i.e., the total increase of $\alpha_{M,j}$'s is less than or equal to 0.4.

Fig. 3.2 shows that xylose consumption rate is accelerated, while glucose consumption rate is reduced. The decrease of glucose consumption rate can be attributed to a combined effect of metabolic burden and cellular regulation. Consequently, simultaneous consumption of glucose and xylose is facilitated, leading to the substantial increase of ethanol productivity from 1.5 to 2.07 g/L/h (i.e., increase of volumetric productivity by 38%), while the ethanol yield is slightly decreased from 0.402 to 0.392. The total conversion of mixed sugars is fixed to 0.99 in this calculation.

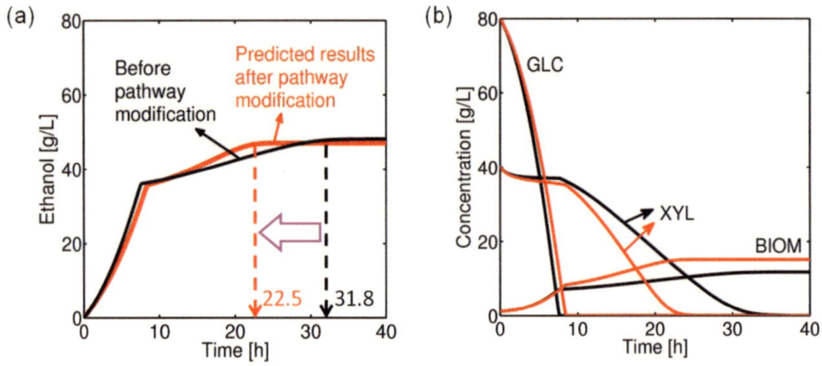

Fig. 3.2. Dynamic fermentation curves in a batch reactor before (black lines) and after (red lines) overexpressing the target mode (EM6). (a) Ethanol, (b) glucose (GLC), xylose (XYL) and biomass (BIOM).

3.3 Implications of amplifying EM6 flux

Comparison of the flux distributions between before (**r**) and after (**r′**) pathway modification suggests an approach to redirect flux distribution for increasing ethanol productivity. Fig. 3.3(a) shows **r** and **r′** at a specific instant when cell density is 3g/L. Metabolic shift caused by the genetic change is also presented by displaying the difference between **r** and **r′** (Fig. 3.3(b)). Then, amplification of the mode throughput flux could be translated as amplification of a set of reactions with positive values of **r′**- **r** which are highlighted in colors in Figs. 3.3(a) and (b). From this analysis, we obtain several interesting findings as follows:

i. First of all, it is observed that none of the reactions in the glycolytic pathway are amplified. It implies that amplification of the glycolytic enzymes may not be a key to increasing ethanol productivity. This is consistent with experimental findings reported in the literature. Overproduction of different glycolytic enzymes of *S. cerevisiae* showed no effect on the rate of ethanol formation (Schaaff et al., 1989). It is because flux control is not inside the glycolytic pathway. Understandably, past efforts for increasing the glycolytic flux by overproduction of glycolytic enzymes have been often unsuccessful (Koebmann et al., 2002). In the *in silico* analysis, flux control is found elsewhere (highlighted in color) which includes xylose utilization pathway, and pentose phosphate (PP) pathway.
ii. While recombinant strain 1400 (pLNH33) efficiently utilizes xylose through the pathway constructed by overexpressing exogenous genes (XR and XDH), as well as endogenous gene (XK), simulation shows that the increase of ethanol productivity requires further overexpression of not only xylose transport reactions (i.e., R31 and R32), but also xylitol conversion to X5P (i.e., R34 and R35).
iii. In addition, it is shown that four reactions in the PP pathway (R19 to R22), i.e., transaldolase (TAL1), transketolase (TKL1), ribulose-5-phosphate 4-epimerase (RPE1) and ribulokinase (RKI1), are possible targets for overexpression. The finding by Johansson & Hahn-Hagerdal (2002) that overexpression of all four genes resulted in better ethanol production than the overexpression of each gene individually is also consistent with the simulation result.
iv. Another interesting aspect that emerges from the model is as follows. Jeppsson et al. (2002) observed that deletion of ZWF1 (i.e., R17), coding for glucose-6-phosphate

dehydrogenase, results in higher ethanol yield but lower productivity. Instead, the hybrid model shows the need to overexpress this oxidative PP pathway to increase ethanol productivity. The calculations show an increase in productivity though there is a small drop in the yield.

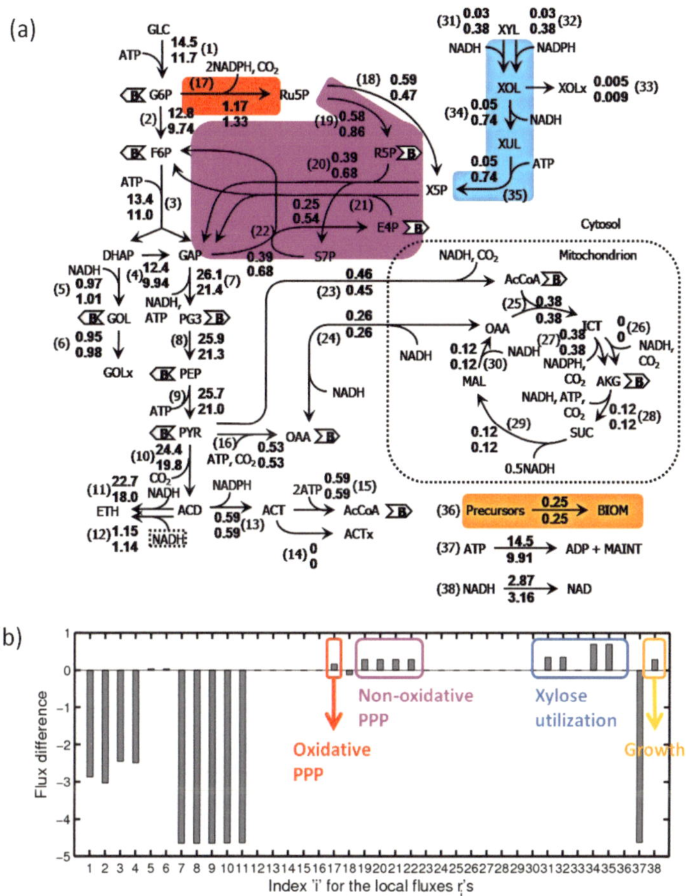

Fig. 3.3. Comparison between before and after amplifying the flux of the target mode (EM6). (a) Flux distributions within the network. The upper and lower numerical values along arrow denote the magnitude of fluxes before (**r**) and after (**r'**) the genetic change. The unit of flux is mmol/gDW/h. (b) Difference between **r'** and **r**.

4. Reactor-level approaches

In this section, we discuss reactor-level strategies towards the enhanced ethanol productivity in two ways. First, we seek optimal ratios of glucose and xylose in batch and continuous cultures to maximize bioethanol productivity. Second, various configurations combining batch, fed-batch and continuous reactors are considered. Their maximum achievable productivities are assessed using the model for the original recombinant strain *S. cerevisiae* 1400 (pLNH33) (i.e., with no amplification of EM6 flux).

The ethanol productivity in a batch and continuous reactor is computed as follows:

$$P_{ETH} \equiv \begin{cases} [x_{ETH}(t_f) - x_{ETH}(0)]/(t_f + t_s) & \text{(batch)} \\ [x_{ETH}(t) - x_{ETH,IN}]D & \text{(continuous)} \end{cases} \tag{9}$$

where $x_{ETH,IN}$ is the ethanol concentration in the feed (which is zero in our case), t_s is the extra time taken for harvesting and preparation for the next batch. The normal range of t_s is from 3 to 10 hours (Shuler & Kargi, 2002) and we set it to 6 hours.

4.1 Effect of sugar composition

We examine the effect of increasing the portion of xylose in the culture medium on ethanol productivity (Fig. 4.1). The total conversion of mixed sugars is set to 0.99 as before. Additional xylose is assumed obtainable by collecting an unconverted sugar from fermentation systems using wild-type yeast which converts glucose only.

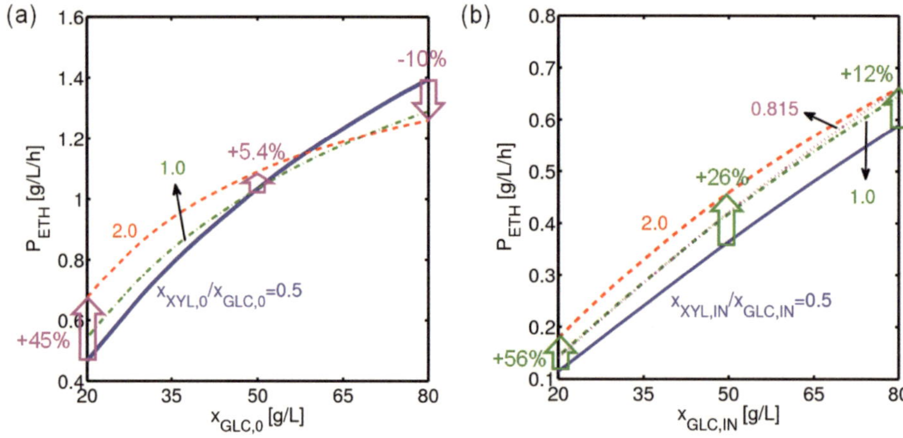

Fig. 4.1. Productivities with different initial sugar concentration in (a) batch and (b) continuous reactors. Adapted from Song and Ramkrishna (2010).

First, the change of ethanol productivity with initial glucose concentration in a *batch* reactor is given in Fig. 4.1 (a). Ethanol productivity may or may not increase with the ratio of xylose to glucose concentration depending on initial glucose concentrations. If, for example, the upper limit of $x_{XYL,0}/x_{GLC,0}$ is 1.0, xylose addition results in increase (or decrease) of productivity when $x_{GLC,0}$ is below (or above) about 50 g/L. If the ratio of initial sugar concentration is allowed to vary up to 2.0, such threshold is extended to $x_{GLC,0}$ = 58 g/L. Higher improvement of ethanol productivity is expected for lower initial concentrations of glucose (e.g., 45% up at $x_{GLC,0}$ = 20 g/L, but 5.4% up at $x_{GLC,0}$ = 50 g/L). Optimal operating conditions correspond to segments of curves above other ones. In Fig. 4.1(a), for example, optimal operating conditions imply that $x_{XYL,0}/x_{GLC,0} = 2$ when $20 \leq x_{GLC,0} \leq 58$, and $x_{XYL,0}/x_{GLC,0} = 0.5$ when $58 \leq x_{GLC,0} \leq 80$.

Next, operating curves in a continuous reactor are presented in Fig. 4.1(b). It is shown that, unlike the batch case, it is always recommendable to increase xylose level in the feed to increase the ethanol productivity. The best productivity is obtained when $x_{XYL,IN}/x_{GLC,IN}$ = 2.0, which increases the productivity by 56%, 26%, and 12% at $x_{GLC,IN}$ = 20, 50, and 80 g/L, respectively.

4.2 Comparison of batch and continuous reactors

Ethanol productivity curves at standard conditions in batch and continuous systems are collected together in Fig. 4.2 for clear comparison. In general, the productivity of growth-associated products in a chemostat is far higher than in a batch reactor. This is not the case with ethanol production because it is suppressed by growth (Shuler & Kargi, 2002). The ethanol productivity from mixed sugars in batch culture is about two to three times higher than in continuous culture (Fig. 4.2(a)). Meanwhile, the foregoing considerations show that chemostats outperform batch fermenters in ethanol production from glucose alone as cells grow relatively fast (Fig. 4.2(b)). Choice of preculture medium affects ethanol productivity of batch fermentation and its effect is more clearly shown for mixed sugars (Fig. 4.2(a)) than a single substrate (Fig. 4.2(b)).

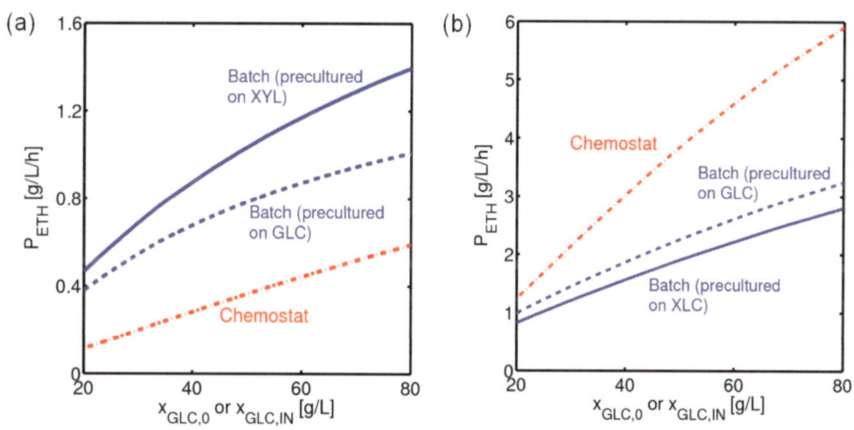

Fig. 4.2. Performance comparison between batch and continuous systems: (a) fermentation of mixed sugars, (b) fermentation of glucose only. Adapted from Song and Ramkrishna (2010).

4.3 Synergistic integration of different type of reactors

In the preceding section, the possibility of improving the productivity in batch reactors was examined by increasing the initial concentration of xylose. Elevation of xylose concentration has both positive and negative effects, i.e., it facilitates simultaneous consumption initially, but prolongs the fermentation time after glucose consumption. Overall, this trade-off resulted in the increase of ethanol productivity only at low sugar concentrations. It was further shown that continuous operation produces significantly more ethanol than batch when only glucose is consumed, but less when mixed sugars are consumed. These findings suggest the investigation of new reactor configurations which may outperform conventional batch fermentation.

We consider the following five configurations (denoted by C1 to C5), each of which combines two different reactor operations (O1 and O2) (Table 4.1). C1 represents a conventional batch operation where mixed sugars are fermented to ethanol by recombinant *S. cerevisiae*. The same is repeated at every batch (i.e., O1 is identical to O2). In C2, O1 is a batch reactor for the growth of the "wild-type" *S. cerevisiae* which can ferment glucose alone. Leftover sugars in O1 are then fed to O2 (i.e., fed-batch operation) where mixed sugars are fermented using the recombinant strain. C3 is the same as C2 except that a chemostat is used

Config.	Operation 1 (O1)			Operation 2 (O2)		
	Reactor	Strain	Sugar	Reactor	Strain	Sugar
C1	Batch	GM	GLC, XYL	Batch.	GM	GLC, XYL
C2	Batch	WT	GLC	Fed-batch	GM	GLC, XYL
C3	Cont.	WT	GLC	Fed-batch	GM	GLC, XYL
C4	Batch	WT	GLC	Fed-batch	GM	GLC, XYL
C5	Cont.	WT	GLC	Fed-batc	GM	GLC, XYL

Table 4.1. Reactor configurations integrating two different types of reactors. Acronyms: C1 to C5 = reactor configurations 1 to 5, GM = genetically modified strain, WT = wild-type strain. GLC = glucose, XYL = xylose. Redrawn from Song et al. (2011).

for O1. C4 and C5 are respective counterparts of C2 and C3, and these two groups are differentiated only by the xylose feeding policy in O2. That is, in C2 and C3, all leftover sugars in O1 are fed into O2 at its start-up (which implies that O2 is a batch system with elevated initial concentration of xylose). In C4 and C5, on the other hand, the xylose feeding rate is optimized such that the ethanol productivity in O2 is maximized.

We introduced a continuous reactor in C3 and C5 in the above. Chemostats have been preferred less than batch reactors in practice. One of the primary reasons for this is the genetic instability of fermenting organisms as continuous operation will impose strong selective pressure of fast growing cells instead of efficient ethanol producers. This will pose a serious problem for recombinant yeast strains, but may not for the wild-type. Thus, we consider C3 and C5 also as practically meaningful configurations.

In Table 4.2, an overall comparison is made for C1 to C5 at three different sugar concentrations with respect to the actual productivity and its relative change (in comparison to C1), respectively. From the comparison of the C2-C3 group and the C4-C5 group, it is clear that the effect of optimizing the feed rate is most significant at high sugar concentration, and appreciable at medium, but least at low. Strangely, at [GLC]/[XYL]=20/10, the productivities of C4 and C5 with optimal feeding policies are lower than those of C2 and C3, respectively, where all extra sugars are dumped into reactors at their start-up without optimization. This is because the initial feeding of C2 and C3 is closer to the "true" optimal than the feed profiles of C4 and C5 obtained from direct methods involving control profile discretization (Song et al., 2011). Other than this exception, C5 exhibits the highest productivity among all other configurations. In

	Productivity (increase or decrease in comparison to C1)		
	[GLC]/[XYL]=20/10	70/35	120/60
C1	0.43	1.04	1.30
C2	0.51 (19%)	1.06 (2%)	1.22 (-6%)
C3	0.66 (52%)	1.29 (23%)	1.40 (8%)
C4	0.51 (18%)	1.13 (9%)	1.41 (9%)
C5	0.64 (48%)	1.34 (29%)	1.60 (23%)

Table 4.2. Total bioethanol productivities of C1 to C5 and relative increase (or decrease) of productivities of C2 to C4 in comparison to C1. The total conversion of mixed sugars in all configurations is fixed to 0.95. Redrawn from Song et al. (2011).

comparison to C1, C5 achieves a substantial increase of the bioethanol productivity, i.e., by 48, 29 and 23 % when [GLC]/[XYL] = 20/10, 70/35, and 120/60, respectively. [GLC]/[XYL] denotes the mass concentration ratio of glucose and xylose.

5. Conclusion

Various possibilities of increasing the productivity of lignocellosic bioethanol at the fermentation step have been discussed, including metabolic pathway modification of fermenting organisms, optimization of reactor operating conditions, and synergistic combination of different types of reactors. Mathematical models play a key role in establishing rational strategies at such diverse levels. The success of the proposed methods, of course, depends on the reliability of the employed mode. We have demonstrated that the cybernetic models are uniquely effective for the *in silico* analysis of fermentation systems in view of their capacity to address productivity.

In regard to strain modification, it is emphasized that increasing the productivity rather than the yield is a more suitable goal as the former is directly related to economic competiveness. Note that emphasis on productivity is not at undue expense of yield since any pronounced drop on yield would also lead to a drop in productivity. On the other hand, sole stress on yield at the expense of productivity (due to a possible drop in growth rate) is not conducive to economics. Therefore, in the course of metabolic engineering undergoing several rounds of analysis and synthesis of strains, the productivity issue must be considered from the very outset. While the HCM framework based on a reduced subset of EMs can be useful in developing basic guidelines for flux redistribution of fermenting organisms, reasonable interpretation should be made under the possible loss of modes with significance for strain improvement. For metabolic engineering application, more sophisticated frameworks such as Lumped HCM (L-HCM) (Song & Ramkrishna, 2010; 2011) or Young's model (Young et al., 2008) represent promising methodologies in the future.

It is also shown that the productivity of lignocellulosic bioethanol can significantly be enhanced by synergistic combination of continuous and fed-batch reactors and optimizing their operating conditions. While experimental verification should follow, our model-based study provides solid proof-of-concept support for the success of the proposed methods.

6. Acknowledgment

The authors acknowledge a special grant from the Dean's Research Office at Purdue University for support.

7. References

Bailey, J. E. (1998). Mathematical modeling and analysis in biochemical engineering: Past accomplishments and future opportunities. *Biotechnology Progress*, Vol. 14, No. 1, pp. 8-20, ISSN 8756-7938

Bentley, W. E.; Mirjalili, N.; Andersen, D. C.; Davis, R. H. & Kompala, D. S. (1990). Plasmid-Encoded Protein - the Principal Factor in the Metabolic Burden Associated with Recombinant Bacteria. *Biotechnology and Bioengineering*, Vol. 35, No. 7, pp. 668-681, ISSN 0006-3592

Binod, P.; Sindhu, R.; Singhania, R. R.; Vikram, S.; Devi, L.; Nagalakshmi, S.; Kurien, N.; Sukumaran, R. K. & Pandey, A. (2010). Bioethanol production from rice straw: An overview. *Bioresource Technology*, Vol. 101, No. 13, pp. 4767-4774, ISSN 0960-8524

Cardona C. A.; Quintero J. A.; Paz I. C. (2010). Production of bioethanol from sugarcane bagasse: Status and perspectives. *Bioresource Technology*, Vol. 101, No. 13, pp. 4754-4766, ISSN 0960-8524

Chu, B. C. H. & Lee, H. (2007). Genetic improvement of Saccharomyces cerevisiae for xylose fermentation. *Biotechnology Advances*, Vol. 25, No. 5, pp. 425-441, ISSN 0734-9750

Glick, B. R. (1995). Metabolic Load and Heterologous Gene-Expression. *Biotechnology Advances*, Vol. 13, No. 2, pp. 247-261, ISSN 0734-9750

Hahn-Hagerdal, B.; Karhumaa, K.; Fonseca, C.; Spencer-Martins, I. & Gorwa-Grauslund, M. F. (2007). Towards industrial pentose-fermenting yeast strains. *Applied Microbiology and Biotechnology*, Vol. 74, No. 5, pp. 937-953, ISSN 0175-7598

Jeppsson, M.; Johansson, B.; Hahn-Hagerdal, B. & Gorwa-Grauslund, M. F. (2002). Reduced oxidative pentose phosphate pathway flux in recombinant xylose-utilizing Saccharomyces cerevisiae strains improves the ethanol yield from xylose. *Applied and Environmental Microbiology*, Vol. 68, No. 4, pp. 1604-1609, ISSN 0099-2240

Johansson, B. & Hahn-Hagerdal, B. (2002). Overproduction of pentose phosphate pathway enzymes using a new CRE-loxP expression vector for repeated genomic integration in Saccharomyces cerevisiae. *Yeast*, Vol. 19, No. 3, pp. 225-231, ISSN 0749-503X

Kadam, K. L. & McMillan, J. D. (2003). Availability of corn stover as a sustainable feedstock for bioethanol production. *Bioresource Technology*, Vol. 88, No. 1, pp. 17-25, ISSN 0960-8524

Keshwani, D. R. & Cheng, J. J. (2009). Switchgrass for bioethanol and other value-added applications: A review. *Bioresource Technology*, Vol. 100, No. 4, pp. 1515-1523, ISSN 0960-8524

Kim, J. I.; Varner, J. D. & Ramkrishna, D. (2008). A Hybrid Model of Anaerobic E. coli GJT001: Combination of Elementary Flux Modes and Cybernetic Variables. *Biotechnology Progress*, Vol. 24, No. 5, pp. 993-1006, ISSN 8756-7938

Klamt, S. & Stelling, J. (2002). Combinatorial complexity of pathway analysis in metabolic networks. *Molecular Biology Reports*, Vol. 29, No. 1-2, pp. 233-236, ISSN 0301-4851

Koebmann, B. J.; Westerhoff, H. V.; Snoep, J. L.; Nilsson, D. & Jensen, P. R. (2002). The glycolytic flux in Escherichia coli is controlled by the demand for ATP. *Journal of Bacteriology*, Vol. 184, No. 14, pp. 3909-3916, ISSN 0021-9193

Kompala, D. S.; Ramkrishna, D.; Jansen, N. B. & Tsao, G. T. (1986). Investigation of Bacterial-Growth on Mixed Substrates - Experimental Evaluation of Cybernetic Models. *Biotechnology and Bioengineering*, Vol. 28, No. 7, pp. 1044-1055, ISSN 0006-3592

Krishnan, M. S.; Xia, Y.; Ho, N. W. Y. & Tsao, G. T. (1997). Fuel ethanol production from lignocellulosic sugars - Studies using a genetically engineered Saccharomyces yeast. *Fuels and Chemicals from Biomass*, Vol. 666, pp. 74-92, ISSN 0097-6156

Lange, J. P. (2007). Lignocellulose conversion: an introduction to chemistry, process and economics. *Biofuels Bioproducts & Biorefining-Biofpr*, Vol. 1, No. 1, pp. 39-48, ISSN 1932-104X

Lee, S. B.; Seressiotis, A. & Bailey, J. E. (1985). A Kinetic-Model for Product Formation in Unstable Recombinant Populations. *Biotechnology and Bioengineering*, Vol. 27, No. 12, pp. 1699-1709, ISSN 0006-3592

Maertens, J. & Vanrolleghem, P. A. (2010). Modeling with a View to Target Identification in Metabolic Engineering: A Critical Evaluation of the Available Tools. *Biotechnology Progress*, Vol. 26, No. 2, pp. 313-331, ISSN 8756-7938

Ramkrishna, D. (1983). A Cybernetic Perspective of Microbial-Growth. *Acs Symposium Series*, Vol. 207, pp. 161-178, ISSN 0097-6156

Ricci, J. C. D. & Hernandez, M. E. (2000). Plasmid effects on Escherichia coli metabolism. *Critical Reviews in Biotechnology*, Vol. 20, No. 2, pp. 79-108, ISSN 0738-8551

Satyagal, V. N. & Agrawal, P. (1989). A Generalized-Model of Plasmid Replication. *Biotechnology and Bioengineering*, Vol. 33, No. 9, pp. 1135-1144, ISSN 0006-3592

Schaaff, I.; Heinisch, J. & Zimmermann, F. K. (1989). Overproduction of Glycolytic-Enzymes in Yeast. *Yeast*, Vol. 5, No. 4, pp. 285-290, ISSN 0749-503X

Schuster, S.; Fell, D. A. & Dandekar, T. (2000). A general definition of metabolic pathways useful for systematic organization and analysis of complex metabolic networks. *Nature Biotechnology*, Vol. 18, No. 3, pp. 326-332, ISSN 1087-0156

Shuler, M. L. & Kargi, F. (2002). *Bioprocess Engineering. Basic Concepts*. Upper Saddle River: Prentice-Hall, Inc., ISBN 978-0130819086.

Song, H.-S. & Ramkrishna, D. (2009). Reduction of a Set of Elementary Modes Using Yield Analysis. *Biotechnology and Bioengineering*, Vol. 102, No. 2, pp. 554-568, ISSN 0006-3592

Song, H.-S. & Ramkrishna, D. (2010). Prediction of Metabolic Function From Limited Data: Lumped Hybrid Cybernetic Modeling (L-HCM). *Biotechnology and Bioengineering*, Vol. 106, No. 2, pp. 271-284, ISSN 0006-3592

Song, H.-S. & Ramkrishna, D. (2011). Cybernetic Models Based on Lumped Elementary Modes Accurately Predict Strain-Specific Metabolic Function. *Biotechnology and Bioengineering*, Vol. 108, No. 1, pp. 127-140, ISSN 0006-3592

Song, H.-S.; Kim, S. J. & Ramkrishna, D. (2011). Synergistic Optimal Integration of Continuous and Fed-Batch Reactors for Enhanced Productivity of Lignocellulosic Bioethanol. Industrial & Engineering Chemistry Research, DOI: 10.1021/ie200879s, ISSN: 0888-5885

Song, H.-S.; Morgan, J. A. & Ramkrishna, D. (2009). Systematic Development of Hybrid Cybernetic Models: Application to Recombinant Yeast Co-Consuming Glucose and Xylose. *Biotechnology and Bioengineering*, Vol. 103, No. 5, pp. 984-1002, ISSN 0006-3592

Talebnia, F.; Karakashev, D. & Angelidaki, I. (2010). Production of bioethanol from wheat straw: An overview on pretreatment, hydrolysis and fermentation. *Bioresource Technology*, Vol. 101, No. 13, pp. 4744-4753, ISSN 0960-8524

von Kamp, A. & Schuster, S. (2006). Metatool 5.0: fast and flexible elementary modes analysis. *Bioinformatics*, Vol. 22, No. 15, pp. 1930-1931, ISSN 1367-4803

Wingren, A.; Galbe, M. & Zacchi, G. (2003). Techno-economic evaluation of producing ethanol from softwood: Comparison of SSF and SHF and identification of bottlenecks. *Biotechnology Progress*, Vol. 19, No. 4, pp. 1109-1117, ISSN 8756-7938

Young, J. D. & Ramkrishna, D. (2007). On the matching and proportional laws of cybernetic models. *Biotechnology Progress*, Vol. 23, No. 1, pp. 83-99, ISSN 8756-7938

Young, J. D.; Henne, K. L.; Morgan, J. A.; Konopka, A. E. & Ramkrishna, D. (2008). Integrating cybernetic modeling with pathway analysis provides a dynamic, systems-level description of metabolic control. *Biotechnology and Bioengineering*, Vol. 100, No. 3, pp. 542-559, ISSN 0006-3592

9

Consolidated Bioprocessing Ethanol Production by Using a Mushroom

Satoshi Kaneko, Ryoji Mizuno, Tomoko Maehara and Hitomi Ichinose
National Food Research Institute
Japan

1. Introduction

Plant cell walls are the most abundant biomass source in nature and are of increasing importance because worldwide attention has now focused on bioethanol production to combat global warming and to safeguard global energy. Because of competition between food and fuel production, lignocelluloses are expected to be utilized for future fuel ethanol production. One of the major problems in producing ethanol from lignocellulosic biomass is the expensive production cost. Consolidated bioprocessing (CBP) is gaining recognition as a potential breakthrough for low-cost biomass processing (Lynd, 1996; Lynd et al., 2002; Lynd et al., 2005; Van Zyl et al., 2007; Xu et al. 2009). CBP of lignocellulose to bioethanol refers to the combination of the 4 biological events required for this conversion process (production of lignocellulose-degrading enzymes, hydrolysis of polysaccharides present in pre-treated biomass and fermentation of hexose and pentose sugars) in one reactor. However, no natural microorganism exhibits all the features desired for CBP. Bacteria and yeast have been the primary candidates for CBP research and some progress has been made in this regard. Traditionally, proponents of CBP processes have identified two primary developmental pathways capable of producing industrially viable CBP microbial strains. These are category I, engineering a cellulase producer, such as *Clostridium thermocellum,* to be ethanologenic; and category II, engineering an ethanologen, such as *Saccharomyces cerevisiae* or *Zymomonas mobilis,* to be cellulolytic (Lynd, 1996; Lynd et al., 2002; Lynd et al., 2005; Van Zyl et al., 2007; Xu et al., 2009). However, the both categories have advantages and disadvantages. Cellulase producer lacks ethanol tolerance, and it is very difficult to coexpress of multiple saccharification enzyme genes in ethanol producer. Especially, heterologous expression of *Trichoderma reesei* cellobiohydrolases (cellobiohydrolase I and cellobiohydrolase II), which play the crucial role in cellulose degradation, are generally poor.

Basidiomycetes, also known as wood-rotting fungi, can achieve the complete breakdown of lignins (Cooke & Rayner, 1984; Cullen, 1997), and are considered primary agents of plant litter decomposition in terrestrial ecosystems (Thorn et al., 1996). Furthermore, some basidiomycetes produce alcohol dehydrogenases, thus allowing the production of wine using a mushroom (Okamura et al., 2000; Okamura et al., 2001). These properties of basidiomycetes appear suitable for use in CBP. In a preliminary study, we screened some edible mushrooms for their ability to produce ethanol and found that *Flammulina velutipes* is a good producer of ethanol. *F. velutipes* is a white-rot fungus that grows from spring through late autumn on a variety of hardwood tree stubs and dead stems and is widely distributed in temperate to subarctic regions. Currently, *F. velutipes* is the most produced mushroom in bed cultivation in Japan, the annual production being 130,000 tons/year. Artificial cultivation of mushrooms in polypropylene bottles is popular in Japan. *F. velutipes* has been characterized as wide adapted strain for various kinds of substance of artificial cultivation media, thus suggesting that the strain may be useful in the conversion of a wide variety of biomass types.

In this study, we investigated the properties of ethanol fermentation by *F. velutipes* to determine its suitability for CBP, because the use of basidiomycetes for bioethanol production is not common and the ethanol fermentation abilities of basidiomycetes are not well characterized. Furthermore, several biomass such as sorghums and rice straw were used as raw material to evaluate the detail conversion from biomass to ethanol by *F. velutipes*.

2. Properties of ethanol fermentation by *F. velutipes*

Because the use of basidiomycetes in bioethanol production is not common, and the ethanol fermentation abilities of basidiomycetes are not well characterized, we investigated the properties of ethanol fermentation by *F. velutipes* to determine its suitability for CBP (Mizuno et al., 2009b). Before the experiment, to obtain a suitable strain for CBP, 10 *F. velutipes* strains, culture stock of the Forest Institute of Toyama Prefectural Agricultural, Forestry, and Fisheries Research Center, were screened for cellulase production and ethanol fermentation. The Fv-1 strain was selected for further study because it not only produces high levels of cellulases, but also because its ability to ferment ethanol is superior to the other strains.

Firstly, fermentation of D-glucose was done by *F. velutipes* Fv-1. Figure 1A shows a conversion of 1% w/v of D-glucose to ethanol by *F. velutipes*. The consumption of D-glucose started gradually after incubation, and it was depleted after 6 d. Ethanol production correlated with sugar consumption, and it reached a maximum after 6 d. Thereafter, the amount of ethanol decreased gradually. Finally, *F. velutipes* converted 10 g/l of D-glucose to 4.5 g/l of ethanol, equivalent to a theoretical ethanol recovery rate of 88%. In the case of ethanol production from 5% w/v D-glucose, ethanol production reached a maximum, and all of the D-glucose was consumed after 18 d of incubation (Fig. 1B), and 50 g/l of D-glucose was converted to 22.4 g/l of ethanol, equivalent to a theoretical ethanol recovery rate of 87%. The conversion rate was the same as the case of 1% w/v of D-glucose. Because the incubation time to ferment 1% w/v sugar is shorter than the case of 5% w/v, we employed 1% w/v of sugar concentration in subsequent experiments.

Secondary, determination of the fermentation specificity of sugars by *F. velutipes* Fv-1 was done using various monosaccharides. As shown in Fig. 2, both D-mannose and D-fructose were converted to ethanol by *F. velutipes*. Consumption of D-mannose occurred slightly faster than that of D-glucose; it started immediately after incubation and was completely depleted after 5 d. Ethanol production from D-mannose was similar to that from D-glucose. It started during the first day of incubation and reached a maximum after 6 d. Furthermore, 4.4 g/l of ethanol was produced from 10 g/l of D-mannose, equivalent to a theoretical ethanol recovery rate of 86% (Fig. 2A). In contrast, consumption of D-fructose was slower than that of D-mannose. It started slowly after incubation and took 7 d to completely consume the D-fructose. Production of ethanol correlated with sugar consumption, and

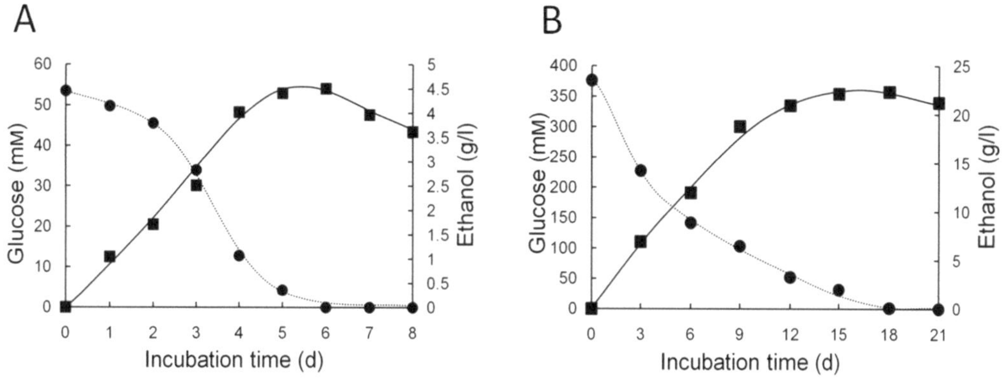

Symbols: closed circle, sugar; closed square, ethanol. The initial D-glucose concentration was (A) 1% w/v and (B) 5% w/v. (Reproduced from Mizuno et al., 2009b)

Fig. 1. Ethanol production from fermentation of D-glucose by *F. velutipes*

maximum conversion of D-fructose to ethanol was observed after 6 d. Upon completion of incubation, 4.0 g/l of ethanol was obtained from 10 g/l of D-fructose (Fig. 2B), yielding a theoretical conversion rate of 77%. In contrast to these sugars, *F. velutipes* did not convert L-arabinose, D-xylose, or D-galactose to ethanol (Figs. 2C, 2D, and 2E). Although there was slight consumption of D-xylose and D-galactose during incubation, ethanol production was not observed. In the case of L-arabinose, little sugar consumption was observed.

Next, we examined the fermentation specificity of *F. velutipes* Fv-1 toward various disaccharides. As shown in Fig. 3, *F. velutipes* possibly converted these sugars to ethanol and produced high yields. The theoretical conversion rates of these sugars were 83% and 77% from sucrose and maltose respectively. Degradation of sucrose was observed immediately after the incubation to import the sugar. The amount of reducing sugars was maximum on day 3 and was completely consumed after 7 d of incubation. Ethanol production was observed 1 d after incubation, and the amount of ethanol reached a maximum after 6 d. Finally, 4.5 g/l of ethanol was produced from 10 g/l of sucrose (Fig. 3A). In the case of maltose, degradation was observed on the first day of incubation, and the amount of reducing sugars reached a maximum after 2 d. Furthermore, the reducing sugars were completely depleted after 7 d of incubation. Ethanol production started during the first day of incubation and reached a maximum after 7 d. At the end of incubation, 10 g/l of maltose was converted to 3.8 g/l of ethanol (Fig. 3B). No conversion of xylobiose to ethanol was detected (data not shown), but a significant amount of ethanol production was observed when cellobiose was used as the carbon source (Fig. 4A). Cellobiose began degrading during the first day of incubation, and both D-glucose and cellobiose were completely depleted after 8 d. β-Glucosidase activity increased gradually during incubation. Ethanol production started after 1 d of incubation, and the amount of ethanol reached a maximum after 8 d. Upon completion of incubation, 10 g/l of cellobiose was converted to 4.5 g/l of ethanol (Fig. 4A). The theoretical conversion rate was 83%, a value similar to that of glucose and significantly higher than that of maltose. A high yield of ethanol was observed also in the higher concentration of cellobiose (Fig. 4D). Finally, 25 g/l of ethanol was produced from 50 g/l of D-glucose, and the theoretical conversion rate was 91%.

Since cellobiose was converted to ethanol at a relatively high rate, the conversions of cello-oligosaccharides to ethanol by *F. velutipes* were also investigated. Figures 4B and 4C show

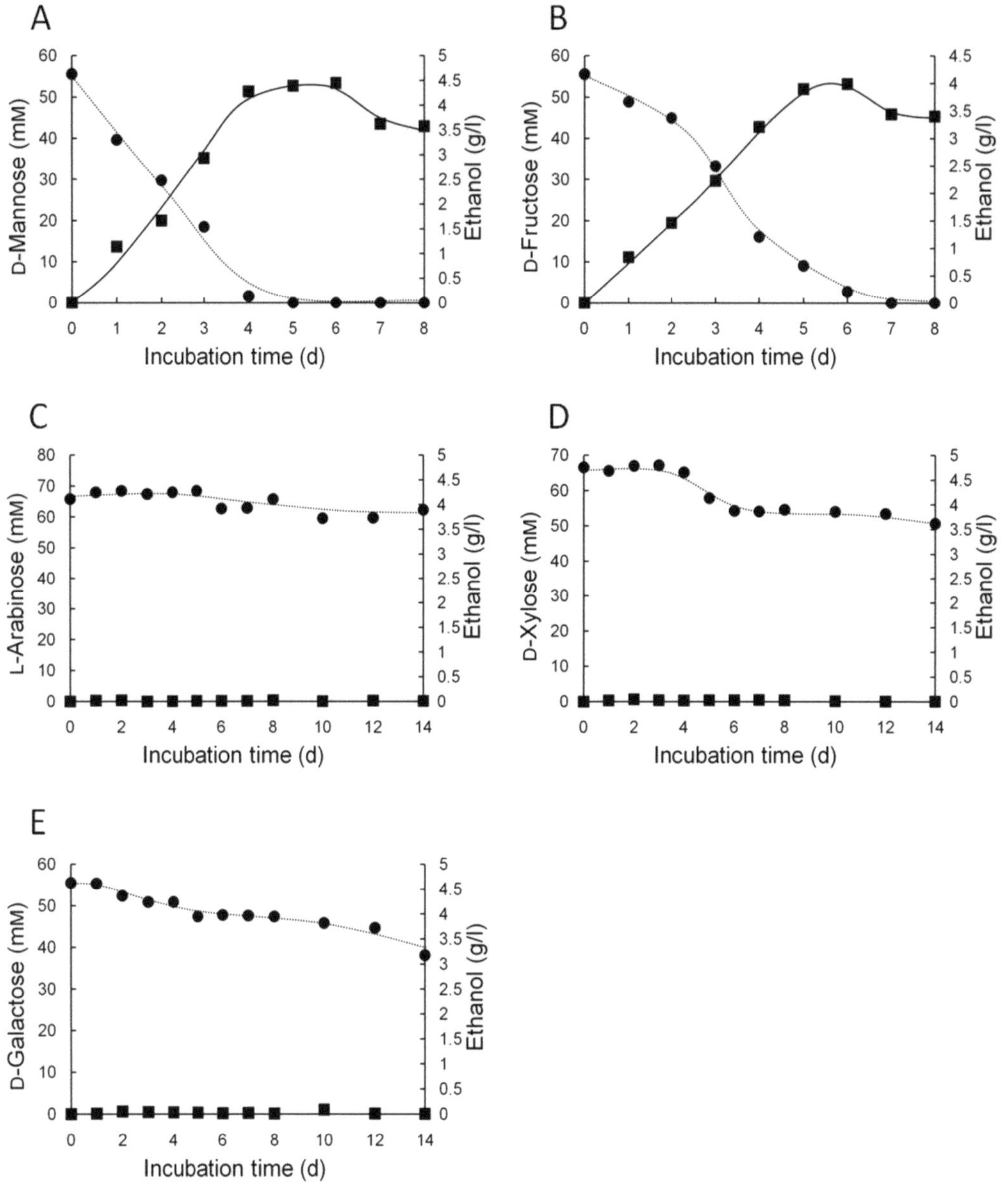

Symbols: closed circle, sugar; closed square, ethanol. The initial sugar concentration was 1% w/v. (Reproduced from Mizuno et al., 2009b)

Fig. 2. Ethanol fermentation from (A) D-mannose, (B) D-fructose, (C) L-arabinose, (D) D-xylose and (E) D-galactose by *F. velutipes*

the results of the conversion of cellotriose and cellotetraose to ethanol. Both cello-oligosaccharides were effectively converted to ethanol by *F. velutipes*. During incubation, cellotriose was initially hydrolyzed to D-glucose and cellobiose, and almost 80% of the initial amount of cellotriose was hydrolyzed by 2 d. Cellotriose was not detected after 5 d of incubation, and D-glucose and cellobiose were completely depleted after 7 d. β-Glucosidase

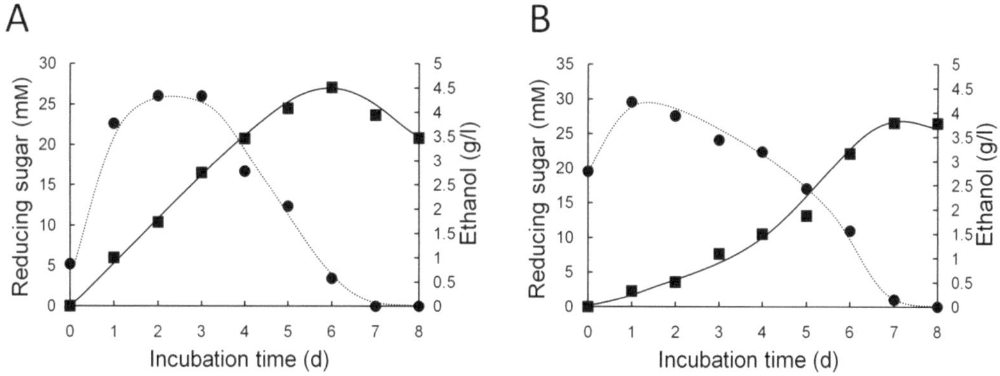

Symbols: closed circle, reducing sugar; closed square, ethanol. The initial sugar concentration was 1% w/v. (Reproduced from Mizuno et al., 2009b)

Fig. 3. Ethanol fermentation from (A) sucrose and (B) maltose by *F. velutipes*

was slightly induced by 6 d, and the activity gradually increased after 6 d. The amount of ethanol increased during incubation and reached a maximum after 7 d of incubation. *F. velutipes* produced 4.2 g/l of ethanol from 10 g/l of cellotriose, equivalent to a theoretical conversion rate of 76% (Fig. 4B). In the case of cellotetraose, it was initially hydrolyzed to cellotriose, cellobiose, and D-glucose, and more than 90% of the cellotetraose was hydrolyzed by 2 d. Cellotetraose was not detected after 3 d of incubation, and cellotriose, cellobiose, and D-glucose were completely depleted after 4, 6 and 7 d respectively. β-Glucosidase activity increased rapidly over 2 d then decreased gradually from 2 d to 5 d, and stabilized at an activity level of about 30 mU/ml. The amount of ethanol increased after incubation, and 4.4 g/l of ethanol was produced from 10 g/l of cellotetraose after 7 d of incubation (Fig. 4C). The ethanol recovery for the theoretical conversion value was 78%.

To date, many microorganisms, including *Saccharomyces cerevisiae*, *Zymmonas mobilis*, *Pichia stipitis*, *Rhizopus oryzae*, and *Clostridium thermocellum*, have been reported to produce ethanol (DeMoss & Gibb, 1951; Maas et al., 2006; Ng et al., 1981; Parekh & Wayman, 1986; Weimer & Zeikus, 1977). In general, *S. cerevisiae* is the most widely used microorganism in the industry and is popular in bioethanol production, because it has high efficiency of ethanol production and high ethanol tolerance. However, we focused on basidiomycetes to develop CBP because these microorganisms have both lignocellulose degradation and ethanol fermentation abilities.

Here, we characterized properties of ethanol fermentation by *F. velutipes* Fv-1. The strain converted D-glucose to ethanol at a theoretical conversion rate of 88%, comparable to those of *S. cerevisiae* and *Zymomonas* (Swings & DeLey, 1977). On the other hand, *F. velutipes* scarcely converted pentose and D-galactose to ethanol (Fig. 2). These properties of *F. velutipes* are similar to those of *S. cerevisiae* (Barnett, 1976). Moreover, *F. velutipes* demonstrated the preferable features for CBP when oligosaccharides were used as starting materials (Figs. 3 and 4). The tested oligosaccharides were converted to ethanol at almost the same rate as that of D-glucose, and β-glucosidase activity increased during fermentation. These features are indispensable in CBP, which requires saccharification and fermentation of cellulose contained in the cell wall. It has been reported that *C. thermocellum* and *P. stipitis* can ferment cellobiose (Parekh & Wayman, 1986). Furthermore, *C. thermocellum* can also convert cellulose to ethanol directly (Ng et al., 1981; Lynd et al., 1989; Weimer & Zeikus,

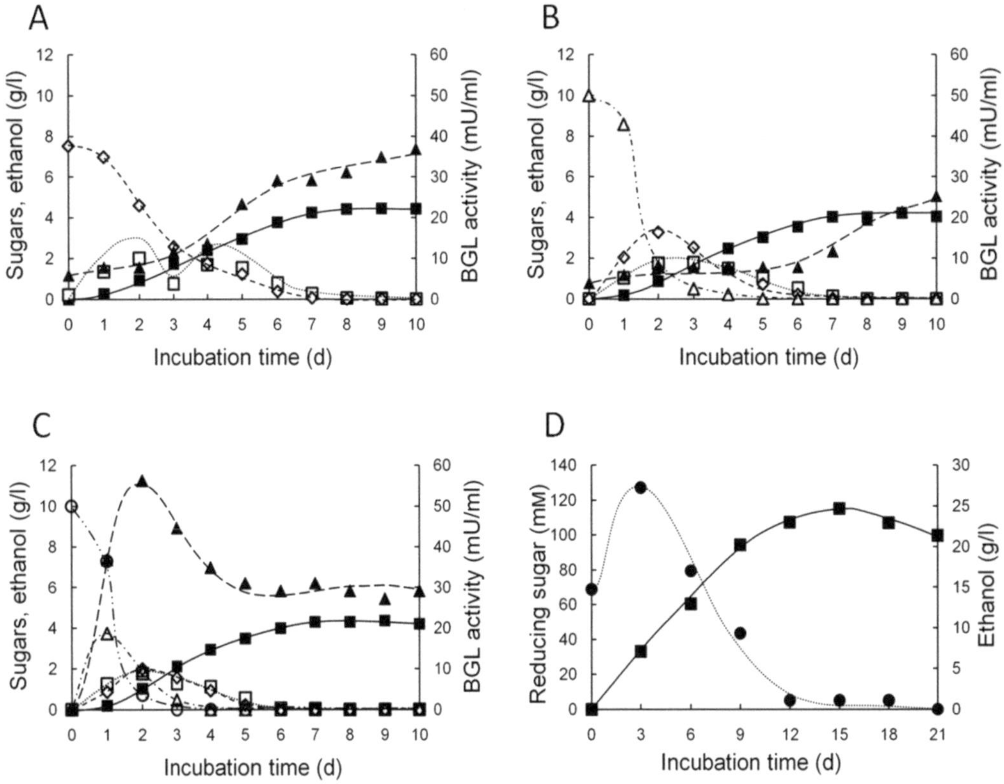

Symbols: open square, D-glucose; open diamond, cellobiose; open triangle, cellotriose; open circle, cellotetraose; closed circle, reducing sugar; closed square, ethanol; closed triangle, β-glucosidase activity. The initial sugar concentration was 1% w/v (A, B, and C) or 5% w/v (D). (Reproduced from Mizuno et al., 2009b)

Fig. 4. Ethanol fermentation from (A) cellobiose, (B) cellotriose, (C) cellotetraose and (D) 5% cellobiose by *F. velutipes*

1997). However, this species cannot be used at the scene of ethanol production because fermentation of *C. thermocellum* is strongly inhibited at relatively low ethanol concentrations (5 g/l) (Herrero & Gomez, 1980). In contrast, it has been reported that basidiomycetes have tolerance of up to 120 g/l of ethanol (Okamura et al., 2001), and therefore basidiomycetes are more suitable for CBP than *Clostridium* strains. From these results, we concluded that *F. velutipes* possesses advantageous characteristics for use in CBP.

3. Properties of ethanol production from biomass by *F. velutipes*

3.1 Use of whole crop sorghums as a raw material in consolidated bioprocessing bioethanol production using *Flammulina velutipes*

The ethanol fermentation abilities of basidiomycetes have not been well characterized, we evaluated the ability of *F. velutipes* in CBP. Preliminary fermentation experiments indicate that *F. velutipes* convert sugars to ethanol much more under the high concentration of biomass which close to solid state cultivation than liquid cultivation condition. Therefore, we employed solid state cultivation which usually performed in artificial cultivation of mushrooms for the coversion of biomass to produce bioethanol. Sorghum is selected as a possible raw material to produce bioethanol by CBP using *F. velutipes*. Sorghum is a C4 crop of the grass family belonging to the genus *Sorghum bicolor* L. It is well adapted to temperate climates and can be cultured from Kyushu to Tohoku area in Japan. The plant grows to a height from about 120 to above 400 cm, depending on the variety and growing conditions, and can be an annual or a short perennial crop. Sorghum is considered to be one of the most drought resistant agricultural crops, as it is able to remain dormant during the driest periods (Xu et al., 2000). These properties of sorghum are suitable as raw material for the ethanol production. We evaluated the ability of *F. velutipes* in CBP using sorghum strains as a raw material, and solid-state CBP of ground sorghum strains (SIL-05 and Kyushukou No. 4) using *F. velutipes* was investigated. The possibility of sorghum strains as a raw material in the CBP ethanol production by *F. velutipes* is also discussed below.

We selected grinding for the pretreatment of sorghum strains. This can be used on both dry and wet materials, and the cost of grinding is one of the cheapest compared to other methods used for milling biomass. The grinding of sorghum was carried out with an ultra-fine friction grinder. Grinding was performed at room temperature, and was repeated twice. To examine the efficiency of grinding as a pretreatment, the degree of saccharification was tested using commercially available enzymes Celluclast 1.5L (Sigma, St. Louis, MO), Novozyme 188 (Sigma) and Multifect xylanase (Genencor Kyowa, Tokyo).

The saccharification yields of SIL-05 and Kyushukou No. 4 by the enzymes were 30.1% and 51.7% respectively (Fig. 5A). Kyushukou No. 4 is one of the sorghum *brown mid-rib* (*bmr*) mutants in which cafferic acid *O*-methyltransferase (COMT), a lignin biosynthetic enzyme, activity is reduced as compared to the wild type (Bout & Vermerris, 2003). This property of *bmr* significantly affected the hydrolysis of polysaccharides in the biomass, but there were no significant differences in the proportions of hydrolysis of the components such as cellulose and hemicellulose (Fig. 5). When the saccharification yields of cellulose and hemicelluloses were compared, degradation of hemicelluloses was slightly higher than for cellulose in both types of sorghum.

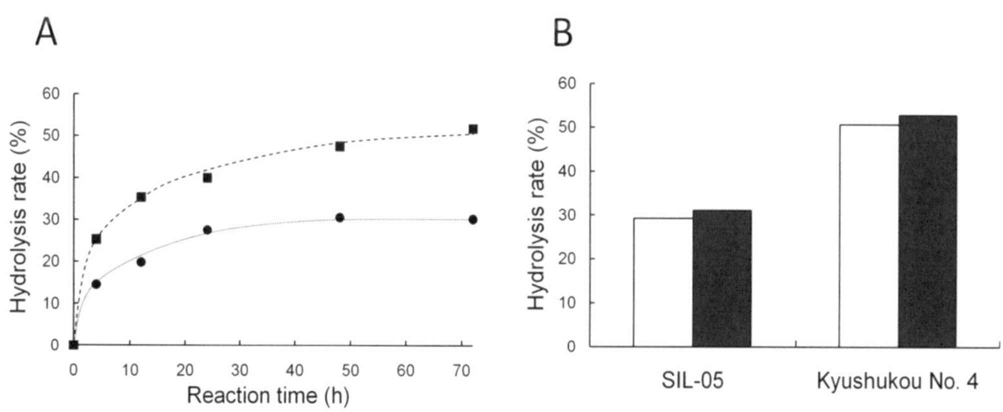

(A) Closed circle, SIL-05; closed square, Kyusyukou No. 4. Broken lines were drawn by roughly following the experimental data points. (B) White, cellulose; black, hemicelluloses. (Reproduced from Mizuno et al., 2009a)

Fig. 5. (A) Time course of sorghum hydrolysis and (B) saccharification yield of cellulose and hemicellulose incubated for 72 h

Next, solid-state ethanol fermentation by *F. velutipes* was performed for both sorghum strains. Solid-state fermentation is advantageous because it carries a low ethanol production cost. Generally, sorghums contain 70–80% v/v water, corresponding to 43–25% w/v. These concentrations are necessary to obtain relatively high final ethanol concentrations.

Furthermore, it is possible to reduce the costs of many procedures, such as amount of water, concentration of biomass, treatment of waste water, and so forth, if the water concentration of the raw materials in the all ethanol production procedures is retained. The Fv-1 strain was selected for further study because it not only produces high levels of cellulases, but also because its ability to ferment ethanol is superior to the other strains. Mycelia of Fv-1 were harvested in the late exponential growth phase by centrifugation at 3,000 × g and washed with sterile water. The prepared wet mycelia (20 mg of dry weight) were mixed with 100 mg of ground sorghum for solid-state fermentation.

A larger amount of ethanol was produced from SIL-05 than from Kyushukou No. 4 (Fig. 6). Because SIL-05 contained a larger amount of soluble sugars than Kyushukou No. 4 (Table 1), it should be advantageous for total ethanol fermentation. The ethanol conversion rates for the soluble sugars contained in SIL-05 and Kyushukou No. 4 were 57.2% and 38.9% respectively. The addition of saccharification enzymes was not effective for SIL-05 (Fig. 6A). This corresponded with the results of enzymatic hydrolysis (SIL-05 just hydrolyzed almost 30%) (Fig. 6). However, the ethanol conversion rate for the degraded cellulose was 85.6%, significantly higher than that for soluble sugars. In contrast, although total ethanol production was not high, ethanol production from Kyushukou No. 4 significantly increased when saccharification enzymes were added to the culture (Fig. 6B). Because the cellulose and hemicellulose in Kyushukou No. 4 were more easily hydrolyzed than SIL-05 by cellulases, significantly more ethanol was produced by the addition of the saccharification enzymes. The ethanol conversion rate for the degraded cellulose of Kyushukou No. 4 (98.3%) was much higher than that of SIL-05 (85.6%). Thus, the *bmr* mutation appears to be useful for CBP because it gives a high yield of glucose from biomass without acid or alkali pretreatment. However, the results indicate that the production of cellulases by *F. velutipes* is not sufficient for CBP, or that the saccharification enzymes are suppressed by carbon

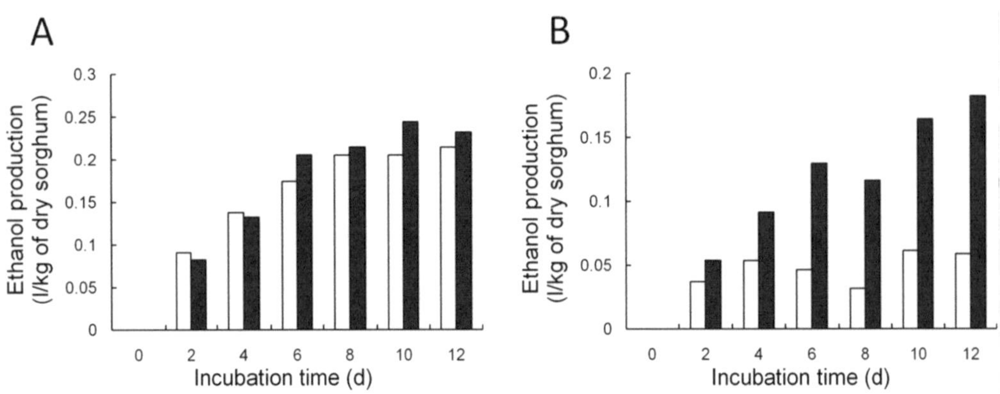

White, no added enzymes; black, 15 µl Celluclast 1.5 L and 10 µl Multifect xylanase added.6 Reproduced from Mizuno et al., 2009a)

Fig. 6. Solid-state ethanol fermentation of (A) SIL-05 and (B) Kyushukou No. 4 by *Flammulina velutipes*

	SIL-05	Kyushukou No. 4
Water content (%)	78 (± 0.7)	80 (± 1.4)
Soluble sugar (%)	12 (± 0.1)	4.5 (± 0.0)
Cellulose content (%)[a]	5.1 (± 0.3)	7.4 (± 0.8)
Hemicellulose content (%)[b]	3.6 (± 0.1)	5.2 (± 0.5)
Other content (%)[c]	1.2 (± 0.2)	2.4 (± 0.5)

[a] The amount of hexose was determined by the anthrone-sulfuric acid method.
[b] The amount of pentose was determined by the orcin-Fe^{3+}-hydrochloric acid method.
[c] All components except for sugars. (Reproduced from Mizuno et al., 2009a)

Table 1. Compositions of SIL-05 and Kyushukou No. 4

catabolites due to the existence of soluble sugars. Therefore, an effective saccharification enzyme inducing method for *F. velutipes* in the CBP is required.

In this work, we demonstrated CBP ethanol fermentation of sorghum strains by *F. velutipes* Fv-1. The procedure is quite simple and cost effective, and can reduce energy consumption, because the raw material is simply ground and then mixed with mycelia. We demonstrated the merit of high concentrations of soluble sugars and the *bmr* mutation in sorghums. Both sorghum strains can be used in CBP. The *bmr* mutation is only found in sorghums, corn, and pearl millet, giving sorghum an advantage over many other crops for ethanol production. Future studies should focus on the improvement of CBP using *F. velutipes* and the selective breeding of novel types of sorghums with high concentrations of soluble sugars and the *bmr* mutation.

3.2 Solid state fermentation of rice straw by *F. velutipes*

The solid state ethanol fermentation by *F. velutipes* was performed for ammonia treated rice straw. Solid state cultivation has a large merit to decrease the ethanol production cost. But it has demerit on the saccharification of biomass. As shown in Fig. 7A, saccharification of biomass at high concentration is quite difficult. Significant amount of cellulase is necessary to obtain enough level of saccharification, and saccharification yield do not increase in proportion to the amount of cellulase if increased the amount of cellulase. Furthermore, saccharification yield will be significantly decreased under the high substrate condition. The hydrolysis rate of 30% w/v biomass was very low (less than 10%). In contrast, ethanol yield was equivalent to 80-90% of hydrolysis rate so that the merit of our process using *F. velutipes* was proven (Fig. 7B). In the case that enzymes were not added, ethanol production by *F. velutipes* was only 0.026 l/kg of dry biomass, equivalent to a theoretical ethanol recovery rate of 5.9% from total hexose. In contrast to no enzymes addition, in the case that 1 and 5 mg/g product of enzymes were added to the fermentation, ethanol production after 15 d by *F. velutipes* was 0.26 and 0.34 l/kg of dry biomass, respectively. The ethanol conversion rates of 1 and 5 mg/g product enzymes addition were 61.6% and 77.8% for total hexose, respectively. The maximum weight loss was approximately 70% in the case that no enzymes were added to the fermentation, while the maximum weight loss for enzyme addition of 1 and 5 mg/g product were approximately 90% and 96% respectively (data not shown).

These results suggest *F. velutipes* has favourable properties for CBP. It could be expected that development of novel bioethanol production process by using *F. velutipes*.

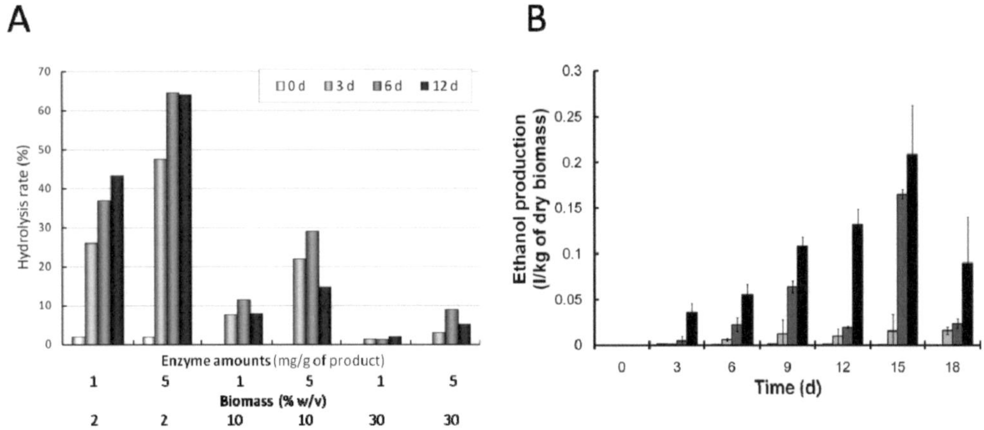

(A) Saccharification yield of ammonia treated rice straw by enzymes. (B) Solid-state ethanol fermentation of ammonia treated rice straw by *F. velutipes*. Right gray, no added enzymes; gray, 1 mg/g product enzymes added; black, 5 mg/g product enzymes added.

Fig. 7. Saccharification and ethanol production at high biomass concentration

4. Development of a gene transfer system for *F. velutipes*

4.1 Development of a gene transfer system for the mycelia of *F. velutipes*

As shown in above, we found the edible mushroom *F. velutipes* Fv-1 strain to be an efficient ethanol producer, and, we demonstrated its preferable properties of ethanol fermentation from various sugars (Mizuno et al., 2009b), whole crop sorghums and rice straw (Mizuno et al., 2009a). However, the strain can only slightly convert pentoses, which account for approximately 20-30% of plant cell walls, into ethanol (Mizuno et al., 2009a). Therefore, genetic engineering of the pentose metabolism is necessary to make possible the ethanol fermentation from pentose. Furthermore, more efficient (low cost) conversion of biomass to ethanol could be expected if saccharification ability was strengthened by expressing cellulases. A transformation method of *F. velutipes* by the electroporation protocol for basidiospores has been reported (Kuo et al., 2004), but it requires a long period to produce basidiospores because it must go through fruiting body formation, and cannot eliminate the risk of contamination in the process of spore harvest. Since screening of many transformants is needed for improvement of the metabolic pathway by genetic engineering, the development of a simpler transformation method is desired to obtain high numbers of transformants.

Therefore, an adequate condition for protoplast preparation from mycelia of *F. velutipes* Fv-1 strain was investigated, and simpler a transformation protocol for this fungus was developed by the calcium-PEG method and the restriction enzyme-mediated-integration (REMI) method.

First, we constructed a pFvT vector for transformation of the *F. velutipes* Fv-1 strain (Fig. 8A). The vector possessed a *F. velutipes* tryptophan synthetase gene promoter and terminator (GenBank no. AB028647) to regulate expression of the constructed genes, and the hygromycin phosphotransferase gene (*hph*) from *Escherichia coli* as selection marker. The *hph* gene was obtained from pCAMBIA1201 vector (CAMBIA; http://www.cambia.org/).

Next, conditions to prepare protoplast from the mycelia of *F. velutipes* were optimized by modifying a method for *Phanerochaete sordida* (Yamagishi et al., 2007). The *F. velutipes* Fv-1

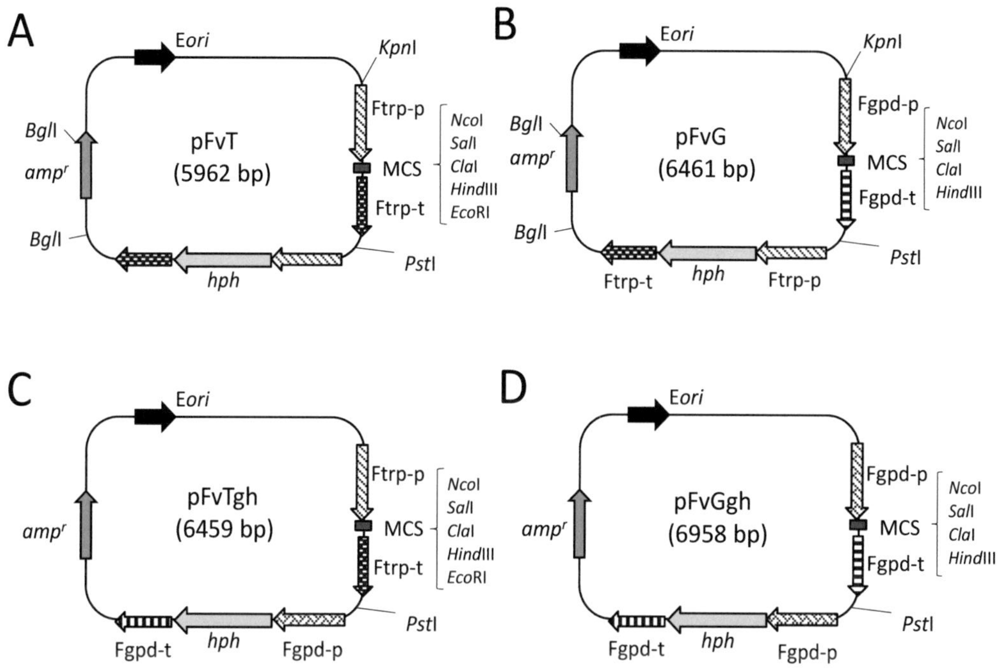

MCS, multiple cloning sites; Ftrp-p, *trp1* promoter from *F. velutipes*; Ftrp-t, *trp1* terminator from *F. velutipes*; Fgpd-p, *gpd* promoter from *F. velutipes*; Fgpd-t, *gpd* terminator from *F. velutipes*; *hph*, hygromycin B phosphotransferase gene; *amp*^{r}, ampicilin resistance gene; E*ori*, pUC19 ori. (Reproduced from Maehara et al., 2010b)

Fig. 8. Structures of the plasmids used in this study

strain was grown in PCMY (1% polypeptone, 0.2% casamino acid, 1% malt extract, and 0.4% yeast extract) medium at 25°C for 3 d. Then the mycelia were collected and incubated in enzyme solution [1.5% cellulase Onozuka-RS (Yakult Pharmaceutical, Tokyo) and 1.5% lysis enzyme (Sigma, St. Louis, MO) in 0.75 M MgOsm (0.75 M $MgSO_4$, 20 mM MES, pH 6.3)] at 30°C for 5 h. The protoplasts were filtered through Miracloth (Cosmo Bio, Tokyo), washed at twice with 1 M SorbOsm (1.0 M sorbitol, 10 mM MES, pH 6.3), and suspended in SorbOsm plus 40 mM $CaCl_2$ solution to a final concentration of approximately 10^8 protoplasts ml^{-1}.

Genetic transformation was investigated using the pFvT vector and the protoplasts prepared as described above. The transformation procedures for *Lentinus edodes* (Sato et al., 1998) and *Schizophyllum commune* (Van Peer et al., 2009) were modified for the transformation of *F. velutipes* Fv-1. In the course of the transformation process, the effect of the structure of the plasmid DNA on transformation was evaluated using circular and linear pFvT plasmids. Approximately 6-fold transformants were obtained when the plasmid DNA was linearized (Table 2).

Because the REMI method is a popular transformation tool for fungi (Hirano et al., 2000; Maier & Schäfer, 1999; Riggle & Kumamoto, 1998; Sato et al., 1998), we evaluated the effect of REMI on the transformation for *F. velutipes* Fv-1. The *F. velutipes* Fv-1 strain was transformed by pFvT with a restriction enzyme, *Bgl*I, *Kpn*I, or *Pst*I. The addition of the restriction enzymes increased the number of transformants by about 1.6- to 5.8-fold (Table 2). The suggests that the addition of restriction enzymes enhanced the transformation efficiency of *F. velutipes*. Therefore, to find the optimum enzyme concentration for REMI, we

No. of transformants*				
Form of pFvT	**Restriction enzyme (50 U)**			
	none	*Bgl*I	*Kpn*I	*Pst*I
Circular	0.7 ± 1.2	7.3 ± 1.2	18.3 ± 6.1	25.7 ± 7.1
Linear	4.4 ± 0.6	12.7 ± 6.4	20.0 ± 8.7	21.7 ± 7.1
Form of DNA	**Plasmid**			
	pFvT	pFvG	pFvTgh	pFvGgh
Circular	11.3 ± 2.6	12.3 ± 2.5	28.3 ± 1.3	33.3 ± 2.9
Linear	11.3 ±6.6	10.7 ± 6.9	24.7 ± 8.4	27.7 ± 11

*The values represent the average and standard deviation of triplicate. (Reproduced from Maehara et al., 2010a & 2010b)

Table 2. Numbers of transformants obtained by the REMI method

performed transformation using circular pFvT plasmid with the presence of various concentrations of *Pst*I (Fig. 9). As for the results, the number of transformants obtained was affected by the amount of restriction enzyme. The efficiency was significantly increased by the addition of *Pst*I at 25 units, by it gradually decreased when the *Pst*I amount was over 25 units, suggesting that the optimal value for transformation mediated by the *Pst*I is 25 units. In conclusion, we found a simple transformation procedure for the mycelia of *F. velutipes* Fv-1 strain by the calcium-PEG method combined with REMI. The transformation method of *F. velutipes* Fv-1 strain does not require a process of spore formation, because the mycelia could be used as starting material. Moreover, a high efficiency of transformation was obtained by the adoption of REMI.

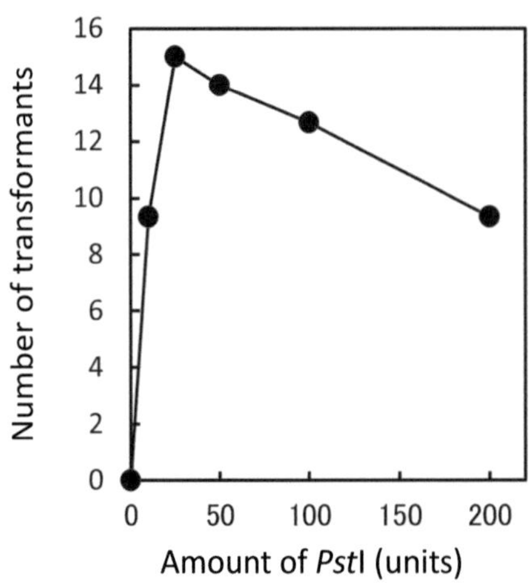

Fig. 9. Effects of the amount of *Pst*I on transformation by REMI method (Reproduced from Maehara et al., 2010a)

4.2 Improvement of the transformation efficiency of *Flammulina velutipes* Fv-1 using the glyceraldehydes-3-phosphate dehydrogenase gene promoter

To make possible genetic manipulation in *F. velutipes*, we constructed the pFvT plasmid containing the hygromycin phosphotransferase gene (*hph*) under the control of the tryptophan synthetase gene (*trp1*) promoter, and developed an easy transformation method for *F. velutipes* by the REMI method (Maehara et al., 2010a). Here, we focused on the promoter of the glyceraldehyde-3-phosphate dehydrogenase (*gpd*) gene because many tools such as promoters and selection markers are desirable for effective metabolic pathway engineering of *F. velutipes* Fv-1. The *gpd* promoters are the most frequently used constitutive promoters in basidiomycetes. GPD constitutes up to 5% of the soluble protein in *Saccharomyces cerevisiae* and other higher eukaryotic organisms (Piechaczyk et al., 1984; Punt et al., 1990), and *gpd* mRNA accounts for 2-5% of the poly (A)+ RNA in yeast (Holland & Holland, 1978).

In this section, we described that construction of new plasmids having the *hph* gene from *Escherichia coli* as a selection marker, which regulated by the *gpd* promoter and the potency of the *gpd* promoter from *F. velutipes* were evaluated.

First we constructed three vectors, pFvG, pFvTgh, and pFvGgh, by modification of pFvT. The pFvT vector possessed a *trp1* promoter and terminator regulating the expression of the constructed genes, and the *hph* gene as selection marker (Fig. 8A, Maehara et al., 2010a). Vectors pFvG (Fig. 8B) and pFvGgh (Fig. 8D) contained the *gpd* promoter and the terminator of *F. velutipes (*Kuo et al., 2004) located upstream and downstream of a multiple cloning site (MCS), and both pFvTgh (Fig. 8C) and pFvGgh (Fig. 8D) contained the *gpd* promoter and the terminator located upstream and downstream of the *hph* gene (Maehara et al., 2010b).

To determine the potency of the *gpd* promoter, we compared transformation efficiency by the *gpd* promoter with that by the *trp1* promoter. Gene integrations were performed by the REMI method. Protoplasts were prepared from mycelia of the *F. velutipes* Fv-1 strain, and then plasmids were transformed into the protoplasts with *Pst*I (25 U). As shown in Table 2, about 10 transformants (10.7 to 12.3) were obtained by the transformation of pFvT and of pFvG, which contain the *hph* gene controlled by the *trp1* promoter. In contrast, as for the results of the transformation of pFvTgh and pFvGgh, the numbers of transformants were significantly increased and about 24.7 to 33.3 transformants were obtained, suggesting that the activity of the *gpd* promoter was higher than that of the *trp1* promoter in *F. velutipes* Fv-1. There is a difference of about 500-bp in the length of pFvT and pFvG, or pFvTgh and pFvGgh, but no significant difference in the number of transformants obtained by pFvT and by pFvG, and by pFvTgh and pFvGgh was not observed. It might suggest, that the difference of the sizes of these plasmids was not affected on transformation efficiency.

To compare the activity of the *gpd* and the *trp1* promoter, the expression levels of the *hph* gene in each transformant were examined by reverse transcription-polymerase chain reaction (RT-PCR). Total RNA was extracted from each set of three transformants and equal amounts of RNAs from each set of three clones were mixed and used as template for RT-PCR. As shown in Fig. 10A, the intensities of the bands of the pFvTgh and pFvGgh transformants were stronger than that of the pFvT and pFvG transformants (upper panel), suggesting that the expression level of the *hph* gene in the pFvTgh and pFvGgh transformants was higher than that in the pFvT and pFvG transformants. The results were corresponded to the transformation efficiency presented in Table 2, and strongly suggest that the *gpd* promoter is functional in the heterologous gene expression system in *F. velutipes* Fv-1 to improve the expression level of the target gene.

Finally, in order to determine whether the plasmid vector was integrated into the genomic DNA by the REMI method, the genomic DNAs of 10 randomly selected pFvGgh transformants were analyzed by Southern blot using the digoxigenin-labeled *hph* gene as a probe (Fig. 10B). Hybridization signals were detected in all the transformants, and multiple

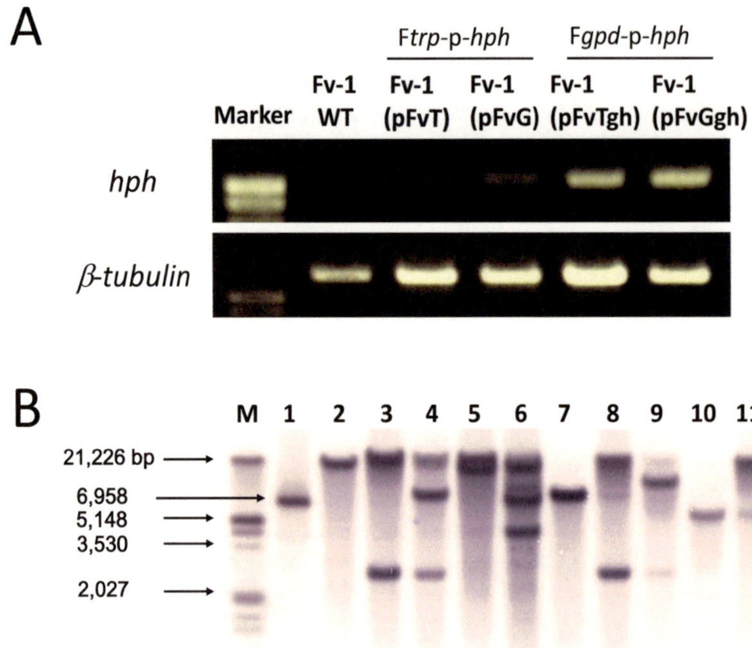

(A) RT-PCR-based evaluation of the expression of the hygromycin phosphotransferase gene: Upper panel, *hph* gene; Lower panel, *β-tubulin* gene as a control. (B) Southern blotting of *Pst*I digested genomic DNA using the *hph* gene probe: Lane M, DNA molecular size markers (values on left); lane 1, pFvGgh as a positive control; lanes 2-11, genomic DNA from transformants. (Reproduced from Maehara et al., 2010b)

Fig. 10. Analysis of the transformants obtained by REMI method

hybridization signals were also detected in some transformants. There was no signal from the genomic DNA of wild-type Fv-1 as a negative control (data not shown). These results indicate that at least a single *hph* gene was introduced into all the transformants, and the *hph* gene is thought to exist as a multicopy in the genomic DNAs of many transformants (Fig. 10B, lanes 3, 4, 5, 6, 8, 9 and 11). The same size bands were detected between 2,027 and 3,530-bp in four transformants (Fig. 10B, lanes 3, 4, 8 and 9). These bands might represent about 2,700-bp of the full-length *gpd* promoter-*hph*-*gpd* terminator region. A 6.9-kb DNA fragment, corresponding to the size of the pFvGgh plasmid, was observed in the genome of only one clone (Fig. 10B, lane 6), indicating that the full length of the plasmid was successfully introduced into the transformant. Consequently, we estimate the probability of integration of full-length pFvGgh vector by the REMI method to be approximately 10%. In our previous study, the probability of integration of the full-length vector was 30% so that the frequency of REMI events of Fv-1 was 10-30% (Maehara et al., 2010a). This value seems to be the comparable level in the case of model mushroom, *Coprinus cinereus* (8-56%) (Granado et al., 1997).

In conclusion, we demonstrated that the *gpd* promoter from *F. velutipes* Fv-1 would be a useful in the transformation system of the strain. The transformation efficiency was about 3 times improved by the use of the *gpd* promoter. The vectors constructed in this study will be available to improve the genetic engineering of *F. velutipes* Fv-1 for ethanol fermentation from pentose.

5. Conclusion

In spite of CBP is gaining recognition of a low-cost biomass processing as it involves enzyme production, completely no enzyme process which does not add the saccharification enzymes have not been established. In this study, we demonstrated that *F. velutipes* can highly convert biomass to ethanol using only small amount of saccharification enzyme even in the quite high concentration of biomass such as 30% w/v. These results suggest *F. velutipes* has favorable properties for CBP. Generally, artificial cultivation of mushrooms in polypropylene bottles is performed under the condition of water content 70 to 80%. The condition must be most suitable condition to cultivate the mushrooms. Therefore, *F. velutipes* will be especially effective in situations that CBP performed under the high concentration of biomass. We believe that this point would be advantage of *F. velutipes* compared with the other microorganisms engineered for CBP and even for fungus which is possible to ferment the both pentose and hexose. In the future, we would like to develop a novel bioethanol production process by using *F. velutipes*.

6. Acknowledgment

This work was financially supported by a grant-in-aid (Development of Biomass Utilization Technologies for Revitalizing Rural Areas) from the Ministry of Agriculture, Forestry, and Fisheries of Japan.

7. References

Barnett, JA. (1976). The utilization of sugars by Yeasts. *Advances in Carbohydrate Chemistry and Biochemistry*, Vol. 32, pp. 125–234, ISSN 0065-2318

Bout , S. & Vermerris, W. (2003). A candidate-gene approach to clone the sorghum Brown midrib gene encoding caffeic acid O-methytransferase. *Molecular Genetics and Genomics*, Vol. 269, pp. 205-214, ISSN 1617-4615

Cooke, RC. & Rayner, ADM. (1984). *Ecology of saprotrophic fungi*, Longman, ISBN 0-582-44260-5, London, United Kingdom

Cullen, D. (1997). Recent advances on the molecular genetics of ligninolytic fungi. *Journal of Biotechnology*, Vol. 53, pp 273–289, ISSN 0168-1656

DeMoss, RD. & Gibb, M. (1951). Ethanol formation in *Pseudomonas lindneri*. *Archives of Biochemistry and Biophysics*, Vol. 43, pp. 478–479, ISSN 0003-9861

Granado, JD.; Kertesz-Chaloupková, K.; Aebi, M. & Kües, U. (1997). Restriction enzyme-mediated DNA integration in *Coprinus cinereus*. *Molecular and Geneneral Genetics*, Vol. 256, pp. 28-36, ISSN 0026-8925

Herrero, AA. & Gomez, RF. (1980). Development of ethanol tolerance in *Clostridium thermocellum*: effect of growth temperature. *Applied and Environmental Microbiology*, Vol. 40, pp. 571–577, ISSN 0099-2240

Hirano, T.; Sato, T.; Yaegashi, K. & Enei, H. (2000). Efficient transformation of the edible basidiomycete *Lentinus edodes* with a vector using a glyceraldehyde-3-phosphate dehydrogenase promoter to hygromycin B resistance. *Molecular and Genneral Genetics*, Vol. 263, pp. 1047-1052, ISSN 0026-8925

Holland, MJ. & Holland, JP. (1978). Isolation and identification of yeast messenger ribonucleic acids coding for enolase, glyceraldehyde-3-phosphate dehydrogenase, and phosphoglycerate kinease. *Biochemistry*, Vol. 17, pp. 4900-4907, ISSN 0006-2960

Kuo, CY.; Chou, SY. & Huang, CT. (2004). Cloning of glyceraldehyde-3-phosphate dehydrogenase gene and use oft he *gpd* promoter for transformation in *Flammulina velutipes. Applied Microbiology and Biotechnology*, Vol. 65, pp. 593-599, ISSN 0175-7598

Lynd, LR.; Grethlein, HG. & Wolkin, RH. (1989). Fermentation of cellulosic substrates in batch and continuous culture by *Clostridium thermocellum. Applied and Environmental Microbiology*, Vol. 55, pp. 3131–3139, ISSN 0099-2240

Lynd, LR. (1996). Overview and evaluation of fuel ethanol production from cellulosic biomass: technology, economics, the environment, and policy. *Annual Review of Energy and the Environment*, Vol.21, pp. 403-465, ISSN 1056-3466

Lynd, LR.; Weimer, PJ.; Van Zyl, WH. & Pretorius, IS. (2002). Microbial cellulose utilization: fundamentals and biotechnology. *Microbiology and Molecular Biology Review*, Vol. 66, pp. 506-77, ISSN 1092-2172

Lynd, LR.; Van Zyl, WH.; McBride, JE. & Laser, M. (2005). Consolidated bioprocessing of cellulosic biomass: an update. *Current Opinion in Biotechnology*, Vol. 16, pp. 577-83, ISSN 0958-1669

Maas, RHW.; Bakker, RR.; Eggink, G. & Weusthuis, RA. (2006). Lactic acid production from xylose by the fungus *Rhizopus oryzae. Applied Microbiology and Biotechnology*, Vol. 72, pp. 861–868, ISSN 0175-7598

Maehara, T.; Yoshida, M.; Ito, Y.; Tomita, S.; Takabatake, K.; Ichinose, H. & Kaneko, S. (2010a). Development of a gene transfer system for the mycelia of *Flammulina velutipes* Fv-1 strain. *Bioscience, Biotechnology, and Biochemistry*, Vol. 74, pp. 1126-1128, ISSN 0916-8451

Maehara, T.; Tomita, S.; Takabatake, T. & Kaneko, S (2010b). Improvement of the transformation efficiency of *Flammulina velutipes* Fv-1 using glyceraldehyde-3-phosphate dehydrogenase gene promoter. *Bioscience, Biotechnology, and Biochemistry*, Vol. 74, pp. 2523-2525, ISSN 0916-8451

Maier, FJ. & Schäfer, W. (1999). Mutagenesis via insertinal- or restriction enzyme-mediated-integration (REMI) as a tool to tag pathogenicity related genes in plant pathogenic fungi. *Biological Chemistry*, Vol. 380, pp. 855-864, ISSN 1431-6730

Mizuno, R.; Ichinose, H.; Honda, M.; Takabatake, K.; Sotome, I.; Maehara, T.; Takai, T.; Gau, M.; Okadome, H.; Isobe, S. & Kaneko, S. (2009a). Use of whole crop sorghums as a raw material in consolidated bioprocessing bioethanol production using *Flammulina velutipes. Bioscience, Biotechnology, and Biochemistry*, Vol. 73, pp. 1671-1673, ISSN 0916-8451

Mizuno, R.; Ichinose, H.; Maehara, T.; Takabatake, K. & Kaneko, S. (2009b). Properties of ethanol fermentation by *Flammulina velutipes. Bioscience, Biotechnology, and Biochemistry*, Vol.73, pp. 2240-2245, ISSN 0916-8451

Nevoigt, E. (2008). Progress in metabolic engineering of *Saccharomyces cerevisiae. Microbiology and Molecular Biology Review*, Vol. 72, pp. 379–412, ISSN 1092-2172

Ng, TK.; Ben-Bassat, A. & Zeikus, JG. (1981). Ehtanol production by themophilic bacteria: fermentation of cellulosic substrates by cocultures of *Clostridium thermocellum* and *Clostridium thermohydrosulfuricum. Applied and Environmental Microbiology*, Vol. 41, pp. 1337–1343, ISSN 0099-2240

Okamura, T.; Ogata, T.; Toyoda, M.; Tanaka, M.; Minamimoto, N.; Takeno, T.; Noda, H.; Fukuda, S, & Ohsugi, M. (2000). Production of Sake by mushroom fermentation. *Mushroom Science and Biotechnology*, Vol. 8, pp. 109–114, ISSN 1348-7388

Okamura, T.; Ogata, T.; Minamimoto, N.; Takeno, T.; Noda, H.; Fukuda, S. & Ohsugi, M. (2001). Characteristics of win produced by mushroom fermentation. *Bioscience, Biotechnology, and Biochemistry*, Vol. 65, pp. 1596–1600, ISSN 0916-8451

Ostergaard, S.; Olsson, L.; Johnston, M. & Nielsen, J. (2000a). Increasing galactose consumption by *Saccharomyces cerevisiae* through metabolic engineering of the *GAL* gene regulatory network. *Nature Biotechnology*, Vol. 18, pp. 1283–1286, ISSN 1087-0156

Ostergaard, S.; Roca, C.; Rønnow, B.; Nielsen, J. & Olsson, L. (2000b). Physiological studies in aerobic batch cultivations of *Saccharomyces cerevisiae* strains harboring the *MEL1* gene. *Biotechnology and Bioengineering*, Vol. 68, pp. 252–259, ISSN 0006-3592

Parekh, S. & Wayman, M. (1986). Fermentation of cellobiose and wood sugars to ethanol by *Candida shehatae* and *Pichia stipitis. Biotechnology Letters,* Vol. 8, pp. 597–600, ISSN 0141-5492

Piechaczyk, M.; Blanchard, JM.; Marty, L.; Dani, C.; Panabieres, F.; El Sabouty, S.; Fort, P. & Jeanteur, P. (1984). Post-transcriptional regulation of glyceraldehyde-3-phosphate-dehydrogenase gene expression in rat tissues. *Nucleic Acids Research,* Vol. 12, pp. 6951–6963, ISSN 0305-1048

Punt, PJ.; Dingemanse, MA.; Kuyvenhoven, A.; Soede, RD.; Pouwels, PH. & Van den Hondel, CA. (1990). Functional elements in the promoter region of the *Aspergillus nidulans gpdA* gene encoding glyceraldehyde-3-phosphate dehydrogenase. *Gene,* Vol. 93, pp. 101-109, ISSN 0378-1119

Riggle, PJ. & Kumamoto, AC. (1998). Genetic analysis in fungi using restriction-enzyme-mediated integration. *Current Opinion in Microbiology,* Vol. 1, pp. 395-399, ISSN 1369-5274

Sato, T.; Yaegashi, K.; Ishii, S.; Hirano, T.; Kajiwara, S.; Shishido, K. & Enei, H. (1998). Transformation of the edible basidiomycete *Lentinus edodes* by restriction enzyme-mediated integration of plasmid DNA. *Bioscience, Biotechnology, and Biochem*istry, Vol. 62, pp. 2346-2350, ISSN 0916-8451

Swings, J. & DeLey, J. (1977). The biology of *Zymomonas. Microbiology and Molecular Biology Review,* Vol. 41, pp. 1-46, ISSN 1092-2172

Thorn, RG.; Reddy, CA.; Harris, D. & Paul, EA. (1996). Isolation of saprophytic basidiomycetes from soli. *Applied and Environmental Microbiology,* Vol. 62, pp. 4288–4292, ISSN 0099-2240

Van Peer, AF.; De Bekker C.; Vinck, A.; Wösten, HAB. & Lugones, LG. (2009). Phleomycin increases transformation efficiency and promotes single integrations in *Schizophyllum commune. Applied and Environmental Microbiology,* Vol. 75, pp. 1243-1247, ISSN 0099-2240

Van Zyl, WH.; Lynd, LR.; Den Haan, R. & McBride, JE. (2007). Consolidated bioprocessing for bioethanol production using *Saccharomyces cerevisiae. Advances in Biochemical Engineering/Biotechnology,* Vol. 108, pp. 205-35, ISSN 0724-6145

Weimer, PJ. & Zeikus, JG. (1977). Fermentation of cellulose and cellobiose by *Clostridium thermocellum* in the absence of methanobacterium thermoautotrophicum. *Applied and Environmental Microbiology,* Vol. 33, pp. 289–297, ISSN 0099-2240

Xu, Q.; Singh, A. & Himmel, ME. (2009). Perspectives and new directions for the production of bioethanol using consolidated bioprocessing of lignocellulose. *Current Opinion in Biotechnology*, Vol. 20, pp. 364-71, ISSN 0958-1669

Xu, W.; Subudhi, PK.; Crasta, OR.; Rosenow, DT.; Mullet, JE. & Nguyen, HT. (2000). Molecular mapping of QTLs conferring stay-green in grain sorghum (*Sorghum bicolor L.* Moench). *Genome*, Vol. 43, pp. 461-469, 0831-2796

Yamagishi, K.; Kimura, T.; Oita, S.; Sugiura, T. & Hirai, H. (2007). Transfomation by complementation of a uracil auxotroph of the hyper lignin-degrading basidiomycete *Phanerochaete sordida* YK-624. *Applied Microbiology and Biotechnology,* Vol. 76, pp. 1079-1091, ISSN 0175-7598

10

SSF Fermentation of Rape Straw and the Effects of Inhibitory Stress on Yeast

Anders Thygesen[1], Lasse Vahlgren[1], Jens Heller Frederiksen[1], William Linnane[2] and Mette H. Thomsen[3]

[1]*Biosystems Division, Risø National Laboratory for Sustainable Energy, Technical University of Denmark*

[2]*Roskilde University*

[3]*Chemical Engineering Program, Madsar Institute*

[1,2]*Denmark*

[3]*United Arab Emirates*

1. Introduction

In 2003 R. E. Smalley (Smalley 2003) made a list of the top 10 problems mankind was going to face in the next 50 years.

1. Energy
2. Water
3. Food
4. Environment
5. Poverty
6. Terrorism and War
7. Disease
8. Education
9. Democracy
10. Population

With the declining amount of fossil fuels, and the increasing energy demand, it is not possible to satisfy our large energy consumption without alternative energy sources, especially sustainable ones that also take care of problem number 4, (Environment). One of the possible ways to produce a liquid sustainable energy source is to replace our large gasoline demand with fermented biomass (bioethanol). To ensure that the food availability does not decrease (problem 3), the biomass for bioethanol fermentation is only gathered from waste materials of the food production, such as the straw of cereals. The fermentation on waste products is commonly referred to as 2nd generation bioethanol.

Biomass of interest for cellulosic produced ethanol includes wheat straw, rape straw and macro algae. In 2009 the production of oilseed rape in EU was 21×10^6 ton together with an even larger amount of rape straw (Eurostat 2010). The most commonly used rape in Europe is a winter rape with a low erucic acid and glucosinolate content (Wittkop et. al. 2009). The rape straw is composed of 32% cellulose and 22% hemicelluloses. In this chapter the focus will be on the fermentation of sugars derived from cellulose.

The enzymatic hydrolysis can be described in three steps: Endoglucanase separates chains of cellulose by breaking down the bonds in amorphous regions of the crystalline cellulose, thereby creating more free ends in the cellulose. Exoglucanase breaks the cellulose down from the non-reducing end into cellubiose (the disaccharide derived from cellobiose) (Teeri and Koivula 1995). This explains the importance of endoglucanase as it creates more "attack points" for exoglucanase. β-glycosidase breaks down cellubiose into glucose. Cellulase enzymes are commonly produced by the fungus *Trichoderma reesei* (Busto et al. 1996). This process has two functions, first it produces the glucose needed for the fermentation, and second it turns the non-soluble cellulose into soluble sugars, which provides the liquid medium required for fermentation.
The fiber structure consists of cellulose microfibrils, bound to each other with hemicellulose and lignin. A model of a plant fibre is shown in Fig. 1.

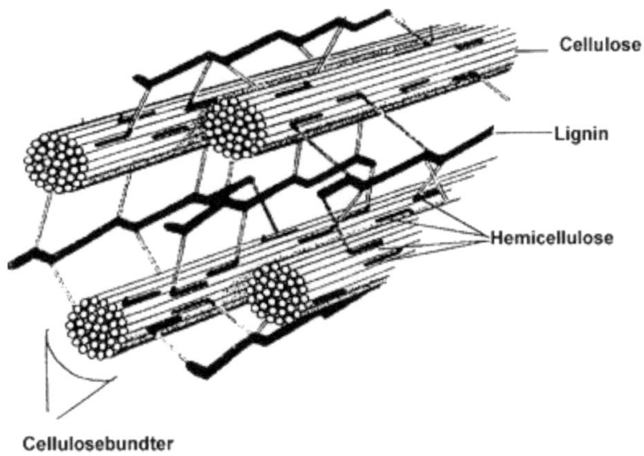

Fig. 1. Structure of cellulosic fibers in e.g. rape straw (adapted from Bjerre & Schmidt 1997)

The hemicellulose content consists of branched and acetylated carbohydrates. These molecules consist of 90% xylan and 10% arabinan in wheat straw (Puls and Schuseil 1993). The lignin content of the straw consists of polymerized molecules with a phenolic structure. The ethanol production process was conducted by simultaneous saccharification of cellulose with cellulase enzymes, and fermentation of the produced glucose with Turbo yeast (*Saccharomyces cerevisiae, Brewer's yeast*) (SSF) (Thomsen et al. 2009).
The production efficiency of the fermentation is strongly dependant on the wellbeing of the yeast. To visualize the health of the yeast, microscopic tests using blue staining were conducted. The bioethanol is produced from wet oxidized rape straw and the effect of the important fermentation inhibitor furfuryl alcohol is tested.

2. Materials and methods

2.1 Furfuryl alcohol effect on glucose fermentation and microscopy

To assess the number of inactive yeast cells "blue staining" is usually done. Methylene blue is the most commonly used color agent. It is a redox indicator that turns colorless in the presence of the active enzymes produced by the yeast *Saccharomyces cerevisiae*. For verification of the results this compound is compared with erioglaucine (E133) (Brilliant blue No. 1).

In order to verify that erioglaucine is equal in quality to methylene blue, a dose response curve for the toxic compound fufuryl alcohol is produced, and the mortality is then examined with the two color agents. 6 fermentations of 100 ml are started with different levels of furfuryl alcohol (0.0, 1.5, 3.5, 6.0, 9.0 and 15.0) (mL/L) and 2 g/L Turbo yeast (from AlcoTec). All the fermentations are completed in blue cap flasks in Millipore water and with 100 g/L glucose monohydrate. The yeast is left to ferment at 25, 32, 40 and 45°C, using 100 rpm of stirring for 24 hours. Weight loss is measured during working hours to monitor the fermentation. After 24 hours samples are taken for HPLC analysis as described by Thomsen et al. (2009). A methylene blue and an erioglaucine solution are produced each of 0.29 g/L. The 6 samples are divided into 12 portions to perform double repetition. Six portions were mixed with methylene blue in a 9:1 ratio and six portions with erioglaucine in a 1:1 ratio. The samples are examined with microscopy at 1000x magnification just after the color agent was added.

2.2 Wet oxidation of rape straw

Rape straw was produced after harvest during cultivation of *Brassica napus Linnaeus* commonly known as rape in Denmark in 2007. The straw was milled to a particle size of 2 mm using a knife mill. The milled straw was then soaked in 80°C hot water for 20 min before wet oxidation in a 2 L loop autoclave (Bjerre et al., 1996). The autoclave setup includes 1 L water, 60 g dry milled rape straw with the canister pressurized to 12 bar oxygen during the reaction. The mixing of oxygen and liquid is obtained by a pumping wheel. The wet oxidation was performed at 205°C for 3 min (Arvaniti et. al 2011). Following the wet oxidation, pressure is released and the prehydrolysate is cooled to room temperature. The filter cake and filtrate are separated and stored at -18° C (Thygesen et. al. 2003).

During wet oxidation, oxygen is introduced in the pre-treatment phase at high pressure and temperature. This causes 50 % (w/w) of the lignin to oxidize into CO_2, H_2O, carboxylic acids and phenolic compounds. It is important that the lignin fraction is low since lignin can denaturize the enzymes involved in the subsequent hydrolysis of cellulose. The majority of the hemicellulose (80 %) is dissolved while 10 % is oxidized to CO_2, H_2O and carboxylic acids. During this process, carboxylic acids, phenolic compounds, and furans are produced, which act as inhibitors in the fermentation process. However, the concentrations are too low to fully hinder microbial growth as wet oxidation also degrades the toxic intermediate reaction products (Thomsen et. al. 2009).

2.3 SSF fermentation and enzymatic hydrolysis

The dry matter content (DM) of the solid straw fraction after wet oxidation (filtercake) is measured by drying at 105°C overnight. The procedure for SSF is as follows: 250 mL flasks are loaded with 100 ml substrate with either 17 g DM or with 8 g DM. The substrate is adjusted to pH 4.8 with 6 mol/L of NaOH. After pH adjustment, 15 FPU/g DM of Celluclast (Novozymes mixture of endo- and exo-glucanase) is added together with 0.20 ml of Novozym 188 (β-glycosidase) per ml Celluclast using sterile conditions. The flasks are shaken at 50 °C with 120 rpm for 24 hours during the pre-hydrolysis. After the pre-hydrolysis, 10FPU/g DM of Celluclast and 0.20 ml of Novozym 188 (β-glycosidase) per ml Celluclast are added together with 1-4 g/L of Turbo yeast and 0.8 g/L of urea and the pH value re-adjusted to 4.8 with addition of NaOH. The flasks are sealed with glycerol yeast locks and incubated at 37°C with shaking at 120 rpm. During SSF, the flasks are weighed to measure ethanol production with respect to time, and after the SSF termination a sample of the supernatant is analyzed by

HPLC (High Performance Liquid Chromatography, Shimadzu) with a Bio-Rad Aminex HPX87H column and 0.6 mL/min flow of the eluent (4 mmol/L of H_2SO_4) at 63°C for ethanol and monosaccharides (Thomsen et al., 2009).

2.4 Analysis of the plant fibers for carbohydrates and lignin

The composition of the raw and the pre-treated straw is measured by strong acid hydrolysis of the carbohydrates. Dried and milled samples (160 mg) are treated with 72 % (w/w) H_2SO_4 (1.5 mL) at 30°C for 1 hour. The solutions were diluted with 42 mL of water and autoclaved at 121°C for 1 hour. After hydrolysate filtration, the Klason lignin content is determined as the weight of the filter cake subtracted the ash content. The filtrate is analysed for sugars on HPLC. The recovery of D-glucose, D-xylose and L-arabinose is determined by standard addition of sugars to samples before autoclavation. The sugars are determined after separation on a HPLC-system (Shimadzu) with a Rezex ROA column (Phenomenex) at 63°C using 4 mmol/L H_2SO_4 as eluent and a flow rate of 0.6 mL/min. Detection is done by a refractive index detector (Shimadzu Corp., Kyoto, Japan). Conversion factors for dehydration on polymerization are 162/180 for glucose and 132/150 for xylose and arabinose (Kaar et al., 1991; Thygesen et al., 2005).

2.5 Calculations

The amount of ethanol produced in the SSF fermentations is calculated based on the weight loss as a result of CO_2 formation in the reaction shown below.

$$C_6H_{12}O_6 \rightarrow 2C_2H_5OH + 2CO_2 \quad (1)$$

The resulting formula is derived as below, using the molar masses of ethanol and CO_2 of 46.07 g/mol and 44.01 g/mol, respectively:

$$wight\ of\ ethanol = wightloss \times \frac{Mw(CH_3CH_2OH)}{Mw(CO_2)} \quad (2)$$

The ECE (ethanol conversion efficiency) is then calculated as such.

$$ECE\% = \frac{Ethanol\ concentration\ after\ SSF}{Ethanol\ potential\ of\ biomass\ in\ SSF} \quad (3)$$

The chemical expression for cellulose and its hydrolysis into glucose is shown in Equation 4.

$$(C_6H_{10}O_5)_nH_2O + (\mathrm{n}-1)H_2O \rightarrow nC_6H_{12}O_6 \quad (4)$$

Where n is in the range 2000 - 10000. The highest ethanol yield is obtained when the cellulose is completely hydrolyzed into glucose and fermented into ethanol as shown in Equation 1 and 4. We now consider the number of monomers (n) in one gram of cellulose. Under complete hydrolysis this amount of matter becomes the amount of glucose. From equation 1 it can be seen that the amount of ethanol produced is 0.512 g per g of glucose fermented. The potential weight of ethanol from 1 g of cellulose is 0.569 g (Equation 5).

$$Yield_{ethanol\ from\ cellulose} = \frac{2Mw_{(C_2H_6O)}}{Mw_{(C_6H_{10}O_5)}} = \frac{2 \times 46.1}{162} = 0.57g/g \quad (5)$$

$$Yield_{ethanol\ from\ biomass} = Yield_{ethanol\ from\ cellulose} \times conc_{cellulose} \times conc_{DM} \quad (6)$$

3. Results and discussion

3.1 Effect of temperature on ethanol fermentation from glucose

Temperature optimizations for glucose fermentation are shown in Fig. 2. The highest rate was obtained at 32°C this fermentation is not limited by hydrolysis of cellulose. A challenge is that the optimal temperature for the yeast is 20°C lower than for the cellulase enzymes.

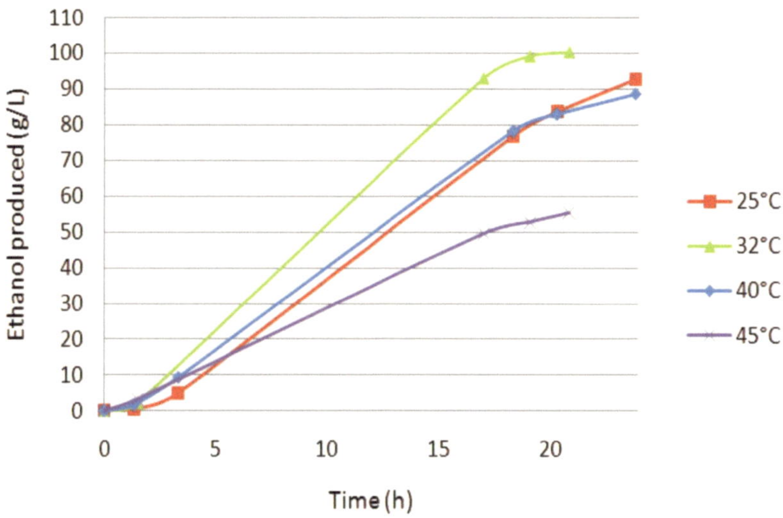

Fig. 2. The effect of temperature on Turbo yeast in ethanol fermentation with glucose.

By further fermentation of the rape straw in which hydrolysis of cellulose is needed it is therefore an advantage to use a higher temperature at which the yeast survives while the enzymes work at a higher reaction rate (Arvaniti et. al 2011).

3.2 Wet oxidation of rape straw

The rape straw used in the fermentation experiments consist of 32 % cellulose, 16 % hemicellulose, 18 % lignin, 5 % ash and 20 % non cell wall material (NCWM). During the wet oxidation process, large contents of hemicellulose and NCWM are extracted resulting in a liquid phase of glucose (1 g/L), xylose (7 g/L) and arabinose (0.5 g/L). The remaining solid phase became thereby enriched in cellulose and contained 54 % cellulose, 14 % hemicellulose, 23 % lignin, 3 % ash and 13 % NCWM. This solid material was subsequently tested by SSF fermentation.

3.3 Effect of yeast concentration on ethanol fermentation from rape straw

Fig. 3. shows the results of SSF on wet oxidized rape straw with 80 g/L DM content using initial yeast concentrations from 2.0 to 8.2 g/L.
As expected, the rate of fermentation dependants on the concentration of yeast, as there is plenty of sugar monomers present in the medium, which are produced in the pre-hydrolysis step of the SSF. At some point the excess sugar from the prehydrolysis is depleted and the rate of fermentation becomes dependent on the amount of sugar released by hydrolysis

Rape straw		Solid phase		Liquid phase	
Compound	g/100 g DM	Compound	g/100 g DM	Compound	g/L
Cellulose	32	Cellulose	54	Glucose	1.1
Hemicellulose	16	Xylan	13	Xylose	6.7
		Arabinan	1	Arabinose	0.5
Lignin	18	Lignin	23	Acetic acid	0.9
Ash	5	Ash	3	Formic acid	0.9
NCWM	20	NCWM	13	Furfural	0.1
				Phenolics	1.3
				pH	3.9

Table 1. Components of rape straw before and after the wet oxidation pretreatment resulting in a solid and a liquid phase. *NCWM is none cell wall material

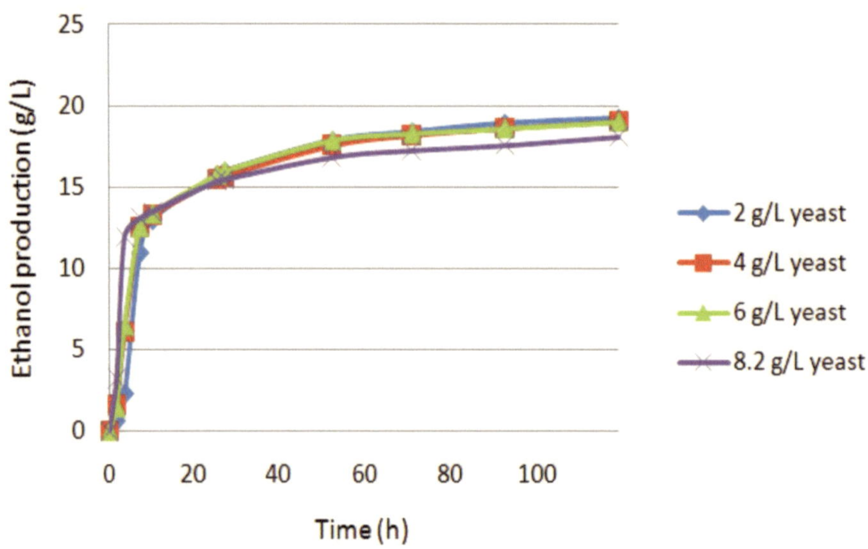

Fig. 3. The amount of ethanol produced on wet oxidized rape straw using 80 g/L DM and yeast concentrations of (2.0 to 8.2) g/L.

(rate of hydrolysis). Since the enzyme loading was the same in all the samples the fermentations produce ethanol at roughly the same rate once the excess sugar has been used. In experiments with high dry matter content such as 170 g/L the difference between the yeast dependant and the enzyme dependant phase is more pronounced than with low DM content (Fig. 4). The viscosity of the sample changes drastically during SSF since hydrolysis changes insoluble cellulose to soluble glucose. This means that the rate of both the yeasts and the enzymes production increases over time, as the production rate of both the enzymes and the yeast is related to the viscosity of the medium. An explanation can be that a high concentration of dry matter gives higher inhibitor concentrations and a larger yeast concentration can make faster detoxification.

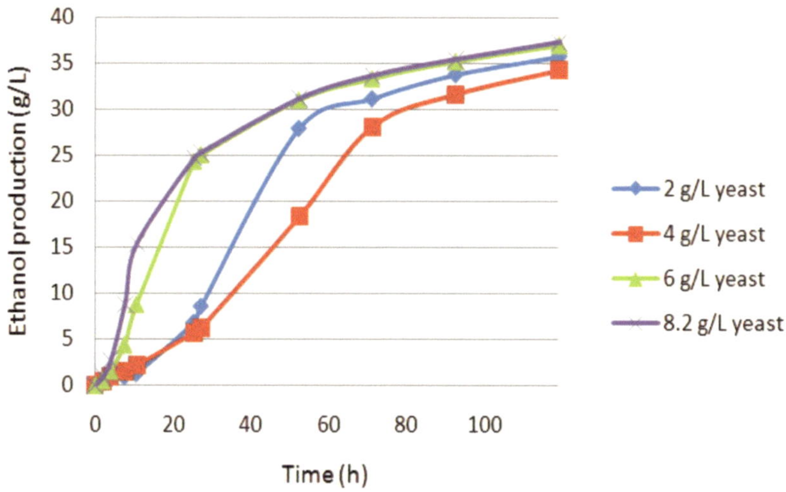

Fig. 4. The amount of ethanol produced during SSF of wet oxidized rape straw using 170 g/L DM and initial yeast concentrations of (2.0 to 8.2) g/L.

Yeast g/L	CO_2 g/L	Ethanol g/L	Ethanol potential g/L	ECE %
170 g/L straw DM		**34.9**	**51.9**	**67.2**
2	32.3	33.8	51.9	65.2
4	30.3	31.7	51.9	61.1
6	33.7	35.3	51.9	68.0
8.2	33.9	35.5	51.9	68.4
80 g/L straw DM		18.4	24.5	75.2
2	16.8	17.6	24.5	71.8
4	17.8	18.6	24.5	76.1
6	17.8	18.6	24.5	76.1
8.2	16.8	17.6	24.5	71.8

Table 2. Ethanol production and ECE (ethanol conversion efficiency) for different DM contents

As shown in table 2, the high DM content result in low ECE% of 67% compared to 75% when using low DM content. This is an essential problem in ethanol production since industrial distillation works best with more than 50 g/L ethanol, which could potentially be achieved with 75% ECE and 220 g/L of rape straw. However, in reality this is a challenge since increased DM contents result in reduced ECE% due to increasing viscosity and difficulties in mixing and higher concentrations of inhibitors. No direct correspondence between the final ECE% and the initial amount of yeast is found in this study. This indicates that the drop in ECE% at high DM can be due to decreasing enzyme performance in high DM pre-hyrolysates of rape straw.

The rate of fermentation is calculated for the time period with the highest fermentation rate and the results are shown in Fig. 5. For 170 g/L DM content that is between 10 and 52 hours, and for 80 g/L DM content it is between 2 and 4 hours. This time period covers the phase

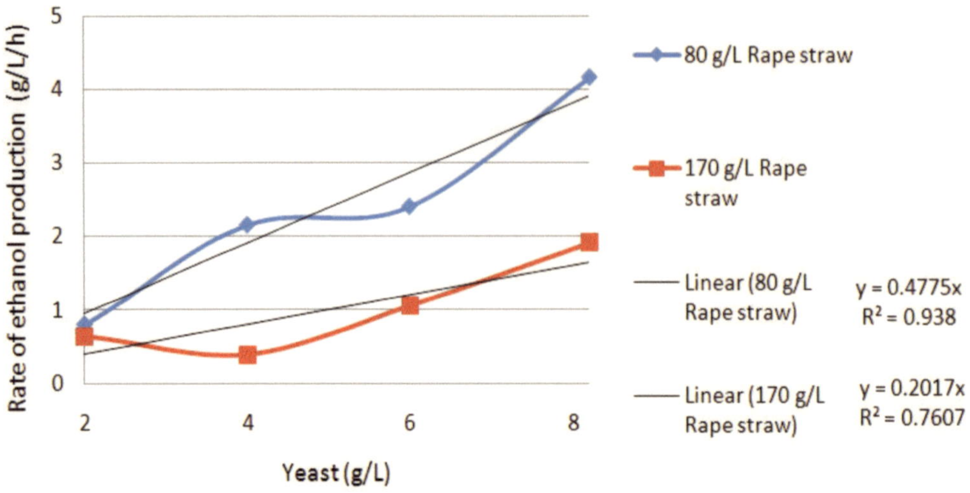

Fig. 5. Rate of ethanol fermentation with respect to the concentration of yeast

where the excess sugars from the hydrolysis is fermented and the positive feedback effect seen in high DM contents is also expressed. As shown in Fig. 5 generally the maximum rate of fermentation increases as a function of the yeast concentration. The slope coefficient of 80 g/L DM was found by linear regression to be 0.48 h^{-1} while it was found to be 0.20 h^{-1} at 170 g/L DM. The amount of yeast therefore contributes strongly to the positive feedback effect as explained in Fig. 4 and in this time period the dependency is also due to the fact that high yeast content can simply ferment the excess sugar from the hydrolysis faster than low contents of yeast.

3.4 The effect of furfuryl alcohol on yeast

The obtained concentration of ethanol versus the concentration of furfuryl alcohol is shown in Fig. 6. The amount of produced ethanol increased from 10 g/L with pure glucose to 24 g/ L at the dose of 1.5 mL/L furfuryl alcohol. At higher concentrations a decrease in ethanol is observed with almost no ethanol being produced at the dose of 15 mL/L. This indicates that with the presence of small doses of furfuryl alcohol, ethanol production can increase.

To provide information about the sharp increase between measurement one and two, another experiment was conducted with the same method as the previous, except the dose of furfuryl alcohol were (0.0, 0.5, 1.0, 1.5, 2.0 and 2.5) mL/L. Samples of this experiment were analyzed by HPLC and the results are shown in table 3.

As shown in table 3, the fermentation will peak with a dose of 1.5 mL/L and at levels above that the amount of ethanol produced will decrease versus the concentration of furfuryl alcohol. It was further shown that there was plenty of leftover glucose for the yeast to ferment. Therefore in none of the cases the loss in fermentation rate was due to lack of glucose.

As shown in Table 3 there is a strong correspondence between the amount of ethanol produced and the dose of furfuryl alcohol. Furfuryl alcohol is a microbial growth inhibitor, but when added in small concentrations it will force the yeast cells to increase their metabolism to survive. This effect should be present for the norm of microbial inhibitors. Despite of the fact that furfuryl alcohol is toxic to the yeast, the stress that it causes can lead

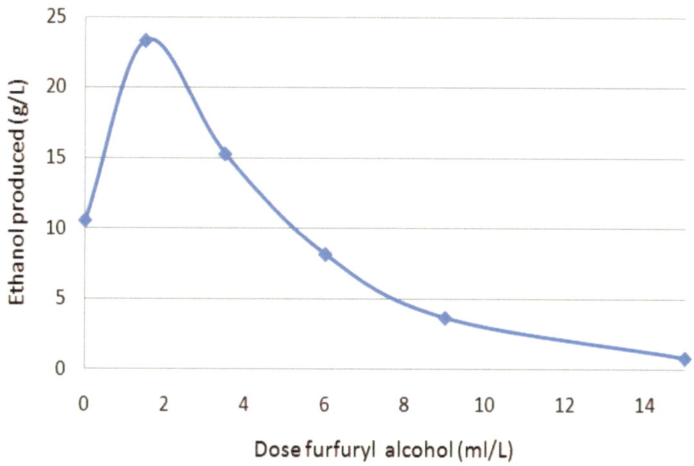

Fig. 6. Relative correspondence between added furfuryl alcohol and ethanol production during 24h of fermentation in a glucose medium.

Dose (mL/L)	Ethanol conc.(g/L)	Glucose (g/L)
0.0	15.0	42.4
0.5	21.0	38.7
1.0	25.7	32.4
1.5	27.2	33.4
2.0	25.0	37.2
2.5	21.6	45.2

Table 3. Final ethanol and glucose concentrations in yeast fermentations with added furfuryl alcohol

to approximately 80% increase in ethanol production given that sufficient amounts of glucose are available. For SSF the amount of glucose needed for this effect to be visible is only present in the start of the SSF which is also where the positive feedback effect of the yeasts fermentation rate is mainly present.

To describe the effect of lacking inhibitory stress on the yeast the maximum fermentation rate with 2 g/L yeast of both 80 g/L DM and 170 g/L DM in the SSF experiments with rape straw are compared and shown in Fig. 7.

The rate of fermentation for the furfuryl alcohol experiment is calculated for the period of 5 to 19 hours, which is estimated to be the highest rate of fermentation for the experiment. As shown in Fig. 7, it seems that even though microbial inhibitors are generally restricting the fermentation process, a medium completely without inhibitors (such as the control from Fig. 8) will have a decreased fermentation rate compared to medium with a small dose of microbial inhibitors.

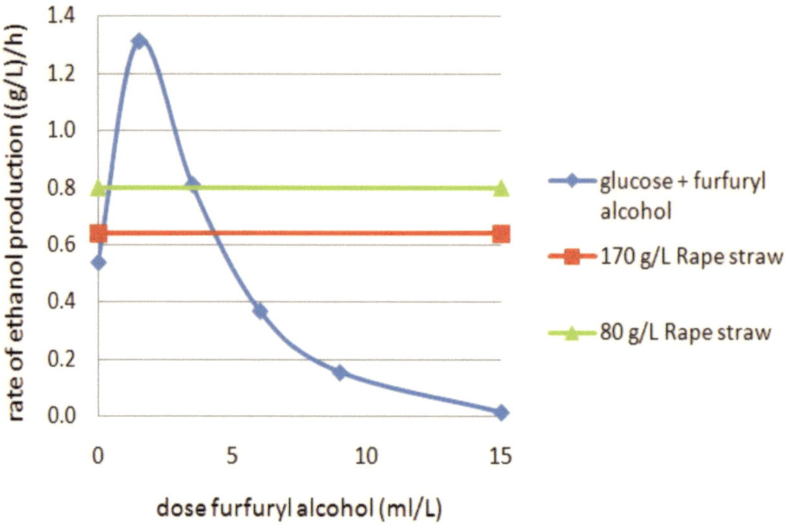

Fig. 7. Fermentation rate with respect to dose of furfuryl alcohol from the glucose experiment from Fig 6. For comparison the fermentation rate with rape straw of both 80 g/L DM and 170 g/L DM content from Fig. 5 is investigated for similar yeast concentrations (2g/L).

3.5 Vitality of the yeast cells

Fig. 8 shows the yeast cells after a glucose fermentation. A large part of the yeast cells have become inactive, even though there is still enough sugar in the substrate to sustain living cells.

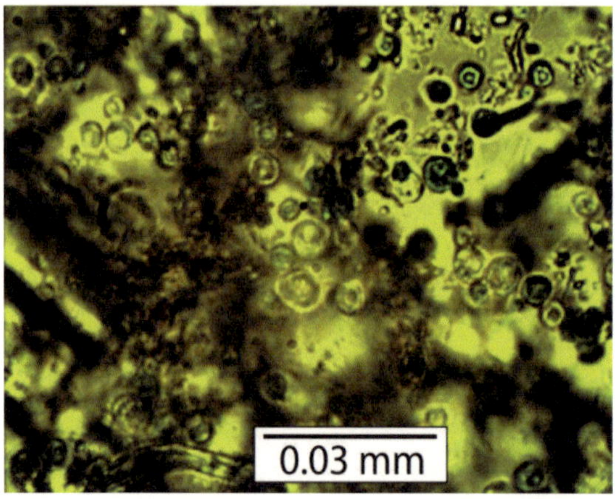

Fig. 8. Methylene blue stained yeast cells after glucose fermentation.

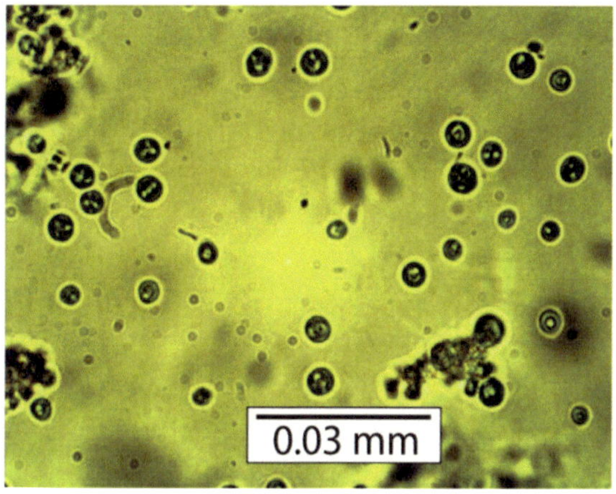

Fig. 9. Methylene blue stained yeast cells after heating to 100 °C.

During the experiments with dose response on furfuryl alcohol in glucose fermentation, a relation was found between cell death and the concentration of furfuryl alcohol (Fig. 8). Yet at this writing, there still exist some inconsistencies between erioglaucine and methylene blue. Overall the results produced by methylene blue appeared slightly more consistent and matching with the data for ethanol production.

3.6 Comparison of feed stocks

In this study the main focus was on production of bioethanol from rape straw, but there are a lot of other possible feedstocks suitable for bioethanol production in general, all cellulosic material can be converted by physical/chemical pretreatment followed by enzymatic hydrolysis into glucose. It is also possible to produce bioethanol from sugar and starch from crops such as corn, wheat, sugarcane, and sugar beet, but since sugar is a food source, using it could decrease food availability for future generations. Using food sources or available agricultural land for pure energy production is generally classified as a 1st generation technology, and is not normally regarded as a sustainable energy source. To compare the different feedstocks Table 4 is produced. However, there are numerous ways to co-produce food and feedstocks for bioenergy when utilizing the lignocellulosic residues from agricultural production as shown in table 4.

As table 4 shows, bast fibers have a very high cellulose content (60-63%) and a low lignin content (3-4 %) which should make them ideal for producing bioethanol, but bast fibers as a feedstock would fall under the category of 1st generation bioethanol, because the production of bast fibers requires land that could otherwise be used for food production. Rape straw has a low cellulose content compared to other straw fibers (32%) but in return the hemicellulose (14%) and lignin (18%) content is also low compared to wheat straw (20%) and corn stover (19 - 21%). Low lignin content is good for the enzymatic hydrolysis, since lignin can denaturize cellulase enzymes (Thygesen et. al 2003). Low hemicellulose content will result in a slightly lower concentration of microbial growth inhibitors derived from oxidation of the hemicellulose. Sugarcane bagasse seems to be the ideal 2nd generation feedstock with its high cellulose content (43%) and low lignin content (11%) but sugarcanes require high temperatures and a lot of rainfall to grow and are therefore only energy efficient when grown in tropical regions, which limits the amount of ethanol produced from sugarcane bagasse worldwide. It is possible to produce bioethanol from wood fibers, like waste wood from carpentry or

Feedstock	Cellulose % w/w	Xylose % w/w	Arabinose % w/w	Lignin % w/w	Ash % w/w	Ref.
Straw fibres						
Corn stover (*Zea mays*)	33	Hemicellulose = 21		19	7	1
Rape straw (*Brassica napus*)	32	14	2	18	5	2
Sugarcane Bagasse (*Saccharum*)	43	Hemicellulose = 31		11	6	3
Winter rye (*Secale cereal*)	41	22	3	16	5	2
Wheat straw (*Triticum*)	39	20	2	20	7	4
Wood fibres						
Norway spruce (*Picea abies*)	49	Hemicellulose = 20		30	0	1
Marine biomass						
Green hairweed (Chaetomorpha linum)	34 - 40	4 - 7	8 - 13	6 - 8	8 - 24	5
Bast fibres						
Flax (*Linum usitatissimum)*	60	8	1	3	4	6
Hemp (*Cannabis*)	63	9	1	4	4	1

Table 4. The composition of cellulose containing and plant -based raw materials including straw, wood, marine biomass and bast fibers. The individual data comes from the following sources: 1. Thygesen et. al. 2005, 2. Petersson et. al. 2007, 3. Martin et. al. 2007, 4. Schultz-Jensen et. al. 2010, 5. Schultz-Jensen et. al. 2011, 6. Hänninen et. al. 2011.

willow, which can grow on land not suitable for agriculture, using pretreatment methods such as steam explosion (Söderström et. al. 2002). The high lignin content in wood fibers increases the amount of enzymes needed and the time period of the fermentation. Furthermore wood fibers have other uses and can easily be burned to produce electricity and heat in a cogeneration plant.

Marine biomass has the advantage that it does not use the same space as agriculture and even though it is not a waste product from food production it is still a viable feedstock because it does not reduce food availability. *Chaetomorpha linum* has very low lignin content (6 – 8 %) and cellulose content similar to straw fibers (34 – 40 %). *C. linum* and other types of useable macroalgae are easy to grow in most of the world, and is therefore a suitable candidate for expanding the bioethanol production to more than what can be obtained from waste products (Schultz-Jensen et. al. 2011).

4. Conclusion

The amount of yeast needed for SSF of pretreated rape straw is dependent on the DM content, despite the fact that enzymes continue to be the primary rate-determining factor. The positive feedback effect from the yeast lowering the sugar concentration can have high relevance when running SSF with high DM content. After prolonged testing of Turbo yeast, the optimal temperature of the SSF is found to be 37°C. Furfuryl alcohol and possibly other growth inhibitors as well, show a positive effect on the rate of fermentation when added in small dosages, since yeast will increase its metabolism under stress. The positive effect of growth inhibitors is so strong that the fermentation rate in sugar media is lower than the fermentation rates in a medium produced from wet oxidized rape straw (filter cake), given the DM concentration does not exceed critical levels.

5. Acknowledgement

The Danish Research Council, DSF is gratefully acknowledged for supporting the research project: Biorefinery for sustainable reliable economical fuel production from energy crops (2104-06-0004). The European Union is acknowledged for supporting the EU-project: Integration of biology and engineering into an economical and energy-efficient 2G bioethanol biorefinery (Proethanol nr. 251151). Efthalia Arvaniti is acknowledged for academic advice. Tomas Fernqvist, Ingelis Larsen and Annette Eva Jensen are thanked for technical assistance and HTX Roskilde for providence of microscope cameras.

6. References

Arvaniti, E.; Thygesen, A.; Kádár, Z. & Thomsen, A.B. (2011). Assessing simultaneous saccharification and fermentation conditions for ethanol production from pre-treated rape straw. *Unpublished data.*

Bjerre, A.B. & Schmidt, A.S. (1997). *Development of chemical and biological processes for production of bioethanol: optimization of the wet oxidation process and characterization of products.* Technical university of Denmark, Risø-R-967, ISBN 87-550-2279-0, Roskilde, Denmark.

Browling, B.L. (1967). *Methods of Wood Chemistry.* Vol.1. Interscience Publishers, A division of John Wiley & Sons, New York, USA.

Busto, M.D.; Ortega, N. & PerezMateos, M. (1996). Location, kinetics and stability of cellulases induced in *Trichoderma reesei* cultures. *Bioresource Technology,* Vol.57, No.2, (August 1996), pp. 187-192, ISSN 09608524.

EUROSTAT, 2010. *Agricultural statistics - Main results 2008-09.* Eurostat. pp. 83-92. ISBN: 978-92-79-15246-7 ISSN: 1830-463X doi:10.2785/44845. Available: http://epp.eurostat.ec.europa.eu/portal/page/portal/product_details/publication?p_product_code=KS-ED-10-001

Hänninen, T.; Thygesen, A.; Mehmood, S.; Madsen, B. & Hughes, M. (2011). Effect of mechanical damage on the susceptibility of flax and hemp fibres to chemical degradation. *Unpublished data.*

Kaar, W.E.; Cool, L.G.; Merriman, M.M. & Brink, D.L. (1991). The complete analysis of wood polysaccharides using HPLC. *Journal of Wood Chemistry and Technology,* Vol.11, No.4, pp. 447-463, ISSN 02773813.

Martin, C.; Klinke, H.B. & Thomsen, A.B. (2007). Wet oxidation as a pretreatment method for enhancing the enzymatic convertibility of sugarcane bagasse. *Enzyme and Microbial Technology,* Vol.40, No.3, (February 2007), pp. 426-432, ISSN 01410229.

Fai, P.B. & Grant, A. (2009). A comparative study of *Saccharomyces cerevisae* sensitivity against eight yest species sensitivities to a range of toxicants. *Chemosphere,* Vol.75, No.3, (April 2009), pp. 289-296, ISSN 00456535.

Petersen, M.Ø.; Larsen, J. & Thomsen, M.H. (2009). Optimization of hydrothermal pretreatment of wheat straw for production of bioethanol at low water consumption without addition of chemicals. *Biomass and Bioenergy*, Vol.33, No.5, (May 2009), pp. 834-840, ISSN 09619534.

Petersson, A.; Thomsen, M.H.; Hauggaard-Nielsen, H. & Thomsen, A.B. (2007). Potential bioethanol and biogas production using lignocellulosic biomass from winter rye, oilseed rape and faba bean. *Biomass and Bioenergy*, Vol.31, No.11-12, (November-December 2007), pp. 812-819, ISSN 09619534.

Puls, J. & Schuseil, J. (1993). Chemistry of hemicelluloses: Relationship between hemicellulose structure and enzymes required for hydrolysis, In: *Hemicellulose and Hemicellulases*, M.P. Coughlan & G.P. Hazlewood, (ed.), pp. 1-27, Portland Press Research Monograph, ISBN 1855780364, Great Britain.

Schultz-Jensen, N.; Leipold, F.; Bindslev, H. & Thomsen, A.B. (2011). Plasma assisted pretreatment of wheat straw. *Applied Biochemistry and Biotechnology*, Vol.163, No.4, (February 2011), pp. 558-572, ISSN 02732289.

Schultz-Jensen, N.; Thygesen, A.; Thomsen, S.T.; Leipold, F.; Roslander, C. & Thomsen, A.B. (2011). Investigation of different pretreatment technologies for the macroalgae *Chaetomorpha Linum* for the production of bioethanol. *Unpublished data.*

Smalley, R.E., 2003-last update, Top Ten Problems of Humanity for Next 50 Years. Energy & NanoTechnology Conference, Rice University, May 3, 2003. Available at http://www.scribd.com/doc/2962283/Richard-Smalleys-energy-talk [15/8-2011, .

Söderström, J.; Pilcher, L.; Galbe, M. & Zacchi, G. (2002). Two step steam pretreatment of softwood with SO_2 impregnation for ethanol production. *Applied Biochemistry and Biotechnology*, Vol.98-100, (Spring 2002), pp. 5-21, ISSN 02732289.

Teeri, T.,T. & Koivula, A. (1995). Cellulose degradation by native and engineered fungal cellulases. *Carbohydrates in Europe*, Vol.12, pp.28-33, ISSN 13850040.

Thomsen, M.H.; Thygesen, A. & Thomsen, A.B. (2009). Identification and characterization of fermentation inhibitors formed during hydrothermal treatment and following SSF of wheat straw. *Applied Microbiology and Biotechnology*, Vol.83, No.3, (June 2009), pp. 447–455, ISSN 01757598.

Thomsen, M.H.; Thygesen, A. & Thomsen, A.B. (2008). Hydrothermal treatment of wheat straw at pilot plant scale using a three-step reactor system aiming at high hemicellulose recovery, high cellulose digestibility and low lignin hydrolysis. *Bioresource Technology*, Vol.99, No.10, (July 2008), pp. 4221-4228, ISSN 09608524.

Thygesen, A.; Oddershede, J.; Lilholt, H.; Thomsen, A.B. & Ståhl, K. (2005). On the determination of crystallinity and cellulose content in plant fibres. *Cellulose*, Vol. 12, No.6, pp. 563-576, ISSN 09690239.

Thygesen, A.; Thomsen, A.B.; Schmidt, A.S.; Jørgensen, H.; Ahring, B.K. & Olsson, L. (2003). Production of cellulose and hemicellulose-degrading enzymes by filamentous fungi cultivated on wet-oxidised wheat straw. *Enzyme and Microbial Technology*, Vol.32, No.5, (April 2003), pp. 606-615, ISSN 01410229.

Thygesen, A.; Possemiers, S.; Thomsen, A. & Verstraete, W. (2010). Intergration of microbial electrolysis (MECs) in the biorefinery for production of ethanol, H_2 and phenolics. *Waste and Biomass Valorization*, Vol.1, No.1, pp. 9-20, ISSN 18772641.

Wittkop, B.; Snowdon, R.J. & Friedt, W. (2009). Status and perspectives of breeding for enhanced yield and quality of oilseed crops for Europe. *Euphytica*, Vol.170, No 1-2, pp. 131-140, ISSN 00142336.

11

Competing Plant Cell Wall Digestion Recalcitrance by Using Fungal Substrate- Adapted Enzyme Cocktails

Vincent Phalip, Philippe Debeire and Jean-Marc Jeltsch
University of Strasbourg
France

1. Introduction

1.1 General needs for energy

General needs for energy are still increasing. In 2000, the energy provided worldwide was 10 Gt of oil equivalent (Gtoe) and the demand is forecasted to be around 15 Gtoe for 2020 (source: Energy Information Administration [EIA], 2002, as cited in Scragg, 2005). During the 20th century, coal proportion in energy supply decreased whereas oil and gas increased drastically. First after the 1973 oil crisis and afterwards periodically depending on oil prices, developments for producing energy by new ways were considered. In the last decade, the depletion of fossil energy sources appeared as a reality although exhaustion time remains highly controversial. Currently, it is clear that considerable efforts to promote alternative sources of energy are driven by both environmental concern (limiting fuel by-products emissions) and economic necessity linked to the fossil fuel depletion.

1.2 Bioethanol among other alternative sources

Ethanol from biomass (Bioethanol) is one of these alternative sources. Despite polemics for biomass uses i.e. biofuels vs food and some alarmist politicians' declarations, this alternative is really promising considering (i) the ability to satisfy a significant part of the demand for energy and (ii) biomass renewability. Polemics and limitations could have been almost rational when first generation of biofuels was concerned, "noble parts" of plants, the same used for food, being transformed. It is not the case anymore for any modern project. Another problem raised is the part of the cultivated surfaces to be reserved to biofuels, but in fact, realistic scenario is not to replace all fossil fuels volumes, but only part of them by using wastes preferentially.

1.3 "Biomass to ethanol" process and review of improvements

The general scheme of "Biomass to ethanol" process is presented elsewhere in this book. Our purpose in this section is to highlight the numerous and various ways to optimize the whole process from biomass to ethanol at different steps: choice of the biomass, pretreatment, enzyme productions, enzymatic hydrolysis, and ethanol fermentation. First of all, as discussed earlier, in the second generation of biofuels, biomass collection should not compete with food plants. Biomass should be abundant and cultural practices as sustainable as possible. Interest was recently focused on plants providing good yields of biomass for a given surface as the tall Miscanthus. Reduction of lignin cell wall content is another interesting approach to enhance sugar recovery from biomass, lignin being an abundant and resistant polymer limiting the digestion of biomass in biofuels processes. With anti-sense technology, tobacco plants lines were obtained with 20% lower lignin content (Kavousi et al., 2010). The modified lines displayed a threefold increase of saccharification efficiency compared to wild type. Of course, the application of such studies in larger scales depends on the acceptance of transgenic plants by the society. Decision to use these plants has to be supported by studies of environmental risks and potential benefits (Talukder, 2006). Literature about pretreatment is very abundant, describing various methods: physical, chemical or combination of both (Soccol et al. 2010). Fine optimization of conditions should be performed individually depending on biomass. Among innovative method proposed, dry wheat straw has been treated successfully with supercritical CO_2. After treatment, 1kg biomass yields to 149 g sugars (Alinia et al., 2010). Another currently emerging feature for bioethanol process amelioration is protein engineering. For instance, a cellulase from the filamentous bacterium *Thermobifida fusca* has been modified both in its catalytic domain and in its carbohydrate binding module (Li et al., 2010). A mutant enzyme displays a two fold increase activity, and a better synergy with other enzymes, leading it to be very useful for biomass digestion. At the next step, i.e. sugar fermentation to ethanol, many efforts have been run to allow yeast to perform both hexoses and pentoses fermentations. Industrial yeast *Saccharomyces cerevisiae* strains, fermenting only hexoses have been modified by addition of xylose degradation enzymes (Hector et al., 2010). Finally the outcome of engineering could be the use of synthetic biology, which is creating cell systems able to convert biomass to sugars and also to ferment them to ethanol. This strategy needs better fundamental knowledge to be developed (Elkins et al., 2010).

As discussed above, the step following the pretreatment of the biomass could be performed via the enzymatic hydrolysis of the cell wall polysaccharides into fermentescible, monomeric sugars. Unfortunately, it is well known that recalcitrance of plant cell wall to enzymatic digestion impairs the process. The behavior and the efficiency of the cell wall degrading enzymes (CWDE) *in situ* and *in vitro* with isolated polysaccharides are completely different. The properties of the CWDE, as conformation, hydrophobicity, capacity of adsorption onto the cell wall, interaction with the lignins, and catalytic efficiency in heterogeneous catalysis, are major parameters which should be considered and studied.

This chapter focuses on biomass degradation enzymes. What is the best strategy to produce the most efficient enzymes? What is the best choice depending of up- and downstream steps: commercial enzyme cocktail, enzymes produced by a given microorganism or heterologous production of individual enzymes? Efficiencies and cost of enzymes, two bottlenecks in the process, will be discussed. For some authors, the improvements of the conversion of biomass to sugar offer larger cost-saving potential than those concerning the step from sugar to biofuels (Lynd et al., 2008). These authors evaluated two scenarios; the first based on current technology and the second one including advanced nonbiological steps. In both cases, conversion of polysaccharides from biomass could be improved by increasing polysaccharides hydrolysis yields combined by lowered enzyme inputs. On-site enzyme production was also identified as beneficial for cost of the whole ethanol production process.

2. Diversity of plant cell wall structures

The plant cell wall structures are highly diverse. Various lignocellulosic species have been used for biofuels production, woods, crop by-products, herbaceous plants, beet pulp, municipal and paper industry wastes. Although all these different biomasses contain typically four major components (i.e. cellulose, hemicelluloses, pectin and lignin), the architecture of the cell wall, the fine biochemical structures of these components and their interactions into the cell wall could be quite different. Nevertheless, cellulose and hemicelluloses leading, with lignins, to the formation of an insoluble, tridimensional network is a constant behavior. A schematic drawing of the plant cell wall polysaccharides is shown in Fig. 1.

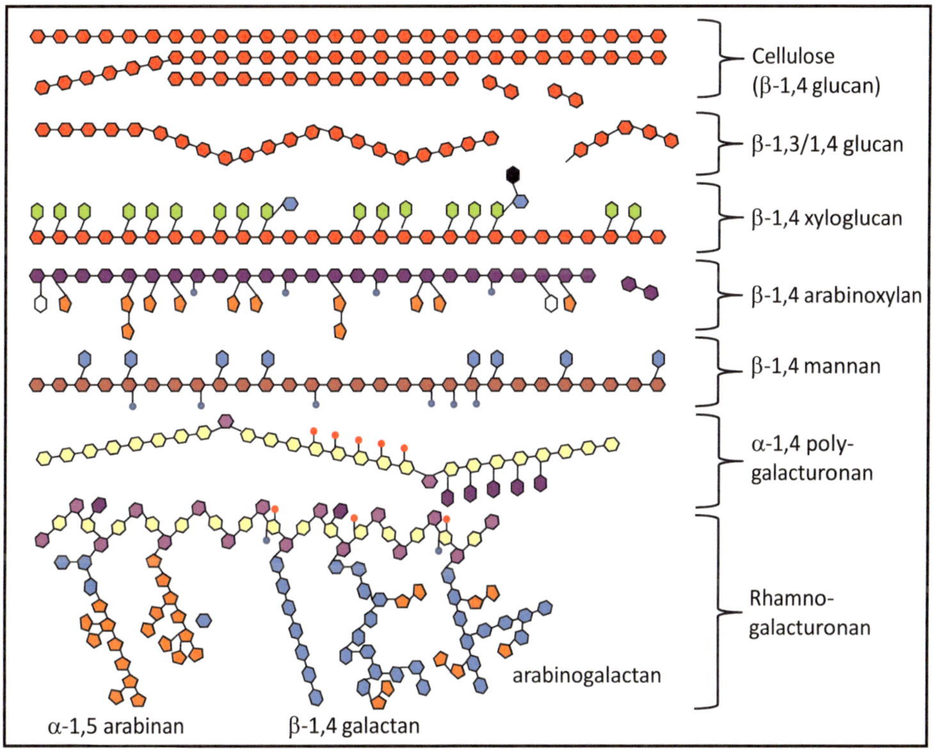

Fig. 1. Schematic representation of plant cell wall polysaccharides. Cellulose and β-1,3/1,4-glucan are composed of glucose residues (red). β-1,4-xyloglucan is a glucose-based polymer substituted by xylose residues (green) themselves possibly linked to galactose (blue) and fucose (black). β-1,4-arabinoxylan is a xylose backbone linked with arabinose (orange), and/or with glucuronic acid acid (white). Xylose could be substituted by acetyl groups (blue circle). β-1,4-mannan is a mannose polymer (dark red) linked to some galactose residues and sometimes acetylated. α-1,4-polygalacturonan is formed of linear chains of galacturonic acid (pale yellow) linked by rhamnose (pink). Galacturonic residues could be either methylated (red circle) or linked to xylose residues. Rhamnogalacturonan is highly ramified and is also called "hairy region" for this reason. The basic backbone is a rhamnose-galacturonic acid motif. Side chain of arabinose, galactose, mixed or not, linear or not, forms a very complex and variable structure. Adapted from Dalboge (1997).

As another example of variable composition of plant cell wall, lignins and sugars in cell walls of different origins were quantified (Table 1). Although this study is not exhaustive, it is clear that every plant has its own characteristic. These biomasses have been used as growth substrates for *Fusarium graminearum* (see paragraph 4).
Regarding cell wall composition variation, it could be postulated that cell wall degradation recalcitrance could be related to cell wall structure and ultra structure variability.

	Hop	Wheat bran	Corn cops	Birch
Lignin	30.9	17.3	6.5	18.1
Total neutral sugars	35.2	50.1	63.5	57.6
Glucose (i.e. glucans)	20.1	14.8	28.8	40.0
Xylose (i.e. xylans)	2.7	15.0	21.0	15.6

Table 1. Lignin, glucose and xylose contents of hop, destarched wheat bran, corn cops and birch. Results are expressed as % of dry matter (unpublished results from the laboratory). Klason lignin was quantified as the acid-soluble residue after sulfuric acid hydrolysis (Rémond et al. 2010) and sugar contents were estimated by enzymatic methods as described in Phalip et al. (2009).

3. A strategy for improving biomass hydrolysis: Studying (and using afterwards) fungi able to degrade plant cell wall components

3.1 Introduction on phytopathogenicity, saprophytism

The primary choice of a microorganism (fungus) potentially providing cell wall degrading enzymes should be directed toward one naturally present in plant environment i.e. a phytopathogen or a saprophyte. Considering the ecology, fungi are qualified as "decomposers" in the opposite to plants, the producers and animals, the consumers. Some fungi, called saprotrophs, get nutrients from dead organisms, especially plants. Some other are pathogen, attacking living organisms. Invasive growth thanks to hyphae gets fungi very adapted to penetrate plants. Hyphae diameter (2–10 µm) permit cell penetration and their hyphal growth in several directions allow them to colonize quickly the plant material with very close contact. Many saprotrophs, phytopathogens and other fungi living in plant environment developed tools for gaining energy from plants during their evolution. Cell wall degrading enzymes (CWDE) are one of these tools which are also efficient for bioethanol production. This is the reason why this chapter focuses on CWDE produced and secreted by some fungi.

3.2 Genome level studies

After the completion of the project of the backer's yeast *Saccharomyces cerevisiae* genome sequencing in 1996, genomes of some fungi have been decrypted. Many reasons drive the decision to sequence one genome and not the other one: some fungi being for a long time scientific models, others displaying industrial relevance, and others acting as saprophytes or pathogens. Backer et al. (2008) propose an interesting concept: let's change our way of thinking and let's consider microorganisms (especially fungi) as reservoirs for sustainable answers to environmental concerns. This point of view fits well with the directing idea of this chapter. To improve the "biomass to ethanol" process, we have to consider many options and enlarge the fields to be prospected rather than being focalized on a single system. As an example for biomass degradation, many efforts have been directed though *Trichoderma reesei* due to historical reasons (discovered during World War II because it degraded uniforms and cotton tents) and to its capacity to produce cell wall degrading enzymes, specially cellulases. But, as it will be described later in this section, this fungus is not -by far for some categories of enzymes- the most equipped in CWDE. Other fungi have to be considered then.

Genome sequence availability offers the scientific community the opportunity to analyze them, deciphering their metabolism, in order to find response to fundamental or applied questions. As valorization of plant biomass arise as an important question to be addressed, several studies attempt to describe the fungal polysaccharide degradation potential. An extensive and complete work leads to a comparison of the genome of 13 fungi (Martinez et al., 2008). The first observation is that the yeast model *Saccharomyces cerevisiae* is poorer in CWDE than filamentous fungi (Fig. 2). This is not surprising regarding their respective lifestyles; all the filamentous fungi shown in Fig. 2 are saprotrophs or pathogens in the opposite of *S. cerevisiae*. This is a first argument for considering the natural habitat of a fungus when examining it for a peculiar application. Here, clearly, fungi living in plant environment displayed many more genes encoding CWDE or associated activities. Note that the model for plant polysaccharide degradation, *Trichoderma reesei*, displays fewer putative glycosyl hydrolases (200) than the pathogens *Magnaporthe grisea* (231) and *Fusarium graminearum* (243). Perhaps even more important is the number of cellulose binding modules (CBM), allowing a better enzyme-substrate binding and then a better efficiency in natural cellulose hydrolysis. *T. reesei* was predicted to have half CBM than the two pathogens (Fig. 2). In the same study, *T. reesei* is shown to be poorer than *M. grisea* and *F. graminearum* in cellulases, hemicellulases and pectinases (Martinez et al., 2008).

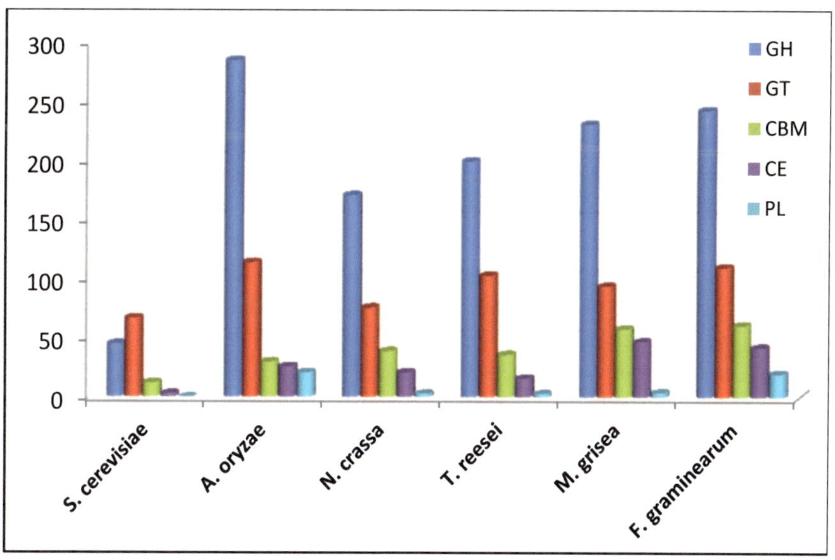

Fig. 2. Number of predicted glycosyl hydrolases (GH), glucosyl transferase (GT), cellulose binding modules (CBM), carbohydrate esterase (CE) and polysaccharide lyases (PL) in the genome of *Saccharomyces cerevisiae, Aspergillus oryzae, Neurospora crassa, Trichoderma reesei,* : *Magnaporthe grisea* and *Fusarium graminearum* (data from Martinez et al., 2008).

The main idea driving genome study is that evolution leads to genomes remodeling: i.e. leading to CWDE diversity for fungi dealing with plants. But through the example of *T. reesei*, it could be concluded that genome study - if available- is useful but not sufficient. Furthermore, obviously, a gene is not a protein; it has to be transcribed and mRNA has to be translated and modified to yield to mature and functional proteins.

3.3 Transcriptome studies

In order to appreciate if the information provided by genome analysis is pertinent, transcriptome studies should also be performed. Our purpose is not to describe the regulation of CWDE in fungi, but it is essential to determinate efficiency of CWDE transcription depending on growth conditions. The goal is of course to optimize conditions leading to high transcription of the required enzymes. This regulation is rather complex, variable and well described in reviews (as an example see Aro et al., 2005). However, global characteristics leading to transcription of hydrolases genes are interesting to point out, since they could be a rational strategy basis for ethanol production. First, and for a long time, CWDE genes were considered generally as being repressed by glucose (catabolic repression) and by other released monosaccharides upon polysaccharide hydrolysis (de Vries & Visser, 2001). On the opposite, CWDE are massively expressed when fungi are grown in presence of polysaccharides and plant material (de Vries & Visser, 2001; Foreman et al., 2003 ; Aro et al., 2005). However, the view of a strict co-regulation of all CWDE is wrong. Induction of a given hydrolase goes on as a function of the polysaccharide in contact with the fungus. An interesting illustration is found in the pea pathogen *Nectria hematococca*. Two pectate lyases were found to be involved in pathology (Rogers et al., 2000). The first one was induced by pectin and repressed *in planta*, whereas the other was induced *in planta* but repressed by pectin. This means that CWDE transcription could be individual and precise. In *Fusarium graminearum*, well known as pathogen of cereals, we performed microarray experiments to test the expression on the whole genome on glucose, cellulose, xylan and hop cell wall (Carapito et al., 2008). Methods and essential findings are summarized in Fig. 3.

First, some genes were actually found to be over-expressed on polysaccharides comparatively to their expression on glucose (Fig. 3.). Their number varies depending on carbon source. CWDE represent also a variable part of overexpressed genes. It is particularly interesting to note that the largest proportion of CWDE was observed when the fungus was grown on plant cell wall (19% of overexpressed genes) i.e. the most diverse substrate. It denotes a strong re-orientation of the metabolism towards cell wall degradation since CWDE correspond to approximately 0.5% of the genome only. Furthermore, cellulases, hemicellulases and pectinases encoding genes are quite equally represented as overexpressed ones when the fungus is grown on plant cell wall, whereas mostly cellulases were shown to be overexpressed on cellulose and mostly hemicellulases were overexpressed on xylan. This data suggest that there is no global response to the presence of plant cell wall, but that the different polysaccharides sent specific signals which are recognized by the fungus and induce various responses.

3.4 Proteomics

As the number of fungal genome sequenced increase, the number of proteomics studies increase the same way. The studies are driven for various reasons, but some of them are devoted to the identification of proteins produced (and most frequently secreted) by fungi

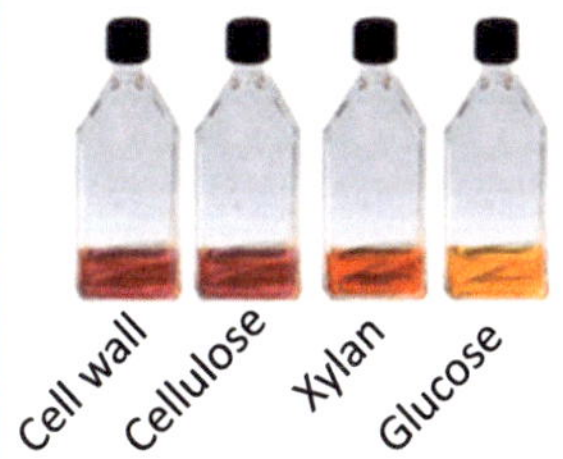

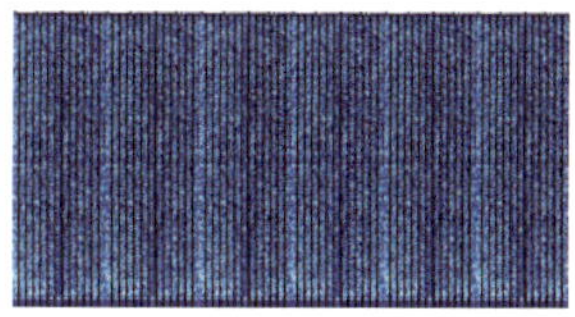

	Cell wall	Cellulose	Xylan
Nb of gene overexpressed	124	150	52
% of CWDE among them	19	13	6

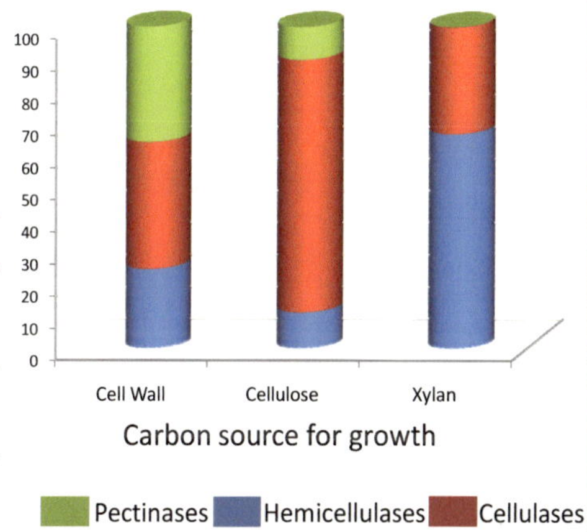

Fig. 3. Expression of the genome of *Fusarium graminearum* when grown on different carbon sources. Microarrays were analyzed and the number of genes overexpressed ($p<0.02$, fold change >2) comparatively that after growth on glucose where determined (Table). The graph represents the repartition of the CWDE (cellulases in yellow, hemicellulases in pink and pectinases in green) as a function of the substrate used for growth.

in response to plant material. This fact fits well with the purpose of this chapter and is perfectly summarized by the title of a recent paper: "Plant-pathogen interactions: what is proteomics telling us" (Mehta et al., 2008). In this paper, is shown that when pathogens are in the presence of plants, their metabolism is changed to secrete proteins, including CWDE, potentially involved in plant cell wall degradation. These findings are in perfect accordance with the conclusion of the transcriptomics studies (see previous section).

We performed a proteomics study with the plant pathogen *Fusarium graminearum* grown either on glucose or on a preparation of plant cell wall as the sole carbon source (Phalip et al., 2005). When it is grown on glucose (Fig. 4.), the fungus secretes a few proteins in small quantities. When the more complex and diverse plant cell wall is used, the fungus reacts by secreting a much higher amount of more diverse proteins. Approximately half (45%) of these are putative CWDE. Furthermore, CWDE identified are able to take in charge the three cell wall layers: cellulose (11 proteins are putative cellulases), hemicellulose (25) and pectin (19). These results are also perfectly correlated with transcriptome studies. The fungus clearly responds to cell wall diversity by enzyme diversity. It could be a good point to keep in mind when looking for an enzyme cocktail for biomass valorization.

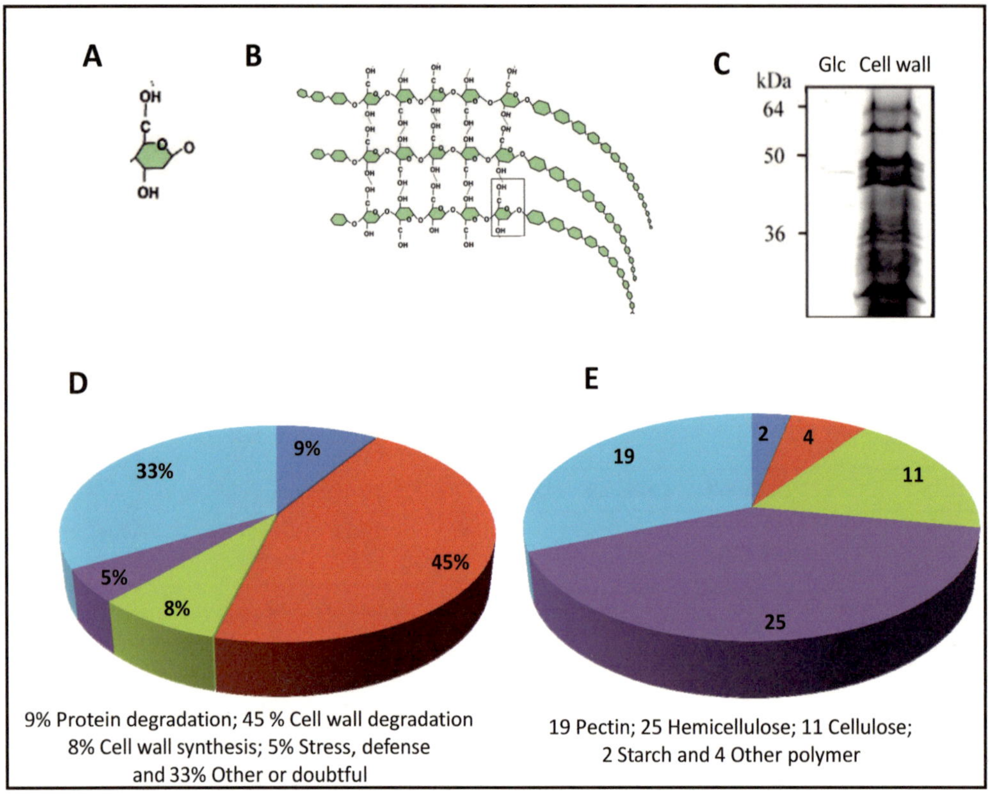

Fig. 4. Proteomics studies of *Fusarium graminearum* grown on glucose (A) and plant cell wall (B). Culture supernatants were concentrated and the equivalent of the same volume of supernatant was loaded on SDS-PAGE (C). The corresponding proteins were by identified by Mass Spectrometry and classified thanks to their homologies with protein in databases (D). The CWDE were further considered as a function of the cell wall layer they degrade (E).

3.5 Enzymatic measurements

Genome, transcriptome and proteome studies lead to interesting but perturbing questions. Several studies indicate that fungi could secrete up to 50 different CWDE in order to degrade cellulose, hemicellulose, pectin and, for some of them, lignin. Among these enzymes, some display the same EC number and/or belong to the same glycosyl hydrolase family (GH). Why fungi use up so much energy to secrete enzymes with quite similar activities? Is it true or apparent redundancy?

Several clues indicate that this is apparent redundancy. Most of the CWDE, still putative, wait for potent substrate specificities characterization. By analogy with the enzymes already characterized, it means that slightly different specificities are likely to be discovered and could be essential to complete plant cell wall degradation. Furthermore, quantitative studies performed on *Fusarium* hemicellulases demonstrate that on hop cell wall, the expression level of the 30 putative enzymes varies greatly from 1 (the less abundant) to 1500 (Hatsch et al., 2006). When another biomass is used for growth, the pattern of secreted enzymes is different, clearly indicating that there is no "general response" to the presence of plant material but specific responses to a given biomass. Taken together these studies mean that the fungus exhibits a large flexibility in its response. Then it could be thought that the observed redundancy actually reflects the multitude of different structures of plant cell wall. Consequently, it is of primary importance that *in silico* studies should be carefully validated by enzymatic measurements. For example, an exponential increase of putative CAZY (Carbohydrate-Active enZYmes) described is observed, but unfortunately only a small proportion of them are biochemically characterized yet (Cantarel et al., 2009). Whereas entries in CAZY database increased exponentially from 1999 to 2007 (a 14-fold increase), less than 10% of them have been enzymatically characterized and less than 1% of the enzyme structures have been solved. This means that there is a real lack in enzyme knowledge regarding to the huge potential of new activities undiscovered yet. It should be noticed that the increasing difference between the number of putative enzymes and well characterized ones also lead to the possibility of mis-annotation and/or false identification. This phenomenon has been known for a long time by molecular biologists and correction of errors and inconsistencies in data bases became an authentic research area (Ghisalberti et al., 2010). In order to thoroughly characterize the enzyme activities and their capacity to degrade the complex structures found in plant cell wall, CWDE substrate specificity should be determined with both artificial and natural substrates. This absolute necessity drives us to perform the characterization of the enzyme cocktail produced by *F. graminearum* on hop cell wall. We used 29 different substrates, poly-or oligosaccharides, natural or artificial (Phalip et al., 2009). Enzyme activities were evaluated by assays of the products (monomers) or by their visualization by polysaccharide analysis using carbohydrate gel electrophoresis (PACE). The conclusion of this study is that the enzymes constituting this cocktail are no more putative but active on each layer of the plant cell wall. On the opposite, the enzyme cocktail produced on glucose displays very tiny activities, furthermore on a small number of substrates. The proof is then provided that to get a large diversity of cell wall degrading enzymes, it is very important to choose the right substrate for a given fungus to grow.

3.6 Synergy

A relevant feature of CWDE activities is the synergy observed between them. Typically an endo-enzyme acts randomly on the polysaccharide to yield oligomers. These oligomers are numerous for a single starting polysaccharide and their extremities are hydrolyzed to di- or monosaccharides by exo-acting enzymes. Accessory enzymes (debranching or desubstituting) proceed if necessary and all three kinds of enzymes work together. Synergy leads to the concept of "Minimal enzyme cocktail", i.e. a few selected enzymes (Sorensen et al., 2007), supposed to be sufficient for the complete digestion of plant cell wall. This paper described efforts performed to digest the more efficiently wheat arabinoxylan to arabinose and xylose. Starting from an "enzymatic base" (a β-xylosidase, and three endo-β-xylanases), they screened three different arabinofuranosidases to enhance substrates digestion. AbfIII was found to be the best enzyme when used alone (Fig. 5). Addition of AbfI significantly increased hydrolysis yields, although in different extent depending on the substrate used.
Addition of the three arabinofuranosidases together did not improve the yields further. The strategy used and the results obtained clearly support the view of the necessity of a rational design of the process leading to ethanol. In other words, such a study has to be repeated when another biomass is concerned. Would the same enzyme be as efficient with the other substrate? Only experiments could address this question. Nevertheless, this study indicates that the starting biomass influence the choice of the enzymes to be used. For the complete

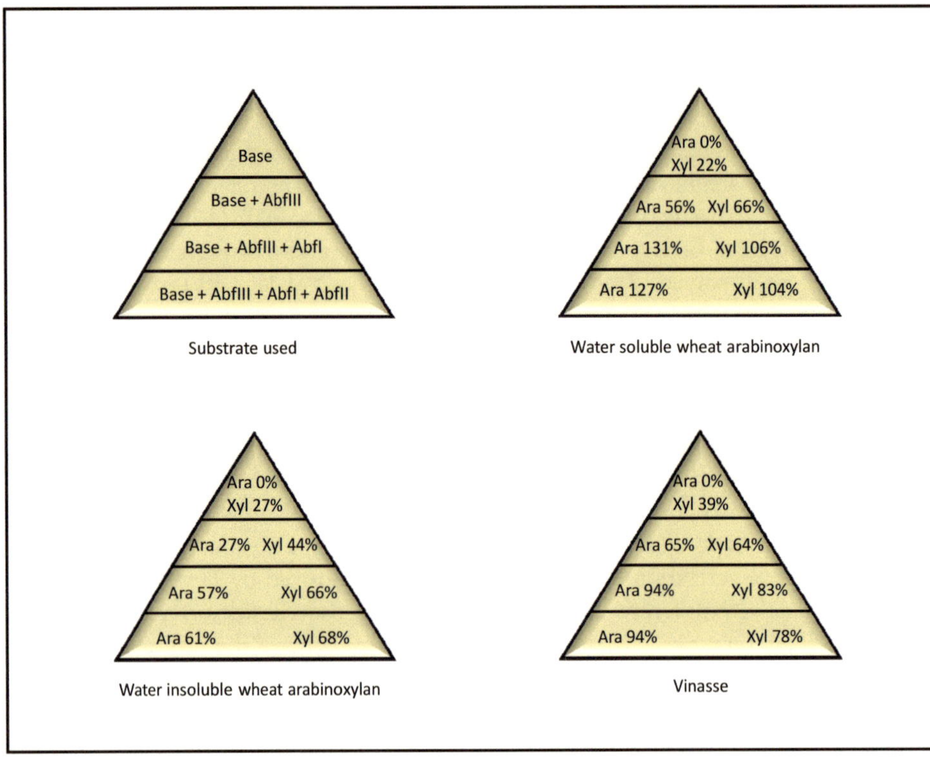

Fig. 5. Yields of arabinose and xylose released from different substrates (substrates are indicated below each pyramid). Base mix of enzyme (top of each pyramid) is constituted of a xylosidase and three endo-β-xylanases. ArfI, AfrI and ArfIII are three arabinofuranosidases belonging to GH (glycosyl hydrolase) families 51, 43 and 51, respectively. As described in the top left pyramid, the more different enzymes were added, the more the result is shown at the bottom of the pyramid. Yield superior to 100% was due to pretreatment. Adapted from Sorensen et al., 2007.

digestion of a plant biomass, minimal enzyme cocktails for all kind of polysaccharide are required. Note that Section2 of this chapter concludes to a great diversity of cell wall fine structures leading to design of much larger and much diverse cocktails.

4. Conclusions and future prospects

Studies of cell wall and CWDE presented in this chapter could be summarized by the same word: diversity. But how the previously mentioned studies inform us about the right strategy driving efficiently from biomass to ethanol? This will be discussed below.

We demonstrated for the first time that the exoproteome of *Fusarium graminearum* grown in presence of plant material was rich in various CWDE: more than 80 different proteins, half of them being putatively involved in cell wall digestion were recovered from culture supernatant (Phalip et al., 2005). It is noticeable that later, rather the same number of proteins was found to be secreted by *Trichoderma reesei* grown on corn cell wall (Nagendran et al., 2009). Commercial preparation Spezyme® used for biomass hydrolysis contains also more than 80 different proteins. All these data corroborate the concept of using complex enzyme cocktails for complete biomass hydrolysis. A thorough analysis of *Fusarium* and *Trichoderma* proteomes reveals some differences between them. For instance, *Fusarium* secretes enzymes belonging to 29 different GH families and 6 pectate lyases whereas *Trichoderma*'s exoproteome exhibits 22 GH families but no pectate lyase. Furthermore, only 14 GH families are present in both proteomes (Fig 6), 15 are recovered in *Fusarium* but not in *Trichoderma* and 8 are only found in *Trichoderma*. These are convincing arguments that in a strategy meant to produce bioethanol from various sources, the design of a biomass to ethanol process should be optimized for each couple biomass/fungus.

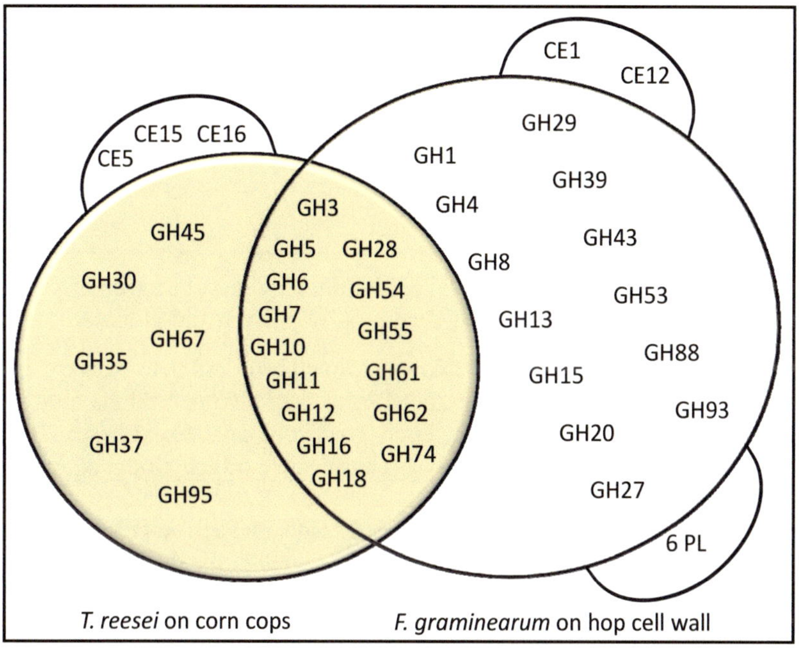

Fig. 6. Nature of putative secreted CWDE found in *Trichoderma reesei* grown on corn cops (left; Nagendran et al., 2009) and in *Fusarium graminearum* grown on hop (right; Phalip et al., 2005).

We studied the patterns of enzymes produced by *F. graminearum* on four different lignocellulosic biomasses, two poorly lignified (wheat bran and corn cobs) and two highly lignified (hop and birch). Each enzyme cocktail was thereafter used for long-term hydrolyses of the ammonia pretreated four biomasses (16 combinations). The oligo- and mono-saccharides (end products) have been characterized. Their patterns showed variations depending on the nature of the biomass used for growth. Accordingly, different enzyme activities were measured on different culture supernatants. For example, enzymes produced when grown on birch, were efficient with pretreated birch and hop in a lesser extent but were also the most efficient for poorly lignified biomasses. Furthermore, the proteins produced in each condition were identified by mass spectrometry. As shown in Fig. 7, the number of unique proteins recovered in supernatants varies greatly from 25 on birch to 72 on hop (14 for the monosaccharide glucose). More interesting is that 80% (20/25) of the proteins recovered on birch corresponds to CWDE whereas only 19% of the proteins are putatively active on polysaccharides with corn cops for growth (0 on glucose). Finally,

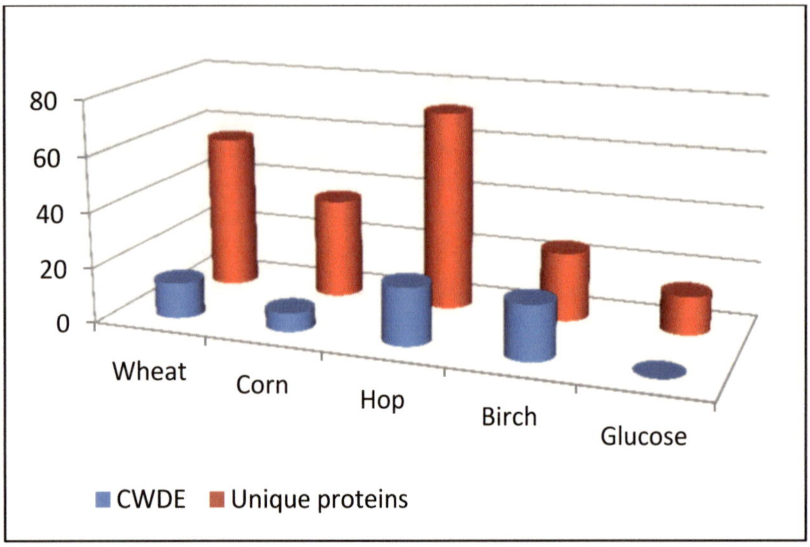

Fig. 7. Number of unique proteins and CWDE identified by mass spectrometry in culture supernatants after growth on wheat bran, corn cops, hop, birch and glucose.

among the 43 CWDE identified during this experiment, not less than 32 were recovered in one culture condition only, 11 were produced in at least two conditions and only 2 were present for each growth conditions (hop, birch, corn cops and wheat bran). Therefore, the fine specificities of the CWDE towards polysaccharides are directly induced by the different lignocellulosic biomasses used for growth.

Of course, even if the couple biomass/fungus is rationally chosen, one could not exclude that the microorganism does not produce a set of keys enzyme(s) for the peculiar biomass to be completely split up. In this case, mixing two (or more) enzymes crude cocktails could be a very good alternative as illustrated by Gottschalk et al. (2010). Enzymes from *Trichoderma reesei* grown on corn steep liquor and *Aspergillus awamori* grown of wheat bran have been obtained. The cocktails display different hydrolytic activities and blends of both led to enhancement of some synergic activities up to a 2-fold factor. It is also observed that their association improved glucose release from steam-treated sugarcane. The authors notice that *T. reesei* cocktail was better for cellulose hydrolysis and *A. awamori* was better for xylan hydrolysis. This is obviously a very good argument for a fungal specific response towards a given biomass. The same kind of concept could also be extended by using one of the rare lignin-degrading fungi, *Phanerochaete chrysosporium* (Kersten & Cullen, 2007) in association with another one providing great amount of CWDE. Lynd et al. (2008) suggest that replacing chemical pretreatment by enzymatic one could be a way to explore to improve the process. Furthermore, proceeding this way is mimicking nature diversity since fungi are often associated in communities acting in synergy for the degradation of plants. Furthermore, Wei et al. (2009) claims rightly that "the plant-microbe-enzyme relationship is the foundation of plant biomass degradation in natural environments". They mean that plant cell wall is naturally degraded by a community of microorganisms. Efficient processes from "biomass to ethanol" could advantageously use this property, of course by means of accelerating the natural process.

Many studies have been performed aiming to get huge quantities of individual enzymes, mostly cellulases. Actually, high quantity of enzyme input is not the panacea since saturation of sugar yields due to increasing enzyme charge is often observed (see Sorensen et al., 2007 as an exemple). After reaching the plateau, adding more enzymes did not improve yields anymore. A real diversity of enzyme responding to that of cell wall is better to overcome cell wall recalcitrance for full degradation. Although fungi naturally secrete small amounts of enzymes in liquid cultures, solid state fermentation (SSF) is preferable as an accurate solution for increasing enzyme production yield at the industrial level. Furthermore, a lot of different biomasses, including wastes, have been proved to support fungal growth and to promote CWDE production. The "waste-to-energy" technology was recently reviewed (Bemirbas et al., 2011) and the authors underlined that as population and urbanization increase, the amount of wastes increased regularly. As described elsewhere in this volume (Verardi et al., 2011; Xavier et al., 2011), wastes included municipal solid waste, paper wastes and also agricultural and forestry by-products, all containing lignocellulosic material.

For sustainable development, we strongly encourage the concept of local small units of bioethanol production. Gnansounou & Dauriat (2010) evoke the necessity of "low-risk biorefineries" in opposition to "complex schemes" production units. Therefore biomass should be easily available, preferentially composed of wastes, transports as limited as possible, wastes almost totally used and co-products fully valorized.

As a conclusion, seeking a universal process for total hydrolysis of all kind of biomasses is utopian. Rather, there is an appropriate methodology to follow, described in this chapter, considering the biomass to be treated and the co-products to be valorized in the respect to sustainable development. This point is perfectly illustrated by Saxena et al., (2009). For every step of biomass conversion, starting by the biomass choice, there are multiple routes for hydrolysis technology, monomers produced, microorganism used for fermentation and by-product formed. In this volume, Xavier and al., 2011 underline the necessity of a specific

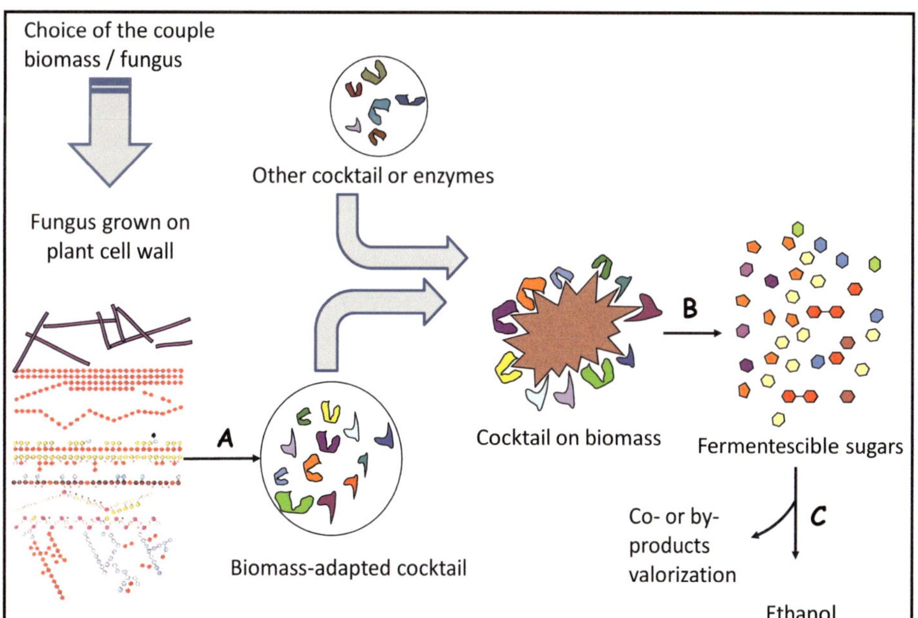

Fig. 8. Schematic representation of the whole process from "biomass to ethanol".

pretreatment for each biomass to be digested. This means that an industrial process should be developed by taking into account the nature of biomass, and consequently the enzymes necessary for its digestion and the down-stream processes.
Fig. 8 summarizes the views developed in this chapter. Taking into account fundamental research, a couple biomass / fungus is chosen. The fungus is grown on the biomass and produced a cocktail rich in CWDE (A). After adding maybe another cocktail (or individual enzymes; see also Verardi et al., 2011), the enzymes are used to digest the pretreated biomass (in a first approach, the same biomass that the one used for growth), yielding diverse fermentescible sugars in quantity (B). The latter are taken in charge by microorganisms to produce ethanol (C).

5. Acknowledgment

We are grateful to Marie-Laurence Phalip for language revision.

6. References

Alinia, R.; Zabihi, S.; Esmaeilzadeh, F. & Kalajahi, J.F. (2010) Pretreatment of wheat straw by supercritical CO_2 and its enzymatic hydrolysis for sugar production. *Biosystems Engineering*, Vol.107, No.1, pp. 61-66, ISSN 1537-5110

Aro, N.; Pakula, T. & Penttilä M. (2005) Transcriptional regulation of plant cell wall degradation by filamentous fungi. *FEMS Microbiology Reviews*, Vol.29, No.4, pp. 719-739, ISSN 1574-6976

Baker, E.; Thykaer, J.; Adney, W.S.; Brettin, T.S.; Brockman, F.J.; D'haeseleer, P. et al. (2008) Fungal genome sequencing and bioenergy. *Fungal Biology Reviews*, Vol.22, No.1, pp. 1-5, ISSN 1749-4613

Cantarel, B.L.; Coutinho, P.M.; Rancurel, C.; Bernard, T.; Lombard, V. & Henrissat, B. (2009) The Carbohydrate-Active EnZymes database (CAZy): an expert resource for Glycogenomics. *Nucleic Acids Research*, Vol.37, pp. D233-D238, ISSN 0305-1048

Carapito, R.; Hatsch, D.; Vorwerk, S.; Petkovski, E.; Jeltsch, J.-M. & Phalip, V. (2008) Gene expression in *Fusarium graminearum* grown on plant cell wall. *Fungal Genetics and Biology*, Vol.45, No.5, pp. 738-748, ISSN 1087-1845

Dalbøge, H. (1997) Expression cloning of fungal enzyme genes; a novel approach for efficient isolation of enzyme genes of industrial relevance. *FEMS Microbiology Reviews*, Vol.21, No.1, pp. 29-42, ISSN 0168-6445

Demirbas, M.F.; Balat, M. & Balat, H. (2011) Biowastes-to-biofuels. *Energy Conversion and Management*, Vol.52, No.4, pp. 1815-1828, ISSN 0196-8904

De Vries, R.P. & Visser, J. (2001) *Aspergillus* enzymes involved in degradation of plant cell wall polysaccharides. *Microbiology and Molecular Biology Reviews*, Vol.65, No.4, pp. 497-522, ISSN 1092-2172

Elkins, J.G.; Raman, B. & Keller, M. (2010) Engineered microbial systems for enhanced conversion of lignocellulosic biomass. *Current Opinion in Biotechnology*, Vol.21, No.5, pp. 657-662, ISSN 0958-1669

Foreman, P.K.; Brown, D.; Dankmeyer, L.; Dean, R.; Diener, S.; Dunn-Coleman, N.S.; Goedegebuur, F. et al. (2003) Transcriptional regulation of biomass-degrading enzymes in the filamentous fungus *Trichoderma reesei*. *Journal of Biological Chemistry*, Vol.278, No.34, pp. 31988-31997, ISSN 0021-9258

Ghisalberti, G.; Masseroli, M. & Tettamanti, L. (2010) Quality controls in integrative approaches to detect errors and inconsistencies in biological databases. *Journal of Integrative Bioinformatics*, Vol.7, No.3, pp. 119-132, ISSN 1613-4516

Gnansounou, E. & Dauriat, A. Techno-economic analysis of lignocellulosic ethanol: A review. *Bioresource Technology*, Vol.101, No.13, pp. 4980-4991, ISSN 0960-8524

Gottschalk, L.M.F.; Oliveira, P.A. & da Silva Bon, E.P. (2010) Cellulases, xylanases, β-glucosidase and ferulic acid esterase produced by *Trichoderma* and *Aspergillus* act synergistically in the hydrolysis of sugarcane bagasse. *Biochemical Engineering Journal*, Vol.51, No.1-2, pp. 72-78, ISSN 1369-703X

Hatsch, D.; Phalip, V.; Petkowski, E. & Jeltsch, J.-M. (2006) *Fusarium graminearum* on plant cell wall: no fewer than 30 xylanase genes transcribed. *Biochemical and Biophysical Research Communications*. Vol.345, No.3, pp. 959-966, ISSN 0006-291X

Hector, R.E.; Dien, B.S.; Cotta, M.A. & Qureshi, N. (2010) Engineering industrial *Saccharomyces cerevisiae* strains for xylose fermentation and comparison for switchgrass conversion. *Journal of Industrial Microbiology and Biotechnology*, Epub ahead of print, pp. 1-10, ISSN 1367-5435

Kavousi, B.; Daudi, A.; Cook, C.M.; Joseleau, J.-P.; Ruel, K.; Bolwell, G.P. & Blee, K.A. (2010) Consequences of antisense down-regulation of a lignification-specific peroxidase on leaf and vascular tissue in tobacco lines demonstrating enhanced enzymic saccharification. *Phytochemistry*, Vol.71, No.5-6, pp. 531-542, ISSN 0031-9422

Kersten, P. & Cullen, D. (2007) Extracellular oxidative systems of the lignin-degrading Basidiomycete *Phanerochaete chrysosporium*. *Fungal Genetics and Biology*, Vol.44, No. 2, pp. 77-87, ISSN 1087-1845

Li, Y.; Irwin, D.C. & Wilson, D.B. (2010) Increased crystalline cellulose activity via combinations of amino acid changes in the family 9 catalytic domain and family 3c cellulose binding module of *Thermobifida fusca* cel19a. *Applied and Environmental Microbiology*, Vol.76, No.8, pp. 2582-2588, ISSN 0099-2240

Lynd, L.R.; Laser, M.S.; Bransby, D.; Dale, B.E.; Davison, B.; Hamilton, R.; Himmel, M.; Keller, M.; McMillan, J.D.; Sheehan, J. & Wyman, C.E. (2008) How biotech can transform biofuels. *Nature Biotechnology*, Vol.26, pp. 169-172, ISSN 1087-0156

Martinez, D.; Berka, R.M.; Henrissat, B.; Saloheimo, M.; Arvas, M. et al. (2008) Genome sequencing and analysis of the biomass-degrading fungus *Trichoderma reesei* (syn. *Hypocrea jecorina*). *Nature Technology*, Vol.26, No.5, pp. 553-560, ISSN 1087-0156

Mehta, A.; Brasileiro, A.C.M.; Souza, D.S.L.; Romano, E.; Campos, M.A.; Grossi-De-Sá, M.F.; Silva, M.S.; et al. (2008) Plant-pathogen interactions: What is proteomics telling us? *FEBS Journal*, Vol.275, No.15, pp. 3731-3746, ISSN 1742-464X

Nagendran, S.; Hallen-Adams, H.E.; Paper, J.M.; Aslam, N. & Walton J.D. (2009) Reduced genomic potential for secreted plant cell-wall-degrading enzymes in the ectomycorrhizal fungus *Amanita bisporigera*, based on the secretome of *Trichoderma reesei*. *Fungal Genetics and Biology*, Vol.46, No.5, pp. 427-435, ISSN 1087-1845

Phalip, V.; Delalande, F.; Carapito, C.; Goubet, F.; Hatsch, D.; Dupree, P.; VanDorsselaer, A. & Jeltsch, J.-M. (2005) Diversity of the exoproteome of *Fusarium graminearum* grown on plant cell wall. *Current Genetics*, Vol.48, pp. 366-379, ISSN 0172-8083

Phalip, V.; Goubet, F.; Carapito, R. & Jeltsch, J.-M. (2009) Plant cell wall degradation with a powerful *Fusarium graminearum* enzymatic arsenal. *Journal of Microbiology and Biotechnology*, Vol.19, No.6, pp. 573-581, ISSN 1017-7825

Rémond, C.; Aubry, N.; Crônier, D.; Noël, S.; Martel, F.; Roge, B.; Rakotoarivonina, H.; Debeire, P. & Chabbert, B. (2010) Combination of ammonia and xylanase pretreatments: Impact on enzymatic xylan and cellulose recovery from wheat straw. *Bioresource Technology*, Vol.101, No.17, pp. 6712-6717, ISSN 0960-8524

Rogers, L.M.; Kim, Y.-K.; Guo, W.; González-Candelas, L.; Li, D. & Kolattukudy, P.E. (2000) Requirement for either a host- or pectin-induced pectate lyase for infection of *Pisum sativum* by *Nectria hematococca*. *Proceedings of the National Academy of Sciences of the United States of America*, Vol.97, No.17, pp. 9813-9818, ISSN 0027-8424

Saxena, R.C.; Adhikari, D.K. & Goyal, H.B. (2009) Biomass-based energy fuel through biochemical routes: A review. *Renewable and Sustainable Energy Reviews*, Vol.13, No.1, pp. 167-178, ISSN 1364-0321

Scragg, A. (2005) *Environmental Biotechnology* (2nd Edition), Oxford University Press, ISBN 978-0-19-926867-2, Oxford, UK

Soccol, C.R.; Vandenberghe, L.P.S.; Medeiros, A.B.P.; (...) Araújo, J.A. & Torres, F.A.G. (2010) Bioethanol from lignocelluloses: Status and perspectives in Brazil. *Bioresource Technology*, Vol.101, pp. 4820-4825, ISSN 0960-8524

Sørensen, H.R.; Pedersen, S.; Jørgensen, C.T. & Meyer, A.S. (2007) Enzymatic hydrolysis of wheat arabinoxylan by a recombinant "minimal" enzyme cocktail containing β-xylosidase and novel endo-1,4-β-xylanase and α-L-arabinofuranosidase activities. *Biotechnology Progress*, Vol.23, No.1, pp. 100-107, ISSN 8756-7938

Talukder, K. (2006) Low-lignin wood - A case study. *Nature Biotechnology*, Vol.24, No.4, pp. 395-396, ISSN 1087-0156

Xavier, A.; Fernandes, D.L.A.; Pereira, R.R.; Serafim, L.S. & Evtyugin, D. (2011). Second generation bioethanol, In: *Bioethanol*, InTech - Open Access Publisher, ISBN: 978-953-307-346-0, Rijeka, Croatia

Verardi, A.; De Bari, I.; Ricca, E. & Calabro, V. (2011) Hydrolysis of lignocellulosic biomass: current status of processes and technologies and future perspectives, In: *Bioethanol*, InTech - Open Access Publisher, ISBN: 978-953-307-346-0, Rijeka, Croatia

Wei, H.; Xu, Q.; Taylor, L.E., Baker, J.O.; Tucker, M.P. & Ding, S.-Y. (2009) Natural paradigms of plant cell wall degradation. *Current Opinion in Biotechnology*, Vol.20, No.3, pp. 330-338, ISSN 0958-1669

12

Heterologous Expression and Extracellular Secretion of Cellulases in Recombinant Microbes

Parisutham Vinuselvi[1] and Sung Kuk Lee[1,2]
[1]*School of Nano-Bioscience and Chemical Engineering, Ulsan National Institute of Science and Technology (UNIST), Ulsan*
[2]*School of Urban and Environmental Engineering, Ulsan National Institute of Science and Technology (UNIST), Ulsan*
Republic of Korea

1. Introduction

Lignocellulose, starch, sucrose, and macroalgal biomass are different forms of plant biomass that have been exploited for bioethanol production. Among them, lignocellulose, found in both agricultural and forest waste, has attracted great attention because of its relative abundance in nature (Lynd et al. 2002). Lignocellulose is a complex polymer made up of cellulose, hemicellulose, and lignin. Efficient conversion of lignocellulose into bioethanol involves a series of steps, namely, the collection of biomass; pretreatment to dissolve lignin; size reduction to reduce the number of recalcitrant hydrogen bonds; enzymatic saccharification to yield simple sugars; and, finally, fermentation of the sugars to ethanol. The main hurdle in this process is the lack of low-cost technology to overcome the recalcitrance associated with lignocellulose (Lynd et al. 2002; Himmel et al. 2007; Xu et al. 2009). Pretreatment is needed to dissolve the lignin, and enzymes such as xylanases are needed to hydrolyze the hemicellulosic fraction that otherwise would prevent cellulases from accessing the cellulose (Wen et al. 2009) (Fig 1A). The half-life of crystalline cellulose at neutral pH is estimated to be one hundred million years (Wilson 2008). A cocktail of saccharification enzymes—with endoglucanases, exoglucanases and β-glucosidases forming the major portion—is needed to disrupt the chemical stability of cellulose. The physical stability of lignocellulose, rendered by hydrogen bonds formed between adjacent cellulose polymers, is still a major obstacle to the efficient hydrolysis of cellulose. An additional challenge in cellulose hydrolysis is the relatively poor kinetics exhibited by cellulases (Himmel et al. 2007). Cellulases have lower specific activities than do other hydrolytic enzymes, because their substrate (cellulose) is insoluble, crystalline, and heterogeneous (Fig 1B) (Zhang and Lynd 2004; Wilson 2008). Activity of each of the cellulases in complex enzyme cocktails is inhibited by intermediates—such as cello-oligosaccharides and cellobiose, produced during cellulose hydrolysis—leading to discontinuity in the process. For example, exoglucanase action yields cellobiose, which inhibits endoglucanase (Fig 1C) (Lee et al. 2010).

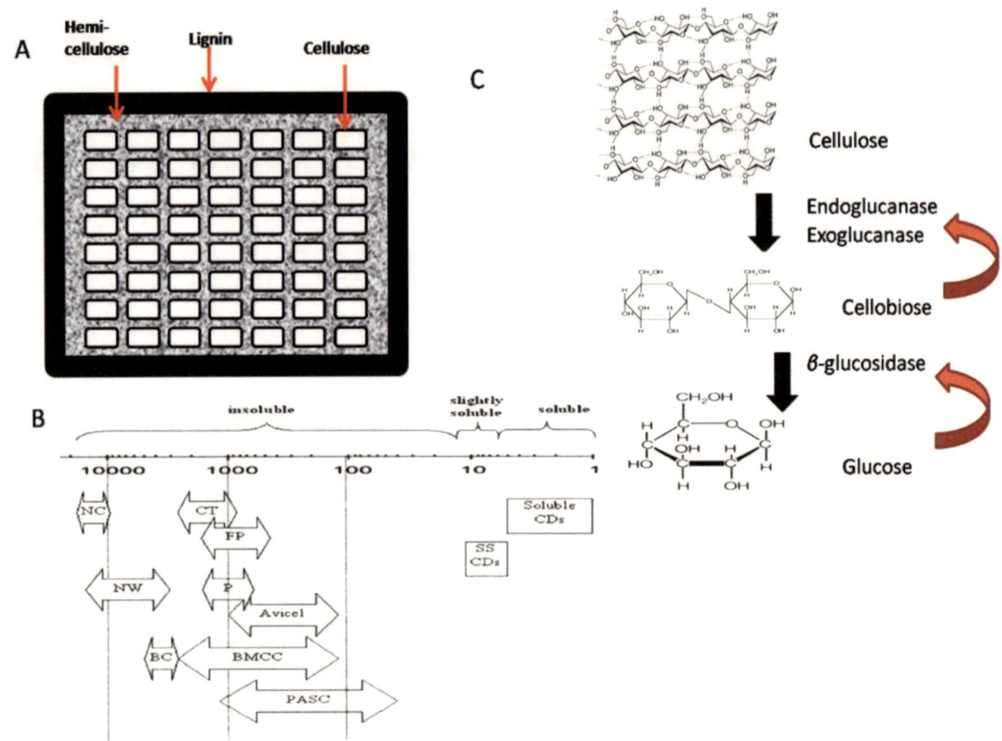

Fig. 1. (A) Schematic representation of the barriers to access the cellulose present in plant cell wall. Adapted from Biotechnology and Bioengineering (Zhang and Lynd 2004). (B) Degree of solubility of various forms of cellulose. CD, cellodextrin; SS CDs, Semi soluble CD; CT, cotton linters; FP, filter paper; P, wood pulp; BC, bacterial cellulose; BMCC, bacterial microcrystalline cellulose; PASC, phosphoric acid swollen cellulose; NW, natural wood; NC, natural cotton. Reproduced with the permission from Biotechnology and Bioengineering (Zhang and Lynd 2004). (C) Schematic representation of enzymatic hydrolysis of cellulose.

Despite these hurdles, several species of *Clostridium*, *Trichoderma*, and *Aspergillus* can efficiently degrade cellulose. Exploitation of the innate potential of the microbial world might be an economical alternative to overcome the recalcitrance associated with lignocellulose (Alper and Stephanopoulos 2009). Two major strategies have been employed to hydrolyze lignocellulose by using microbial consortia. In the first strategy, native cellulolytic organisms like *Clostridium* spp. are engineered to produce bioethanol. In another approach, cellulolytic ability is imposed on efficient ethanol producers such as *Escherichia coli*, *Saccharomyces cerevisiae*, and *Zymomonas mobilis* (Xu et al. 2009). This chapter focuses mainly on the cellulolytic systems that have been engineered into recombinant microorganisms.

2. Enzymatic hydrolysis of lignocellulose

In native cellulolytic organisms, enzymes needed for cellulose hydrolysis—xylanase, endoglucanase, exoglucanase, and β-glucosidase—are expressed either separately or in complexes called cellulosomes (Fig 2). Noncomplexed cellulase systems are characteristic of cellulolytic aerobic bacteria (such as *Bacillus* spp.) and fungi (such as *Trichoderma* spp.) (Lynd et al. 2002). Endoglucanase hydrolyzes amorphous cellulose randomly, leading to the formation of cello-oligosaccharides of varying chain length. Exoglucanases are highly selective enzymes and act on either the reducing or the nonreducing end of cello-oligosaccharides to liberate glucose or cellobiose, respectively. β-Glucosidase hydrolyzes cellobiose into its glucose monomers (Lynd et al. 2002). Cellobiose inhibits both exoglucanase and endoglucanase. Hence, β-glucosidase plays an important role in the overall process, because it prevents the accumulation of cellobiose (Shewale 1982).

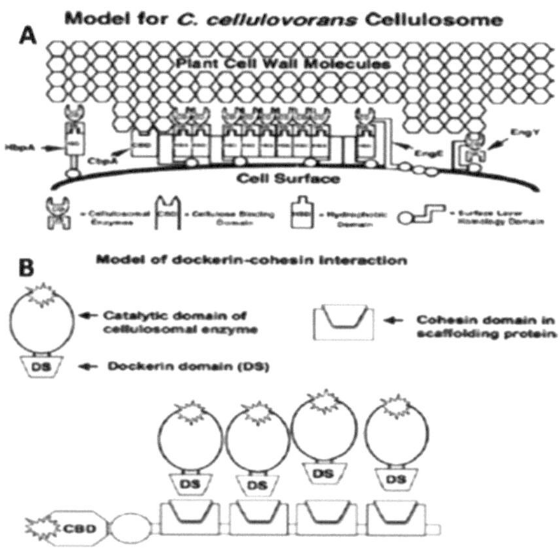

Fig. 2. (A) A model of cellulosome. (B) Synthetic scaffoldin favors arrangement of cellulases with higher activity in close proximity and hence would favor a proper synergy. Reproduced with a permission from Annals of New York Academy of Science (Doi 2008).

Anaerobic bacteria such as *Clostridium* spp. usually produce complexed cellulases called cellulosomes. In cellulosomes, individual enzymes attach to a scaffoldin with their dockerin domains, while exposing the cellulose-binding domain. This complex enables proper synergy among endoglucanase, exoglucanase, and β-glucosidase. (Bayer et al. 1998). Several chimeric scaffoldins have been engineered to position enzymes of higher activity together, and thereby increase the overall hydrolysis efficiency (Fig 2B) (Wen et al. 2009). Even though the large size of the cellulosomes restricts them to only the most readily accessible regions of cellulose, cellulosomes can hydrolyze cellulose more efficiently than free cellulases can (Wilson 2009).

Engineering efforts to increase the efficiency of cellulases and to enhance their kinetic properties have focused mainly on improving the specific activity by improving the thermal or the pH stability of the enzymes (Wen et al. 2009). However, a more important parameter to consider is the efficiency of access to the cellulose interior. While the active-site plays an essential role in other hydrolytic enzymes, the cellulose-binding domain constitutes the key module for cellulases (Bayer et al. 1998). In fact, the cellulose-binding domain determines the type of cellulase. Several efforts to establish a kinetic model for cellulose hydrolysis have failed because of the heterogeneous nature of the cellulosic substrate and the need for multiple enzyme activities (Kadam et al. 2004). In addition to enzyme-substrate proximity, enzyme-enzyme synergy should be considered as a factor for the efficient hydrolysis of cellulose. Whether any relationship or correlation between the crystallinity of lignocellulose and the rate of enzymatic hydrolysis exists remains unclear (Zhang and Lynd 2004). Moreover, the mechanism of cellulose hydrolysis remains incompletely understood, because some groups of cellulases have both exoglucanase and endoglucanase activities.

The low processivity of cellulases demands that the enzymes be replenished several times during the saccharification process. The economic feasibility of enzymatic hydrolysis of lignocellulose to simple sugars is limited by the poor kinetic properties of the enzymes. The use of cellulase-secreting microbes could be an economical alternative to the enzymatic saccharification process. With microbes, the enzymes can be continuously produced, secreted, and used to hydrolyze cellulose into simple sugars that could be directly fermented to ethanol (Fig 3). Thus, microbial fermentation of lignocellulose offers greater promise for economical bioethanol production.

3. Native cellulolytic organisms

The quest for cellulolytic organisms has recently gained increased interest because of the potential to circumvent the cost of enzymes used for cellulose hydrolysis. An ideal host for cellulosic ethanol production should possess certain traits, such as a broad substrate range (utilizing both pentoses and hexoses), high productivity, and tolerance to both ethanol and toxic compounds of lignin (Fischer et al. 2008). In order to identify desirable organisms for cellulosic ethanol production, naturally evolved cellulose-degrading microbes have been characterized from several sources, including the rumen of cattle and the gut of insects, and even from marine environments (Hess et al. 2011). However, most of these microbes cannot be cultivated with synthetic media in the laboratory. Hence, DNA isolates were directly sequenced and putative carbohydrate-hydrolyzing genes were identified (Hess et al. 2011). With this metagenomic approach, identification of microbes suitable for cellulosic fuel production has not been possible, because our current knowledge of the genes is limited.

Well-characterized native cellulolytic organisms include *Cellulomonas fimi*, *Fibrobacter succinogenes*, *Ruminococcus albus*, and *C. thermocellum*. Among these, *C. thermocellum* is of considerable importance, because it is recognized as a "cellulose-using specialist" (Zhang and Lynd 2005). Cellulolytic organisms produce many isoforms of the three different cellulases. *T. reesei,* for example, can secrete five endoglucanases, two cellobiohydrolases, and two β-glucosidases. Apart from cellulases, these organisms also secrete adhesion proteins like glycocalyx, which enables strong adhesion of the cellulolytic organisms to cellulose (Lynd et al. 2002).

Despite the diversity of cellulolytic organisms, none of these organisms are known to produce ethanol efficiently (Xu et al. 2009). Even as the search for a cellulolytic organism with the ability to produce ethanol continues, another strategy would be to engineer efficient ethanol production into cellulolytic organisms such as *Clostridium* spp. (Lynd et al. 2005). However, a lack of proper genetic tools for manipulating these uncommon laboratory strains and very limited knowledge of their genotypes have resulted in a need to engineer the cellulolytic ability into efficient ethanol producers such as *S. cerevisiae, E. coli* and *Z. mobilis*.

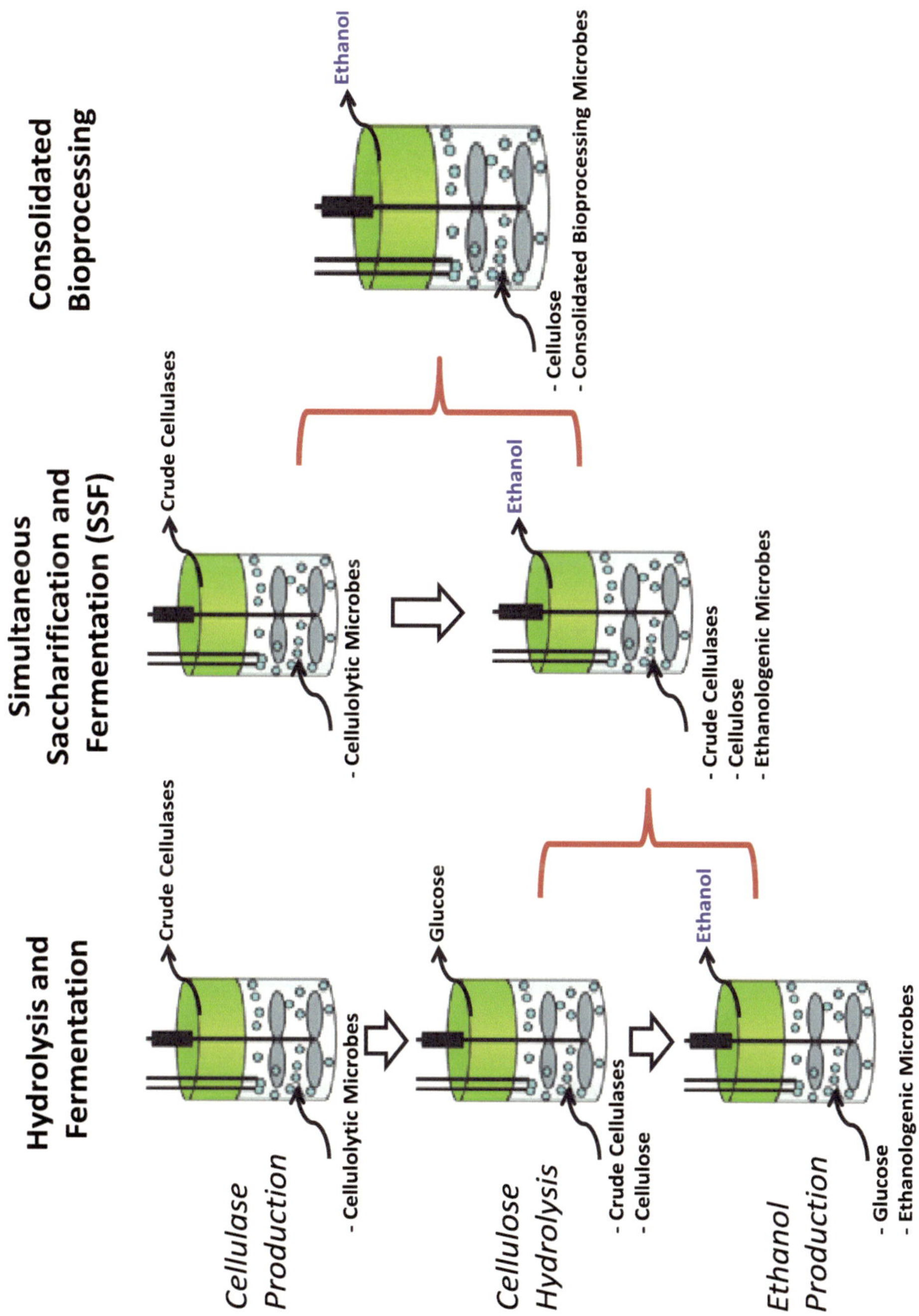

Fig. 3. Schematic representation of the benefits of consolidated bioprocessing over simultaneous saccharification and fermentation.

4. Recombinant cellulolytic organisms

Because the specific activity of cellulase enzymes is at least two-fold lower than that of other hydrolytic enzymes such as starch-hydrolyzing enzymes (Zhang and Lynd 2004; Wilson 2008), even native cellulolytic organisms must produce a high titer of cellulase to efficiently hydrolyze cellulose. The need for synthesis of a large quantity of cellulases is a "metabolic burden," even to native cellulolytic organisms (Zhang and Lynd 2005). Thus, heterologous expression of cellulolytic enzymes in industrial ethanol-producing hosts such as *S. cerevisiae, E. coli* and *Z. mobilis* is especially challenging. Despite these obstacles, several recombinant strains have been engineered for efficient cellulosic ethanol production.

4.1 *S. cerevisiae*

Yeast is an efficient industrial host with a high productivity of ethanol and with well-developed genetic tools. However, yeast does not possess endogenous cellulolytic ability. Several heterologous cellulases have, therefore, been expressed in yeast for direct conversion of cellulose into ethanol. Endoglucanase genes from *Bacillus* spp. were successfully integrated (randomly, at approximately 44 sites) into the chromosome of yeast, resulting in the direct conversion of cellodextrin into ethanol (Cho et al. 1999).

With the advent of cell-surface display technologies, it has become possible to express artificial cellulosomes (rather than free cellulases) in yeast. Cellulosomes facilitate the assembly of different cellulolytic enzymes in close proximity, and thereby favor a proper synergy between the enzymes (Tsai et al. 2010). Surface display of endoglucanase from *T. reesei* and β-glucosidase from *A. aculeatus* in yeast helped in the successful conversion of barley β-glucan into ethanol with 93% of the theoretical yield and without any pretreatment (Fujita et al. 2002). Co-displaying the exoglucanase from *Aspergillus* spp. along with the endoglucanase and β-glucosidase in yeast has resulted in the direct conversion of amorphous cellulose into ethanol (Fujita et al. 2004; McBride et al. 2005). Very recently, recombinant yeast has been further modified to express β-glucosidase within the cell. A high-affinity transporter for cellobiose and cellodextrin has also been cloned into the recombinant yeast. This strain co-metabolizes xylose and cellobiose more efficiently (Ha et al. 2011).

Although several studies have demonstrated efficient ethanol production from amorphous cellulose, attempts to engineer yeast to hydrolyze crystalline cellulose have been unsuccessful because of low exoglucanase activity (la Grange et al. 2010). The exoglucanase and β-glucosidase activities in recombinant cellulolytic yeast strains are insufficient to support growth with cellulose as a sole carbon source. Hence, a synthetic yeast consortium has been developed with four engineered yeast strains, each expressing either the scaffoldin from *Clostridium* spp. and *Ruminococcus* spp. or the three enzymes, namely, exo- and endoglucanases from *Clostridium* spp. and β-glucosidase from *Ruminococcus* spp. (Fig 4) (Tsai et al. 2010). However, investigators have been unable to completely decipher the efficiency of the synthetic consortium, because the ratio of the different cellulases needed for a proper synergy has not been established. A cocktail δ-integration tool has been developed in yeast to predict the optimum ratio of different cellulases, but with little success (Yamada et al. 2010).

Another major problem with recombinant cellulase expression is that heterologous cellulases are made to function at a suboptimal temperature. The optimal temperature for the growth of recombinant hosts is 37°C, but cellulases are more active at temperatures

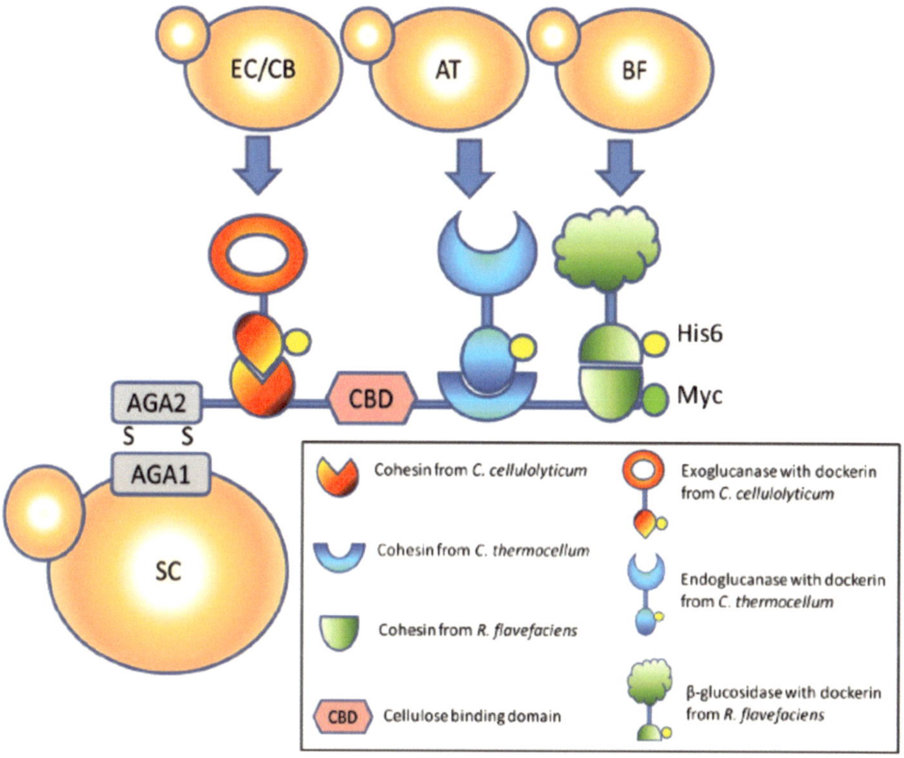

Fig. 4. Schematic representation of the synthetic yeast consortium developed for efficient cellulose utilization. Reproduced with a permission from Applied and Environmental Biotechnology (Tsai et al. 2010). CBD, Cellulose Binding Domain; SC, trifunctional scaffoldin; EC/CB, Exoglucanase; AT, Endoglucanase; BF-β-glucosidase;

above 50°C. Therefore, the thermotolerant yeast *Kluyveromyces marxianus* has been engineered to display thermostable endoglucanase and β-glucosidase on its surface. This engineered, thermostable yeast ferments β-glucan directly to ethanol at 48°C (Yanase et al. 2010).

4.2 *E. coli*

The broad substrate range of *E. coli*, together with its ample genetic tools and its substantial fermentation capacity, renders the species to be a potential candidate for bioethanol production. *E. coli*, with chromosomally integrated genes encoding pyruvate decarboxylase and alcohol dehydrogenase, is an efficient ethanol producer (Ohta et al. 1991). Several attempts have been made to engineer cellulolytic ability in ethanologenic *E. coli*. The species also has endogenous cryptic genes for cellobiose metabolism and an endoglucanase for the hydrolysis of soluble cellulose (Park and Yun 1999; Kachroo et al. 2007; Vinuselvi and Lee 2011).

Achieving a higher extracellular titer of cellulases is a bottleneck in the development of a recombinant cellulolytic *E. coli* for ethanol production. *E. coli* does not have a proper protein secretion system (Shin and Chen 2008). Because *E. coli* is a gram-negative bacterium, it has an outer membrane rich in peptidoglycan, which acts as a barrier for protein secretion. The extracellular protein concentration observed with *E. coli* is 0.0088 g/L, one hundred-fold less than that observed with native cellulolytic organisms (Qian et al. 2008; Xu et al. 2009;

Vinuselvi et al. 2011). Gram-negative bacteria possess five different protein-export pathways (Types I–V), two of which are found in *E. coli* (Type I and Type II).
Several attempts have been made to increase the extracellular titer of recombinant proteins in *E. coli*: by exploiting the Sec/TAT signal sequence (Zhou et al. 1999; Angelini et al. 2001), by fusion of recombinant proteins with extracellular proteins such as OsmY (Qian et al. 2008), or by increasing membrane permeability (Shin and Chen 2008). Cellulase secretion in *E. coli* has been achieved through the expression of endoglucanase, along with the *out* genes of *Erwinia chrysanthemi*, under the control of a surrogate promoter (Zhou et al. 1999). Deletion of *lpp* weakens the outer membrane, allowing any proteins targeted to the periplasmic space to be secreted into the medium. Approximately 70% of the cellulases produced were secreted into the medium in an *lpp* knockout *E. coli* strain (Shin and Chen 2008). Several studies have used OsmY as a fusion partner for recombinant protein secretion in *E. coli*. However, this technique has not been exploited for cellulase secretion because of the large size of cellulases (Aristidou and Penttilä 2000; Qian et al. 2008) (Fig 5).

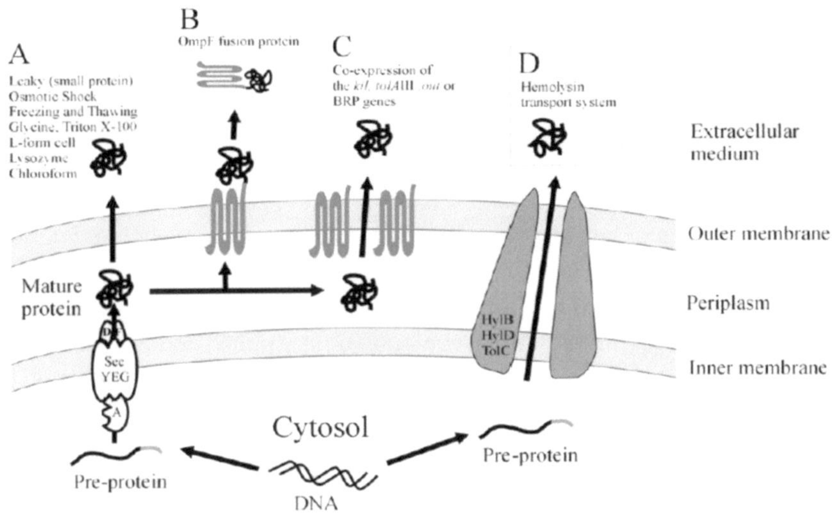

Fig. 5. Schematic representation of the strategies used for extracellular secretion of recombinant proteins in *E. coli*. (A) Membrane disruption using detergents or through lpp deletion increases membrane permeabilization and the periplasmic proteins are leaked into the extracellular space. (B) Use of OmpF fusion proteins helps in the secretion of small proteins. (C) *out* gene of *Erwinia* encodes for a bacteriocin release protein pore which helps in the secretion of the periplasmic proteins. (D) Use of SEC/TAT pathway signal sequence favors direct secretion of cellulases into the medium. Reprinted with a permission from Applied Microbial Biotechnology (Choi and Lee 2004).

The cellobiose metabolic operon from *Klebsiella oxytoca* has been introduced into *E. coli*, but the expression level of the cellobiose transporter and metabolic genes was poor, and hence could not support the growth of *E. coli* on cellobiose (Moniruzzaman et al. 1997). Cellulases from several species of *Clostridium*, *Bacillus*, *Cellulomonas*, and *Ruminococcus* have been expressed and characterized in *E. coli* (Hinchliffe 1984; Zappe et al. 1986; Fierobe et al. 1991; ReverbelLeroy et al. 1996; Lam et al. 1997; ReverbelLeroy et al. 1997; Lee et al. 2008; Li et al. 2009). Co-expression of endoglucanase from *B. pumilus* and *β*-glucosidase from

Fervidobacterium spp. in *E. coli* favored growth of the recombinant strain, with soluble carboxymethyl cellulose (CMC) as the sole carbon source (Rodrigues et al. 2010).

4.3 *Z. mobilis*

Zymomonas is an efficient ethanol producer, together with a higher tolerance to ethanol and to several inhibitory substances of lignin. *Zymomonas* species also possess a gene that codes for cellulase (Rajnish et al. 2008). While protein secretion is not a hurdle in *Zymomonas* spp., a major difficulty arises with the lack of amenable genetic tools for the introduction or modification of a gene (Linger et al. 2010). Two cellulases from *Acidothermus cellulolyticus* have been expressed in *Z. mobilis*, and a significant amount of secretion was observed when they were fused with predicted N-terminal signal peptides of *Z. mobilis* (Linger et al. 2010). Endoglucanases from different cellulolytic organisms such as *Cellulomonas* spp., *Enterobacter cloacae*, *Pseudomonas fluorescens*, and *Erwinia* spp., have been expressed in *Z. mobilis*. However, none of these cellulases were secreted efficiently (Lejeune et al. 1988; Misawa et al. 1988; Brestic-Goachet et al. 1989; Thirumalai Vasan et al. 2011).

5. Future of cellulosic ethanol

An Ideal Biofuel Producing Microorganism (IBPM) should possess four important traits: it should be able to carry out (1) biomass degradation and (2) product formation; (3) it should show tolerance to solvents, and (4) it should serve as a chassis organism for rapid growth in the bioreactor (French 2009). Chassis organisms, such as yeast and *E. coli*, are well characterized. Commercial bioethanol has been produced from sugarcane by yeast. In addition, *E. coli* and *Z. mobilis* are progressing as efficient ethanol producers. A current challenge is to engineer biomass degradation (cellulolytic) ability. Further, investigators seek to enhance tolerance to harsh conditions that arise during cellulose fermentation, such as substrate and product toxicity. In particular, the chassis organism should have enhanced tolerance to toxic compounds of lignin. Classical strain improvement through long-term adaptation and mutagenesis may be an effective way to increase the tolerance to harsh environments, such as ethanol or lignin, because the mechanisms of toxicity and tolerance are largely unknown (Fischer et al. 2008).

Engineering cellulolytic ability into recombinant hosts has long been a challenge. The number of cellulase genes that should be cloned into the recombinant host remains unclear (Vinuselvi et al. 2011). The main obstacle to developing a recombinant cellulolytic host is the inability of hosts to support expression and secretion of a sufficient quantity of cellulases. Although cellulase expression is well established in yeast, there is no known study demonstrating direct conversion of plant biomass into ethanol. Despite the characterization of several cellulases in *E. coli*, a cellulolytic cassette containing all three cellulases has not been established for *E. coli*. Furthermore, efficient genetic tools are still lacking for *Zymomonas*, limiting its potential to be engineered with a heterologous gene.

One way to address the problems associated with heterologous cellulase expression and to reduce the metabolic burden imposed by the expression of cellulolytic enzymes in recombinant hosts would be the development of a well-defined synthetic consortium with two efficient players—native cellulolytic and solventogenic organisms—acting together. A high level of expression of multiple heterologous proteins would impose a heavy metabolic burden on the host. With a synthetic consortium, this burden could be shared by different species or by different strains of the same species. A co-culture of these strains to produce a cellulase cocktail would, therefore, reduces the overall metabolic burden and increase the ethanol yield (Brenner et al. 2008). Synthetic biology also offers superior inducible systems, such as light-inducible promoters and the *fim* inversion system, which are capable of providing spatiotemporal changes in gene expression (Levskaya et al. 2005; Ham et al. 2006). With such systems, it is possible to regulate the expression of genes with time and, potentially, to help reduce the metabolic burden imposed on the recombinant host (Drepper et al. 2011). Using metabolic engineering and synthetic biology, Steen et al. (2010) have developed a promising way of causing *E. coli* to produce more complex biofuels—fatty esters and fatty alcohols—directly from hemicellulose, a major component of plant-derived biomass (Fig 6). This study is representative of the recent progress in cellulosic fuel production. However, the possibility of increasing the productivity of such advanced biofuels remains a significant challenge.

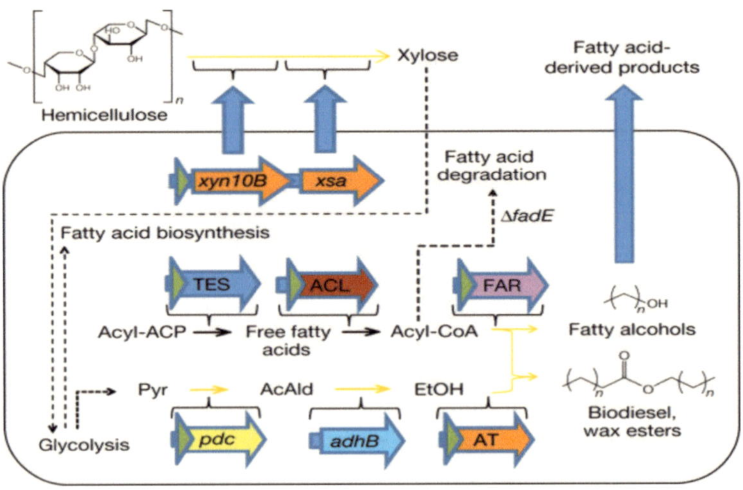

Fig. 6. Schematic representation of the new pathways engineered into recombinant *E. coli* for the production of advanced biofuels from hemicellulosic fraction of plant biomass. This recombinant strain is a representative candidate proving the potency of synthetic biology and metabolic engineering to develop a cellulosic ethanol producer. TES, thioesterase; ACL, acyl-CoA ligase; FAR, fatty acyl-CoA reductase; AT, acyltransferase; *pdc,* pyruvate decarboxylase; *adhB,* alcohol dehydrogenase; AcAld, acetaldehyde; EtOH, ethanol; pyr, pyruvate; *xyn10B* & *xsa,* xylanase. Overexpressed genes or operons are indicated; green triangles represent the *lacUV5* promoter. Reproduced with a permission from Nature (Steen et al. 2010).

6. Conclusions

Cellulosic bioethanol is gaining importance to circumvent the oil crisis and climate change. However, two major problems remain to be solved, in order to produce cellulosic ethanol economically. One problem is the high price of the cellulolytic enzymes used in the saccharification of lignocelluloses. The other problem is that the traditional saccharification and fermentation for bioethanol requires huge initial capital investment and operational cost. Consolidated bioprocessing presents a desirable way to produce bioethanol economically from lignocellulose. Microorganisms such as *Trichoderma* spp. and *C. thermocellum* effectively challenge the recalcitrance of lignocellulose, whereas microbes such as yeast and *Z. mobilis* can produce ethanol more efficiently. Several attempts have been made to combine these two abilities into a single organism, but with little success. Recent progress in synthetic biology, metabolic engineering, and protein engineering gives hope that the goal of generating cellulosic ethanol with a single organism may not be far from reality.

7. Acknowledgements

This work was supported by the National Research Foundation of Korea (NRF) through grants funded by the Ministry of Education, Science and Technology (NRF-2009-C1AAA001-2009-0093479, NRF-2009-0076912, NRF-2010-0006436) and UNIST (Ulsan National Institute of Science and Technology) research grant.

8. References

Alper, H. and G. Stephanopoulos (2009). "Engineering for biofuels: exploiting innate microbial capacity or importing biosynthetic potential?" *Nature Reviews Microbiology* 7(10): 715-723, 1740-1526

Angelini, S., R. Moreno, et al. (2001). "Export of *Thermus thermophilus* alkaline phosphatase via the twin-arginine translocation pathway in *Escherichia coli*." *FEBS Letters* 506(2): 103-107, 0014-5793

Aristidou, A. and M. Penttilä (2000). "Metabolic engineering applications to renewable resource utilization." *Current Opinion in Biotechnology* 11(2): 187-198, 0958-1669

Bayer, E. A., H. Chanzy, et al. (1998). "Cellulose, cellulases and cellulosomes." *Current Opinion in Structural Biology* 8(5): 548-557, 0959-440X

Brenner, K., L. You, et al. (2008). "Engineering microbial consortia: a new frontier in synthetic biology." *Trends in Biotechnology* 26(9): 483-489, 0167-7799

Brestic-Goachet, N., P. Gunasekaran, et al. (1989). "Transfer and expression of an *Erwinia chrysanthemi* cellulase gene in *Zymomonas mobilis*." *Journal of General Microbiology* 135(4): 893-902, 0022-1287

Cho, K. M., Y. J. Yoo, et al. (1999). "δ-Integration of endo/exo-glucanase and β-glucosidase genes into the yeast chromosomes for direct conversion of cellulose to ethanol." *Enzyme and Microbial Technology* 25(1-2): 23-30, 0141-0229

Choi, J. H. and S. Y. Lee (2004). "Secretory and extracellular production of recombinant proteins using *Escherichia coli*." *Applied Microbiology and Biotechnology* 64(5): 625-635, 0175-7598

Doi, R. H. (2008). "Cellulases of mesophilic microorganisms." *Annals of the New York Academy of Sciences* 1125(1): 267-279, 1749-6632

Drepper, T., U. Krauss, et al. (2011). "Lights on and action! controlling microbial gene expression by light." *Applied Microbiology and Biotechnology* 90(1): 23-40, 0175-7598

Fierobe, H. P., C. Gaudin, et al. (1991). "Characterization of endoglucanase-A from *Clostridium cellulolyticum*." *Journal of Bacteriology* 173(24): 7956-7962, 0021-9193

Fischer, C. R., D. Klein-Marcuschamer, et al. (2008). "Selection and optimization of microbial hosts for biofuels production." *Metabolic Engineering* 10(6): 295-304, 1096-7176

French, C. E. (2009). "Synthetic biology and biomass conversion: a match made in heaven?" *Journal of The Royal Society Interface* 6(Suppl 4): S547-S558, 1742-5662

Fujita, Y., J. Ito, et al. (2004). "Synergistic saccharification, and direct fermentation to ethanol, of amorphous cellulose by use of an engineered yeast strain codisplaying three types of cellulolytic enzyme." *Applied Environmental Microbiology* 70(2): 1207-1212, 0099-2240

Fujita, Y., S. Takahashi, et al. (2002). "Direct and efficient production of ethanol from cellulosic material with a yeast strain displaying cellulolytic enzymes." *Applied Environmental Microbiology* 68(10): 5136-5141, 0099-2240

Ha, S.-J., J. M. Galazka, et al. (2011). "Engineered *Saccharomyces cerevisiae* capable of simultaneous cellobiose and xylose fermentation." *Proceedings of the National Academy of Sciences* 108(2): 504-509, 0027-8424

Ham, T. S., S. K. Lee, et al. (2006). "A tightly regulated inducible expression system utilizing the *fim* inversion recombination switch." *Biotechnology and Bioengineering* 94(1): 1-4, 1097-0290

Hess, M., A. Sczyrba, et al. (2011). "Metagenomic discovery of biomass-degrading genes and genomes from cow rumen." *Science* 331(6016): 463-467, 0036-8075

Himmel, M. E., S.-Y. Ding, et al. (2007). "Biomass recalcitrance: engineering plants and enzymes for biofuels production." *Science* 315(5813): 804-807, 0036-8075

Hinchliffe, E. (1984). "Cloning and expression of *Bacillus subtilis* endo-1,3-1,4-beta-D-glucanase gene in *Escherichia coli* K12." *Journal of General Microbiology* 130: 1285-1291, 0022-1287

Kachroo, A. H., A. K. Kancherla, et al. (2007). "Mutations that alter the regulation of the *chb* operon of *Escherichia coli* allow utilization of cellobiose." *Molecular Microbiology* 66: 1382-1395, 0950-382X

Kadam, K. L., E. C. Rydholm, et al. (2004). "Development and validation of a kinetic model for enzymatic saccharification of lignocellulosic biomass." *Biotechnology Progress* 20(3): 698-705, 1520-6033

la Grange, D. C., R. den Haan, et al. (2010). "Engineering cellulolytic ability into bioprocessing organisms." *Applied Microbiology and Biotechnology* 87(4): 1195-1208, 0175-7598

Lam, T.-L., R. S. C. Wong, et al. (1997). "Enhancement of extracellular production of a *Cellulomonas fimi* exoglucanase in *Escherichia coli* by the reduction of promoter strength." *Enzyme and Microbial Technology* 20(7): 482-488, 0141-0229

Lee, S. M., L. H. Jin, et al. (2010). "Beta-glucosidase coating on polymer nanofibers for improved cellulosic ethanol production." *Bioprocess and Biosystems Engineering.* 33(1): 141-147, 1615-7591

Lee, Y. J., B. K. Kim, et al. (2008). "Purification and characterization of cellulase produced by *Bacillus amyoliquefaciens* DL-3 utilizing rice hull." *Bioresource Technology* 99(2): 378-386, 0960-8524

Lejeune, A., D. E. Eveleigh, et al. (1988). "Expression of an endoglucanase gene of *Pseudomonas fluorescens var. cellulosa* in *Zymomonas mobilis*." *FEMS Microbiology Letters* 49(3): 363-366, 0378-1097

Levskaya, A., A. A. Chevalier, et al. (2005). "Synthetic biology: engineering *Escherichia coli* to see light." *Nature* 438(7067): 441-442, 0028-0836

Li, W., X. J. Huan, et al. (2009). "Simultaneous cloning and expression of two cellulase genes from *Bacillus subtilis* newly isolated from Golden Takin (*Budorcas taxicolor Bedfordi*)." *Biochemical and Biophysical Research Communications* 383(4): 397-400, 0006-291X

Linger, J. G., W. S. Adney, et al. (2010). "Heterologous expression and extracellular secretion of cellulolytic enzymes by *Zymomonas mobilis*." *Applied and Environmental Microbiology* 76(19): 6360-6369, 0099-2240

Lynd, L. R., W. H. van Zyl, et al. (2005). "Consolidated bioprocessing of cellulosic biomass: an update." *Current Opinion in Biotechnology* 16(5): 577-583, 0958-1669

Lynd, L. R., P. J. Weimer, et al. (2002). "Microbial cellulose utilization: fundamentals and biotechnology." *Microbiology and Molecular Biology Reviews* 66(3): 506-577, 1092-2172

McBride, J. E., J. J. Zietsman, et al. (2005). "Utilization of cellobiose by recombinant β-glucosidase-expressing strains of *Saccharomyces cerevisiae*: characterization and evaluation of the sufficiency of expression." *Enzyme and Microbial Technology* 37(1): 93-101, 0141-0229

Misawa, N., T. Okamoto, et al. (1988). "Expression of a cellulase gene in *Zymomonas mobilis*." *Journal of Biotechnology* 7(3): 167-177, 0168-1656

Moniruzzaman, M., X. Lai, et al. (1997). "Isolation and molecular characterization of high-performance cellobiose-fermenting spontaneous mutants of ethanologenic *Escherichia coli* KO11 containing the *Klebsiella oxytoca casAB* operon." *Applied and Environmental Microbiology* 63(12): 4633-4637, 0099-2240

Ohta, K., D. S. Beall, et al. (1991). "Genetic improvement of *Escherichia coli* for ethanol production: chromosomal integration of *Zymomonas mobilis* genes encoding pyruvate decarboxylase and alcohol dehydrogenase II." *Applied and Environmental Microbiology* 57(4): 893-900, 0099-2240

Park, Y. W. and H. D. Yun (1999). "Cloning of the *Escherichia coli* endo-1,4-D-glucanase gene and identification of its product." *Molecular and General Genetics* 261(2): 236-241, 0026-8925

Qian, Z.-G., X. -X. Xia, et al. (2008). "Proteome-based identification of fusion partner for high-level extracellular production of recombinant proteins in *Escherichia coli*." *Biotechnology and Bioengineering* 101(3): 587-601, 1097-0290

Rajnish, K., G. Choudhary, et al. (2008). "Functional characterization of a putative endoglucanase gene in the genome of *Zymomonas mobilis*." *Biotechnology Letters* 30(8): 1461-1467, 0141-5492

ReverbelLeroy, C., A. Belaich, et al. (1996). "Molecular study and overexpression of the *Clostridium cellulolyticum celF* cellulase gene in *Escherichia coli*." *Microbiology-UK* 142: 1013-1023, 1350-0872

ReverbelLeroy, C., S. Pages, et al. (1997). "The processive endocellulase CelF, a major component of the *Clostridium cellulolyticum* cellulosome: purification and characterization of the recombinant form." *Journal of Bacteriology* 179(1): 46-52, 0021-9193

Rodrigues, A. L., A. Cavalett, et al. (2010). "Enhancement of *Escherichia coli* cellulolytic activity by co-production of beta-glucosidase and endoglucanase enzymes." *Electronic Journal of Biotechnology* 13(5), 0717-3458

Shewale, J. G. (1982). "β-Gucosidase: its role in cellulase synthesis and hydrolysis of cellulose." *International Journal of Biochemistry* 14(6): 435-443, 0020-711X

Shin, H. D. and R. R. Chen (2008). "Extracellular recombinant protein production from an *Escherichia coli lpp* deletion mutant." *Biotechnology and Bioengineering* 101(6): 1288-1296, 0006-3592

Steen, E. J., Y. Kang, et al. (2010). "Microbial production of fatty-acid-derived fuels and chemicals from plant biomass." *Nature* 463(7280): 559-562, 0028-0836

Thirumalai Vasan, P., P. Sobana Piriya, et al. (2011). "Cellulosic ethanol production by *Zymomonas mobilis* harboring an endoglucanase gene from *Enterobacter cloacae*." *Bioresource Technology* 102(3): 2585-2589, 0960-8524

Tsai, S.-L., G. Goyal, et al. (2010). "Surface display of a functional minicellulosome by intracellular complementation using a synthetic yeast consortium and its application to cellulose hydrolysis and ethanol production." *Applied and Environmental Microbiology* 76(22): 7514-7520, 0099-2240

Vinuselvi, P. and S. K. Lee (2011). "Engineering *Escherichia coli* for efficient cellobiose utilization." *Applied Microbiology and Biotechnology* 92(1): 125-132, 0175-7598

Vinuselvi, P., J. M. Park, et al. (2011). "Engineering microorganisms for biofuel production." *Biofuels* 2(2): 153-166, 1759-7269

Wen, F., N. U. Nair, et al. (2009). "Protein engineering in designing tailored enzymes and microorganisms for biofuels production." *Current Opinion in Biotechnology* 20(4): 412-419, 0958-1669

Wilson, D. B. (2008). "Three microbial strategies for plant cell wall degradation." *Incredible Anaerobes: from Physiology to Genomics to Fuels* 1125: 289-297, 0077-8923

Wilson, D. B. (2009). "Cellulases and biofuels." *Current Opinion in Biotechnology* 20(3): 295-299, 0958-1669

Xu, Q., A. Singh, et al. (2009). "Perspectives and new directions for the production of bioethanol using consolidated bioprocessing of lignocellulose." *Current Opinion in Biotechnology* 20(3): 364-371, 0958-1669

Yamada, R., N. Taniguchi, et al. (2010). "Cocktail delta-integration: a novel method to construct cellulolytic enzyme expression ratio-optimized yeast strains." *Microbial Cell Factories* 9(1): 32, 1475-2859

Yanase, S., T. Hasunuma, et al. (2010). "Direct ethanol production from cellulosic materials at high temperature using the thermotolerant yeast *Kluyveromyces marxianus* displaying cellulolytic enzymes." *Applied Microbiology and Biotechnology* 88(1): 381-388, 0175-7598

Zappe, H., D. T. Jones, et al. (1986). "Cloning and expression of *Clostridium acetobutylicum* endoglucanase, cellobiase and amino-acid biosynthesis gene in *Escherichia coli*." *Journal of General Microbiology* 132: 1367-1372, 0022-1287

Zhang, Y.-H. P. and L. R. Lynd (2004). "Toward an aggregated understanding of enzymatic hydrolysis of cellulose: noncomplexed cellulase systems." *Biotechnology and Bioengineering* 88(7): 797-824, 1097-0290

Zhang, Y.-H. P. and L. R. Lynd (2005). "Cellulose utilization by *Clostridium thermocellum*: bioenergetics and hydrolysis product assimilation." *Proceedings of the National Academy of Sciences of the United States of America* 102(20): 7321-7325, 0027-8424

Zhou, S., L. P. Yomano, et al. (1999). "Enhancement of expression and apparent secretion of *Erwinia chrysanthemi* endoglucanase (encoded by *celZ)* in *Escherichia coli* B." *Applied and Environmental Microbiology* 65(6): 2439-2445, 0099-2240

Part 3

Bioethanol Use

13

Catalytic Hydrogen Production from Bioethanol

Hua Song
RTI International
USA

1. Introduction

Along with the maturity of the production technology (i.e., fermentation) for a long history, bioethanol has become one of the most significant chemicals and energy carriers in large quantity derived from biomass. Although ethanol production from non-food resources remains challengeable for scientists, how to utilize ethanol in an efficient and economical way opens more space for all researchers both from industry and academia to play with.

Hydrogen is likely to play an important role in the energy portfolio of the future due to its high gravimetric energy density. Especially when it is used in fuel cells, it is an ideal energy carrier that can offer clean and efficient power generation. In the United States, ~95 % of hydrogen is produced using a steam reforming process [1]. Over 50% of world's hydrogen production relies on natural gas as the feedstock [2]. As the concern for a sustainable energy strategy grows, replacing natural gas and other fossil fuels with renewable sources is gaining new urgency. In this context, producing hydrogen from bio-derived liquids such as bio-ethanol has emerged as a promising technology due to the low toxicity, ease of handling and the availability from many different renewable sources (e.g., sugar cane, switchgrass, algae) that ethanol has to offer. An added advantage of producing hydrogen from bio-derived liquids is that it is quite suitable for a distributed production strategy.

This chapter is aimed to provide a big overview of the current technologies for catalytic hydrogen production from bioethanol while focusing the discussion on the hydrogen production through steam reforming of bioethanol over non-precious metal based catalysts, more specifically, cobalt-based catalysts. By combing the work performed at the author' laboratories, this chapter will also provide the professional insights on the future development direction of such technologies. Through the estimated economic analysis of this process simulated at industrial scale, the ways of final commercialization of the developed catalyst system specially tailored for central and distributed hydrogen production from steam reforming of bioethanol will be suggested.

2. Production technology overview

Multiple techniques have been developed during the past decades to convert bioethanol to hydrogen by following the reaction (1).

$$C_2H_5OH_{(l)} + 3\ H_2O_{(l)} \quad 2\ CO_2 + 6\ H_2 \qquad (\Delta H_{r,298K} = 348\ kJ/mol) \qquad (1)$$

It is clearly observed that 6 moles of hydrogen can be produced per mole of ethanol fed. However, the highly endothermic feature of this reaction requires external energy supply. Depending on the type of energy input, the current hydrogen production technologies can be categorized into two areas: non-thermal including bio, photo, plasma, and thermal-chemical processes. Besides, several hybrid systems have also been recently developed to produce hydrogen relying on the energy supply of more than one source (e.g., photo-fermentation and thermal plasma). Compared to thermochemical conversion, non-thermal hydrogen production can take place at much mild conditions with minimal thermo-energy input requirement from surroundings. However, the biological or photo hydrogen production efficiency is much lower than acceptable scale for industrial application. Unlike biological or photo process, thermochemical conversion can happen at much higher reaction rate, but under relatively severe conditions (e.g., high temperature and pressure) with notable amount of thermo-energy input. In addition to water, CO_2 (dry reforming) and O_2 (partial oxidation or oxidative reforming) can also act as oxidant to oxidize ethanol for hydrogen production. Among all the available techniques described in details in this section, steam reforming might possess the highest potential to be commercialized in the near term.

2.1 Fermentative hydrogen production

In this process, metabolically engineered microorganisms such as bacteria convert ethanol to hydrogen under the facilitation of hydrogenase enzymes which are metalloproteins, containing complicated metal active centres that catalyze the interconversion of protons and electrons with dihydrogen. According to literature reporting [3-5], two major classes of hydrogenases are recognized based on their metal active sites: [FeFe] and [NiFe]. Depending on whether light will be involved, this biological hydrogen production process can be simply classified as photo- and dark-fermentation processes [6].

During the photo-fermentation process, the hydrogenase enzyme synthesized and activated under dark anaerobic condition is used to convert ethanol to biohydrogen under light anaerobic condition. Since the light acts as the energy source, the consumption rate of substrate is less than that required for dark fermentation. However, the hydrogen efficiency will be dramatically reduced in the presence of oxygen concurrently produced through photosynthesis by bacteria, which has been evidenced by many researchers [7]. Furthermore, the ultra-violet wavelength radiation requirement and relatively slower production rate limit its industrial application at large scale.

Under the dark operation environment, there is no risk for hydrogenases exposed to oxygen, which makes the hydrogenase enzymes remain active throughout the whole process, leading to more efficient hydrogen production. Compared to photo-fermentation, the inherent continuous and fast production feature makes dark anaerobic digestion economically promising for industrial scale practice. In recent years, many publications have reported their efforts spent on optimization of operation parameters, development of genetically modified microorganism, metabolic engineering, improvement of reactor designs, use of different solid matrices for cell immobilization, etc. to maximize hydrogen yield. Among many considerations, the blockage of methanogenesis in the anaerobic pathway is crucial to improve hydrogen selectivity through the inhibition of methane formation.

2.2 Photocatalytic hydrogen production

In addition to biological process, photocatalytic oxidation of ethanol provides alternative interesting approach to generate hydrogen. Similar to photo-fermentation where enzyme is used to catalyze the conversion, solar energy is again utilized to offer sufficient power to produce hydrogen from ethanol under the facilitation of inorganic catalyst. Among many catalysts documented in the literature, TiO_2 [8-10] is the most commonly used catalyst base due to its excellent photoreactivity which has a suitable band gap for efficient light photon absorption. Upon radiation, the electron contained in a semiconductor such as TiO_2 will be excited and transferred from valence band to conduction band, resulting in the creation of an electron-hole pair and in turn providing an active site for redox reaction. As shown in Figure 1, reaction (1) is a typical redox reaction where H_2O serves as the oxidant to oxidize ethanol while itself being reduced to H_2. The adsorbed ethanol and water species will react with each other on the surface of the active sites of the synthesized photocatalyst to produce H_2. Usually, certain amount of active metal (noble metal or transition metal) will be loaded to the TiO_2 support to promote its photoactivity. According to the publications, Cu, Ni, V, Pt, Pd, Rh, Au, Ir, and Ru have been tested [11-14], among which Pt doped TiO_2 exhibits the highest photoactivity toward hydrogen production from bioethanol. Various synthesis methods have been successfully demonstrated to get TiO_2 supported catalyst with desirable particle size and morphology for hydrogen generation maximization. Besides TiO_2 supported catalyst, there are multiple other novel semiconductors being developed recently for effective hydrogen production including CdS [15], VO_2 [16], WO_3 [17], and $ZnSn(OH)_6$ [18]. Nevertheless, the hydrogen production efficiency from catalytic ethanol oxidation still remains at very low level probably due to two facts: the fast recombination rate of the created electron-hole pairs and the low photon absorption efficiency at visible light range. Although hydrogen evolution rate of 21 mmol/g_{cat}/h has been reported and is the fastest rate claimed so far in the literature [19], it is still significantly lower than that obtained from thermochemical ethanol conversion. Therefore, the technical breakthrough is required in the field of photocatalysis before the commercialization of this technique can be seriously considered.

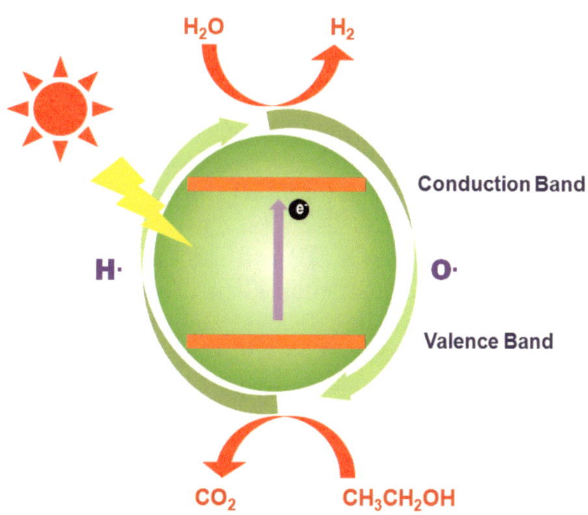

Fig. 1. The schematic diagram of photocatalytic ethanol reforming

2.3 Aqueous phase reforming

As a low temperature alternative to steam reforming, Aqueous Phase Reforming (APR) has emerged as a valuable means of converting organic compounds of biological origin to value-added chemicals and fuel components. Due to its feature of low temperature operation, the energy required for water and oxygenated hydrocarbon evaporation is eliminated, leading to the notable reduction of overall energy input, which overcomes the evaporation difficulty of some organic compounds with high boiling point required for steam reforming. In order to keep all reactants in the liquid phase at operation temperature (typically ~500 K), certain pressure (typically 15~50 bar) has to be applied to the whole reactor system. Such operation temperature and pressure benefit the happening of water-gas shift reaction, making it possible to produce hydrogen with low amounts of CO in a single reactor. Undesirable organic compound decomposition can also be minimized under such low reaction temperature. Furthermore, the relatively high pressure operation will also favour the downstream gas separation and purification, and even subsequent gas compression, storage, and delivery. This process is exclusively suitable for the biomass derived organic compounds with relatively longer carbon chain such as sorbitol, which has been comprehensively reviewed by the researchers in Dumesic's group [20]. For smaller organic compounds like ethanol discussed in this chapter, APR process for hydrogen generation is less favourable from the overall energy utilization viewpoint, which is concluded by Tokarev, et al. in their recent publication [21]. Moreover, the relatively high pressure requirement raises the concerns on safety and operation cost. Hydrogen selectivity is another big challenge APR has to face, because H_2 and CO_2 produced are thermodynamically unstable and methane formation is favourable at such low temperature.

2.4 CO_2 dry reforming

In addition to H_2O, CO_2 can also acts as oxidant to reform ethanol to generate gaseous products. The reaction involved in this process is depicted in Reaction (2).

$$C_2H_5OH_{(l)} + CO_2 \quad 3\,H_2 + 3\,CO \qquad (\Delta H_{r,298K} = 338\ kJ/mol) \tag{2}$$

Compared to Reaction (1), although only 3 moles of hydrogen are produced per mole of ethanol by using dry reforming process, it is still a valuable approach to utilize CO_2 for hydrogen or syngas production beneficial for reducing greenhouse gas emission. The process feasibility and optimal operation parameters have been investigated by W. Wang, et al. thermodynamically, which is valuable for desirable product yield maximization. According to the calculations performed in [22], higher temperature, lower pressure, addition of inert gas, and lower CO_2 to ethanol ratio benefit the improvement of hydrogen yield. Several catalysts such as Ni/Al_2O_3 [23] and Rh/CeO_2 [24] have been developed in recent years for hydrogen or syngas production. Generally speaking, CO_2 is less active than water in oxidizing ethanol. Therefore, more active catalysts are critical for making ethanol dry reforming more attractive to industrial investors. Similarly to methane dry reforming, coke can be formed with high possbility at certain reaction conditions on the catalyst surface, resulting in catalyst deactivation. Carbon tends to form at low temperature and low CO_2/ethanol ratio based on thermodynamic prediction, which should be avoided to prevent catalyst deactivation. However, sometimes as a preferable byproducts, production of various types of carbon nanofilaments is desired by following Reaction (3), which has been found to be effectively catalyzed by stainless steel or carbon steel catalysts [25, 26].

$$C_2H_5OH_{(l)} + CO_2 \quad 2\,H_2 + 2\,CO + 2\,C + H_2O_{(l)} \quad (\Delta H_{r,298K} = 163\ kJ/mol) \tag{3}$$

2.5 Plasma reforming

The energy required for ethanol reforming can also be provided by the electrical discharge powered by high voltage transformer. The ethanol solution fed can thereafter be ionized to plasma state under such discharge, leading to the creation of a variety of chemically active species and energetic electrons which will quickly react with each other to form product gases. Depending on their energy level, temperature, and electronic density, plasma state can be generally classified as thermal and non-thermal plasma. Compared to thermal plasma, the hydrogen production under non-thermal plasma condition has much lower energy consumption. The features of low temperature operation, rapid reaction start-up, no involvement of catalyst handling, and non-equilibrium properties make non-thermal plasma technique very promising for energy conversion and fuel gas treatment [27]. Comparable performance has been reported through non-thermal plasma process toward hydrogen production, which is very close to the ones obtained from catalytic reactors [28]. However, its relatively high energy requirement, complicated reaction network, and low selectivity remain the main obstacles preventing it from industrial application at current stage.

2.6 Partial oxidation

Compared to H_2O and CO_2, O_2 is much active in partially oxidizing ethanol for hydrogen production by following a representative Reaction (4) which is a slightly endothermic reaction, indicating that much less external energy is needed for reaction proceeding.

$$C_2H_5OH_{(l)} + 0.5\,O_2 \quad 3\,H_2 + 2\,CO \quad (\Delta H_{r,298K} = 56\ kJ/mol) \tag{4}$$

As a result, the ethanol partial oxidation can take place at much lower temperature (200 ~300 ^{o}C) in the presence of catalyst than those required for steam or dry reforming (typically 450 ~650 ^{o}C). Depending on the reaction conditions and catalyst used, in addition to CO, various ethanol oxidation products with different oxidation states have been observed including acetaldehyde, acetone, acetic acid, and CO_2. Plenty of catalyst systems have been extensively studied for catalyzing ethanol oxidation at low temperature. Among them, Ni-Fe alloy [29] from transition metal group and Pt from noble metal group based catalyst [30] have drawn special attentions. According to literature reporting, 51% ethanol conversion and 97% hydrogen selectivity has been successfully achieved at temperature as low as 370 K over Pt/ZrO_2 [31]. Although O_2 usage significantly improves the ethanol reactivity and lowers down the energy input, it reduces the hydrogen production by half, referring to Reaction (1). Moreover, the likelihood of hot-spot formation makes the control of this reaction difficult.

2.7 Steam reforming

As mentioned earlier in this chapter, hydrogen production can be maximized per fed ethanol through pure steam reforming. However, the highly endothermic feature of this reaction limits its widely industrial application for hydrogen production. In order to lessen its heavy dependence on external energy supply, part of ethanol is sacrificed to provide required energy for steam reforming through the introduction of oxygen, which is named as oxidative steam reforming (Reaction 5). Depending on the value of δ, the enthalpy change of

Reaction (5) will become less positive, indicating less energy requirement from surroundings. The reaction will finally become autothermal at the point where little or no energy is needed from external sources (e.g., if δ=0.6, $\Delta H_{r,298K}$ =4.4 kJ/mol).

$$C_2H_5OH_{(l)} + \delta\, O_2 + (3\text{-}2\,\delta)\, H_2O_{(l)} \quad (6\text{-}2\,\delta)\, H_2 + 2\, CO_2 \tag{5}$$

Although the products from the desired reactions are only CO_2 and H_2, in reality, depending on the reaction conditions and catalysts used, the product distribution can be governed by a very complex reaction network. Possible reactions involved can be as follows.

$$CH_3CH_2OH \quad CH_4{+}CO{+}H_2 \quad \text{(ethanol decomposition)} \tag{6}$$

$$CH_3CH_2OH \quad CH_3CHO{+}H_2 \quad \text{(ethanol dehydrogenation)} \tag{7}$$

$$CH_3CH_2OH \quad C_2H_4{+}H_2O \quad \text{(ethanol dehydration)} \tag{8}$$

$$CH_3CH_2OH{+}H_2O \quad 2\,CO{+}4H_2 \quad \text{(ethanol incomplete reforming)} \tag{9}$$

$$2\,CH_3CH_2OH \quad (C_2H_5)_2O{+}H_2O \quad \text{(ethanol dehydrative coupling)} \tag{10}$$

$$CH_3CH_2OH{+}H_2O \quad CH_3COOH{+}2\,H_2 \quad \text{(acetic acid formation)} \tag{11}$$

$$CH_3CHO \quad CH_4{+}CO \quad \text{(acetaldehyde decomposition)} \tag{12}$$

$$2CH_3CHO \quad CH_3COCH_3{+}CO{+}H_2 \quad \text{(acetone formation)} \tag{13}$$

$$CO{+}3\,H_2 \quad CH_4{+}H_2O \quad \text{(methanation)} \tag{14}$$

$$C_2H_4 \quad \text{coke} \quad \text{(polymerization)} \tag{15}$$

$$CH_4{+}2\,H_2O \quad CO_2{+}4\,H_2 \quad \text{(methane steam reforming)1} \tag{16}$$

$$CH_4 \quad C{+}2\,H_2 \quad \text{(methane cracking)} \tag{17}$$

$$CO{+}H_2O \leftrightarrows CO_2{+}H_2 \quad \text{(water-gas shift)} \tag{18}$$

$$2\,CO \quad CO_2{+}C \quad \text{(Boudouard reaction)} \tag{19}$$

There are many side reactions that might take place during ethanol steam reforming, complicating the product distribution. To get the highest possible H_2 yield for industrial applications, it is essential to investigate the effects of temperature, reactants ratio, pressure, space velocity as well the catalytic parameters. A thermodynamic analysis was performed using the software HSC® Chemistry 5.1. All possible products, including solid carbon were included among the possible species that could exist in the equilibrium state. In the thermodynamic analysis, the following definitions are used.

$$H_2 \text{ Yield } \% = \frac{\text{moles of } H_2 \text{ produced}}{6 \times (\text{moles of ethanol fed})} \times 100$$

$$\text{Selectivity } \% = \frac{\text{mol of a certain product}}{\text{mol of total products}} \times 100$$

$$\text{EtOH Conv. } \% = \frac{\text{moles of ethanol converted}}{\text{moles of ethanol fed}} \times 100$$

The thermodynamic analysis in Fig.2 shows ethanol conversion, yield and selectivity of main products starting from a reactant composition similar to a bio-ethanol stream from biomass fermentation (ethanol-to-water ratio of 1:10). Ethanol conversion is not thermodynamically limited at any temperature. The methanation reaction, which is exothermic, is thermodynamically favored at lower temperatures (below 400 °C). At higher temperatures (above 500 °C) the reverse of this reaction, i.e., steam reforming of methane to CO_2 and H_2 becomes favorable. This would suggest that, if operated in a thermodynamically controlled regime, in order to minimize CH_4 concentration in the product stream, the reaction temperature should be kept as high as possible. However, as shown in Fig.2, once the temperature is increased above 550 °C, the reverse-water-gas shift reaction takes off, i.e., CO formation becomes significant and hydrogen yield decreases. At this ethanol-to-water ratio, there is no solid carbon at the equilibrium state.

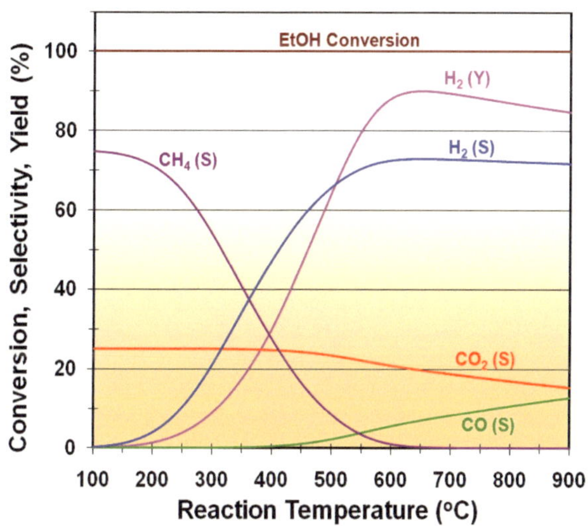

Fig. 2. Product distribution from ethanol steam reforming at thermodynamic equilibrium with EtOH:Water=1:10 (molar), C_{EtOH}=2.8%, and atmospheric pressure

Fig.3 shows the effect of ethanol-to-water molar ratio on H_2 yield. Lower molar ratios of ethanol-to-water can increase the hydrogen yield, since both water gas shift reaction and CH_4 reforming reactions would shift to the left with increased water concentration. In Fig.3, solid carbon selectivities for the lowest water concentrations are also included. At high ethanol-to-water ratios, solid carbon deposition becomes thermodynamically favorable, especially at lower temperatures.

The effect of dilution with an inert gas on the equilibrium H_2 yield is shown in Fig.4. The addition of inert gas increases the equilibrium hydrogen yield at low temperatures and has no effect at high temperatures. At low temperatures, the dominant reaction is the methanation/methane steam reforming. Diluting the system favors the methane steam reforming, and hence we see a difference at low temperatures. At high temperatures, the main reaction is the reverse water gas shift reaction, which is not affected by dilution, since there is no change in the number of moles with the extent of this reaction. Increased pressure has a negative influence on hydrogen yield at lower temperatures and no effect at higher temperatures (Fig.5).

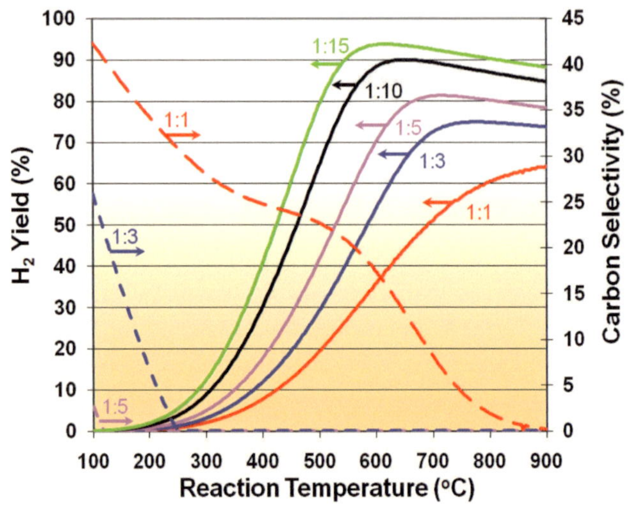

Fig. 3. Effect of EtOH-to-water molar ratio on equilibrium H_2 yield and C selectivity at (no dilution)

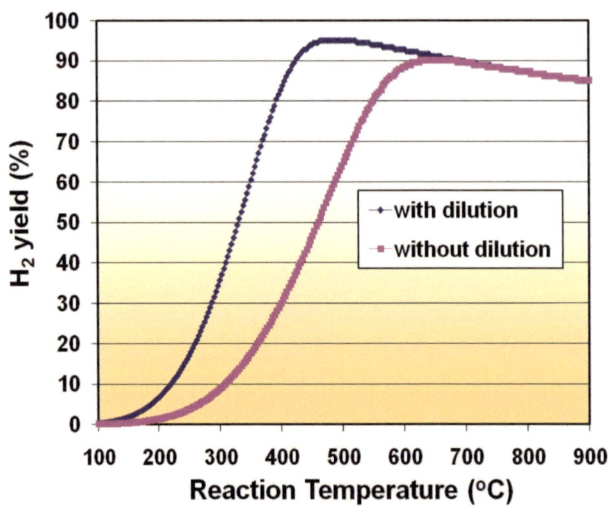

Fig. 4. Effect of dilution on equilibrium hydrogen yield (Dilution ratio used: Inert:EtOH:H_2O = 25:1:10)

Although it is important to be aware of the thermodynamic limitations, these analyses do not provide any information about the product distribution that would be obtained under kinetically controlled regimes. However, the study is still meaningful for guiding the choice of the desirable reaction parameters such that reaction is always controlled by kinetics under thermodynamically favorable conditions.

Due to its simplicity, flexibility, maturity, and high hydrogen yield, thermal bioethanol steam reforming has been extensively studied and a variety of technical improvements and researches directions have been proposed and implemented over the past several decades. The discussions of the following sections will focus on this technique.

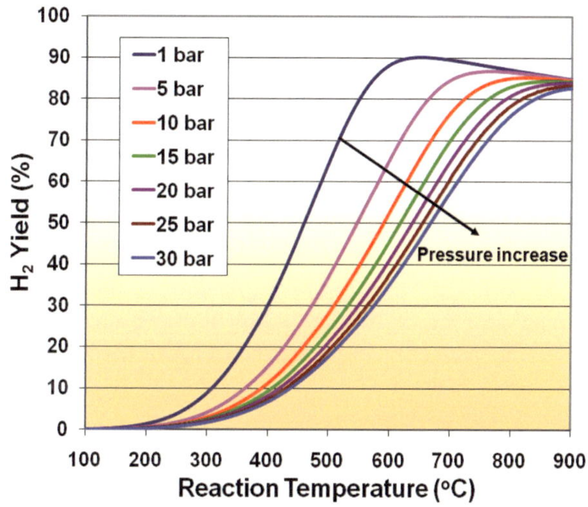

Fig. 5. Effect of pressure on equilibrium hydrogen yield (EtOH:Water=1:10 (molar ratio), no dilution)

3. Catalyst overview

In order to achieve equilibrated or even higher hydrogen yield especially at lower temperatures, catalytic bio-ethanol steam reforming (BESR) has been studied increasingly in recent years. More than three hundreds papers have been devoted to this field within the last two decades. The catalyst systems developed in these studies can be generally classified into two categories, i.e., supported noble and non-noble metal catalysts [32, 33]. However, based on the results reported in the literature, there is no commonly accepted optimal catalyst system which has excellent performance as well as low cost.

The noble metal catalysts such as Rh, Ru, Pd, Pt, Re, Au, and Ir [34-39] have been extensively investigated for BESR, which exhibit promising catalytic activity within a wide range of temperatures (350 °C~800 °C) and gas hourly space velocities (GHSV: 5,000~300,000 h^{-1}). The outstanding catalytic performance experienced by noble metal catalysts might be closely related to its remarkable capability in C-C bond cleavage [40]. Among the noble metal catalysts reported so far, it is evidenced [41-44] that Rh is generally more effective than other noble metals in terms of ethanol conversion and hydrogen production. Diagne et al. [45] claimed that up to 5.7 mol H_2 can be produced per mol ethanol (equal to 95 % H_2 yield) at 350 °C–450 °C over CeO_2–ZrO_2 supported Rh catalyst. However, although the metal loading is relatively low (1~5 wt.%) compared with its non-noble counterparts (10~15 wt.%), the extremely high unit price still limits its wide-scale industrial applications.

As a less expensive alternative way to address the cost issue, increasing attention has been focused on the development of non-noble metal catalysts. According to the publications documented so far, the efforts are mainly focused on the Cu, Ni, and Co based catalyst systems, especially supported Ni catalysts. As typical transition metals, the active outer layer electrons and associated valence states determine their identities as the candidates for BESR. Similar with noble metals, Ni also works well as it favors C-C rupture. Based on the observations reported by several authors [38, 43, 46], the non-precious metals are less reactive than noble metal supported samples. Specifically, Rh sites resulted to be 3.7 and 5.8 times more active than Co and Ni, respectively, supported by MgO under the reaction conditions used in [43]. For obtaining the same reactivity (H_2 yield > 95 %), much higher temperatures (650 °C) have to be employed [43, 47] over Ni catalysts. Furthermore, the non-noble metals are more prone to be deactivated due to sintering and coking compared with Rh. In order to achieve the comparable catalytic performance with noble metals, the formulation modifications of non-noble metal catalyst systems are worth studying for future commercialization. After summarizing the papers dedicated to investigation of various supports, ZnO and La_2O_3 seem more promising than MgO, Y_2O_3, and Al_2O_3 in terms of activity and stability [48, 49]. The basicity of sample surface has been evidenced crucial to improve its stability by adding La_2O_3 into the Al_2O_3 support aiming to neutralize the acidic sites present on the Al_2O_3 surface [50]. The addition of alkali metals (e.g., Na, K) to Ni/MgO has been observed to depress the deactivation occurrence by preventing Ni sintering [51]. It is worth noting that the recent interests on Ni catalysts seem to be transferred to CeO_2 and ZrO_2 supported samples, which could be ascribed to its well-known oxygen mobility, oxygen storage capability (OSC), and thermal stability [52-55], in turn improving coke-resistance. In addition, the synergetic effects become notable leading to better catalytic performance (activity, selectivity, and stability) when the second component (Cu) is incorporated into the Ni catalysts indicated by the work performed by Fierro et al., Marino et al., and Velu et al. [56-58]. They believe that the introduction of Cu might favor the dehydrogenation of ethanol to acetaldehyde, one of the important surface reaction intermediates during BESR. Compared with Ni based catalysts, cobalt samples have been less studied as catalysts for BESR. However, recent years have witnessed a significant increase in publications focusing on the development of Co-based catalysts, among which is the pioneering work by Haga et al. [59, 60]. Then Llorca et al. reported the promising results that 5.1 mol of H_2 can be produced per mol of reacted ethanol over Co/ZnO sample [61]. Although the reaction condition is slightly unrealistic for industrial applications, this study proved that cobalt is also efficient in C-C bond breakage [62]. Neither copper nor nickel alone supported on zinc oxide appears to have as good reactivity and stability as that of its Co counterpart for hydrogen production under the same reaction conditions [63, 64]. After thorough investigation of the product distribution at various temperatures, it was indicated that the copper sample prefers dehydrogenation of ethanol into acetaldehyde but the reforming reaction does not further progress significantly into H_2 and CO_x. On the other hand, the nickel sample favors the decomposition reaction of ethanol to CH_4 and CO_x, accounting for the lower H_2 yield at lower temperatures. Only at high temperatures can the methane production be lowered through steam-reforming. Moreover, Co catalysts have been applied in the Fischer-Tropsch to generate liquid hydrocarbons for more than 80 years. The knowledge accumulated during the study of Co based catalyst systems provides a good starting point. With these encouraging initial data, cobalt catalysts merit to be studied extensively as an alternative solution for reducing the cost from usage of noble metals.

4. Catalyst optimization strategies

In order to acquire competitive catalytic performance with noble metals, a series of optimization procedures need to be carried out over cobalt based catalysts. The significance of support was first explored by Haga et al. [59] indicating that Co/Al_2O_3 shows more promising activity than SiO_2, C, ZrO_2, and MgO. A relatively systematic investigation of the effect of supports was performed by Llorca and his coworkers [65]. Among the supports of CeO_2, Sm_2O_3, MgO, Al_2O_3, SiO_2, TiO_2, ZnO, La_2O_3, V_2O_5 reported in this study, ZnO was ranked the best.

Recently mixed metal oxides have been employed as the support to improve the behavior of single metal oxides by doping one or more additional components into the original support lattice. For instance, in the implementation of $Ce_{1-x}Zr_xO_2$, as the washcoat material in three-way catalysts, support combines the oxygen mobility of CeO_2 and thermal tolerance of ZrO_2 [66-69]. The introduction of Ca creates oxygen vacancies, which is beneficial for the enhancement of oxygen mobility [70, 71]. Besides, the perovskite-type oxides such as $LaAlO_3$, $SrTiO_3$, and $BaTiO_3$ have been used as the support for BESR catalysts due to their highly labile lattice oxygen [72, 73].

The cobalt precursor was proved by several authors [60, 74, 75] to have prominent effect on catalytic performance, which was proposed to be related to the cobalt dispersion. From the comparison between several precursor candidates, the one complexed with organic functional groups gave higher dispersion, which could be attributed to its isolation effect on the nearby Co atoms from agglomeration. It has been accepted that the active site during bio-ethanol steam reforming is related to the metal cobalt [76], that is, the higher the percentage of the cobalt that is available, the better the catalytic performance for BESR. Therefore, the improvement of cobalt dispersion will benefit the enhancement of corresponding catalytic activity.

It is expectable that cobalt loading has direct impact on the cobalt dispersion in the final catalyst. From the studies performed over Ni-based catalysts [53, 77], there exists an optimal loading, which can obtain the highest metal dispersion, through increasing the metal loading while avoiding metal sintering occurring at high loading due to the agglomeration of nearby metal atoms during thermal treatment. To the best of our knowledge, there is no systematic research of the effect of cobalt loading on its catalytic performance during BESR. Therefore, executing such a study can provide us better control of the catalyst optimization.

The impregnation medium is expected to have influence on the diffusion of cobalt precursor during impregnation and redistribution of cobalt atoms during the subsequent thermal treatment, which is shown by the experimental observations over Co/SiO_2 [78]. The smaller Co_3O_4 crystallite size obtained for samples using ethanol rather than water as impregnation solvent is attributed to the formation of ethoxyl groups on silica and/or Co_3O_4 surface during impregnation which hindered the sintering of Co_3O_4 by physically interfering during the thermal decomposition of nitrates. As a result, a higher percentage dispersion of cobalt metal was achieved from reduction of smaller crystallites of Co_3O_4. In addition, further sintering of cobalt metal during reduction might be hindered by ethoxyl groups as well. Since the cobalt dispersion is closely correlated to the activity during BESR as described above, this effect needs to be further investigated.

It was reported by Enache et al. [79] and Ruckenstein [80] in their studies of cobalt-based catalysts for Fischer-Tropsch reaction that the parameters used in the sample heat treatment before being charged for reaction play a significant role on the cobalt dispersion and in turn catalytic activity. Thus the synthesis parameters during calcination and reduction need to be explored to optimize the catalytic performance.

The promotion effect of alkali metal addition has been observed separately by Llorca et al., and Galetti et al. [63, 64, 81]. The hydrogen yield enhancement and carbon deposition inhibition showed the improvement of catalytic performance even when a small amount of Na and K (~0.7 wt.%) was introduced. As an inexpensive additive, this promising modification should be further explored.

Similar to Ni catalysts, promotion effect has also been evidenced over the samples with the formation of metallic alloy. According to the results published so far, the second active metal in addition to Co can be generally categorized as noble metals (e.g., Rh [82] and Ru [83-85]) and non-noble metals (e.g., Ni, Cu [63, 86], Fe, and Mn [87]). The integration of each metal specialized in different functions might be responsible for the synergetic interaction on the improvement of catalytic performance. The non-noble metal additives also merit further investigation.

Not only the modifications to the formulation of catalyst system, but also the preparation methods can impact the catalytic performance. Versatile synthesis strategies have been developed for obtaining catalysts with high performance during BESR. Incipient wetness impregnation (IWI) [88-91], wet impregnation [84, 92, 93], sol-gel (SG) [94, 95], and co-precipitation (CP) [63, 64, 86, 87] are the most commonly utilized methods, each of which has its own advantages and disadvantages. Impregnation is the most convenient method to be scaled up, for manufacturing. However, nonhomogeneous distribution of the metal precursor is the biggest issue associated with the impregnation method, leading to metal agglomeration, one of the reasons which contribute to catalyst deactivation. On the contrary, it is easier for SG and CP to achieve homogeneous dispersion of active metal. However, the synthesis procedure of SG and CP is more complicated compared with that of impregnation, leading to poor reproducibility between various batches. Also, since most of the active metal atoms are embedded in the matrix of support, resulting in less exposure of active metal on the sample surface, SG and CP prepared samples are more stable but less active than those prepared by impregnation. In addition, several novel preparation protocols such as hydrothermal [96], solvothermal [97], and microemulsion [98] have been developed to control the sample particle size and morphology which have been shown to be highly relevant to catalytic activity. On the other hand, most of the newly developed methods mentioned involve the employment of organic solvents, which could be harmful to the surroundings. Although all the preparation techniques documented up to now supply abundant resources to start with, the establishment of an appropriate method balancing low cost, easy operation, and environmental benignancy is important to be researched.

4.1 Cobalt based catalyst performance optimization

A series of catalyst optimization efforts have been carried out in the past several years aiming to enhance the catalytic performance during BESR. Studies on cobalt-based catalysts supported on γ-Al_2O_3, TiO_2, ZrO_2 supports have indicated that ethanol conversion correlates closely with metal dispersion and hence, the metallic Co sites. Among the supports studied, zirconia is shown to provide the highest metal dispersion and the highest H_2 yield. H_2 yields as high as 92% (5.5 mol of H_2 per mole of ethanol fed) are achieved over a 10% Co/ZrO_2 catalyst at 550 ^oC [69].

Investigation of the evolution of the Co–ZrO_2 catalysts through different stages of the synthesis process showed that catalyst precursors start out with Co existing primarily in a nitrate phase and transforming into a Co_3O_4 phase in the fully calcined state. The reduction proceeds in two distinct steps as in $Co_3O_4 \rightarrow CoO$ and $CoO \rightarrow Co$. There is an optimum in each of the synthesis parameters, which gives the highest metallic Co surface area. The maximum in metallic Co area is often determined by a series of competing processes, such as transformation from a nitrate to an oxide phase and onset of crystallinity versus reaction with the support at higher calcination temperatures, reduction to metallic state versus sintering at higher reduction temperatures. The maximum in metallic Co area was seen to coincide with the maxima in both ethanol adsorption capacity and H_2 yield in the BESR reaction, suggesting a strong correlation between metallic Co sites and BESR activity [99].

Although promising activity toward hydrogen production is observed over Co/ZrO_2, steady-state reaction experiments coupled with post-reaction characterization experiments showed significant deactivation of Co/ZrO_2 catalysts through deposition of carbon on the surface, mostly in the form of carbon fibers, the growth of which is catalyzed by the Co particles. The addition of ceria appears to improve the catalyst stability due to its high OSC and high oxygen mobility, allowing gasification/oxidation of deposited carbon as soon as it forms. Although Co sintering is also observed, especially over the ZrO_2-supported catalysts, it does not appear to be the main mode of deactivation. The high oxygen mobility of the catalyst not only suppresses carbon deposition and helps maintain the active surface area, but it also allows delivery of oxygen to close proximity of ethoxy species, promoting complete oxidation of carbon to CO_2, resulting in higher hydrogen yields. Overall, oxygen accessibility of the catalyst plays a significant role on catalytic performance during BESR [100].

the effect of impregnation medium on the activity of Co/CeO_2 catalysts was also systematically investigated under the environment of BESR. The significant catalytic performance improvement has been observed over ethanol impregnated Co-CeO_2 catalyst, especially at lower temperature (300-400 ^{o}C), compared with its counterpart with aqueous impregnation. This promotion effect is considered to be closely related to the cobalt dispersion amelioration through cobalt particle segregation under the facilitation of surface carbon oxygenated species derived from ethanol impregnation. Moreover, even better catalytic performance is achieved using ethylene glycol as impregnation medium in our recent study, which might be closely related with the achievement of even smaller cobalt particle size due to its superior ability in preventing cobalt agglomeration probably originating from the presence of organic surface species [101].

In order to further improve the oxygen mobility within the catalyst, the effect of Ca doping on CeO_2 support has been intensively studied. According to the observations obtained from the various characterization techniques employed, the introduction of calcium into the CeO_2 lattice structure leads to the unit cell expansion and creation of oxygen vacancies due to lower oxidation state of Ca (2+) compared to Ce (4+), which facilitates the improvement of oxygen mobility. As a result, the catalytic performance has been significantly enhanced when Ca is present, leading to larger amount of final product formations (H_2 and CO_2) from BESR reaction [102].

The influence of cobalt precursor on catalytic performance was also systematically investigated. Multiple cobalt precursors including inorganic salts and organometallic compounds were used to prepare Co/CeO_2 catalysts. The steady-state reaction experiments show much higher H_2 yields and fewer side products over the catalysts prepared using organometallic precursors. Among these, the catalyst prepared using cobalt acetyl acetonate has the highest H_2 yield, most favorable product distribution, and best stability. The superior performance is verified by the transient data. Characterization results point to an improved dispersion on the surface. It is possible that the organic ligands surrounding Co ions provide a spatial barrier effect, keeping the particles segregated and leading to better dispersion [103].

In the interest of figuring out the impact of catalyst preparation method on its performance during BESR, in addition to conventional Incipient Wetness Impregnation (IWI) method, solvothermal, hydrothermal, colloidal crystal templating, and reverse microemulsion methods have also been employed to prepare CeO_2 support and CeO_2 supported Co catalysts with various morphologies. All of the novel preparation techniques led to superior behavior in ethanol steam reforming reaction compared to IWI method. Among the catalysts studied, the one prepared with the reverse microemulsion technique showed the best performance, giving higher H_2 yields at much higher space velocities. The catalyst also showed good stability, with no sign of deactivation when it was kept on-line at 400 °C for 120 h. The superior performance is likely to be related to the improved cobalt dispersion, enhanced metal-support interaction and increased metal-support interphase facilitated by the reverse microemulsion technique. In addition, the hydrothermal method has also been employed to prepare the Co/CeO_2 catalyst. The CeO_2 particles with various shapes and size distribution have been successfully achieved in our laboratories by controlling the parameters during preparation process. The morphological effect on the catalytic performance will be evaluated in the future [104].

5. Reaction mechanism and kinetic studies

As can be seen in Section 2.7, the reaction network which would possibly occur during BESR is fairly complicated and heavily dependent on the catalyst system employed. In order to obtain maximum amount of hydrogen out of ethanol used, the side reactions should be effectively suppressed, leading to the minimization of byproducts such as methane, carbon monoxide, acetaldehyde, acetone, acetic acid and so on. For controlling the reaction proceeding along the desired pathway which will give us the highest hydrogen yield, it is critical to gain a comprehensive understanding of the reaction mechanisms involved, which will in turn guide the rational design of catalyst system. There are two approaches we can follow to achieve our final goal, that is, theoretical and experimental directions. The theoretical approach (reaction mechanism study through computational chemistry) is still at its initial stage referring to the papers published in this area and will be covered in detail in Section 6. However, the experimental route has been widely adopted to study the catalytic behaviors present during BESR.

As an interfacial phenomenon, any heterogeneous catalytic reaction takes place involving three basic steps: reactants adsorption, surface reaction, and products desorption. To be a gas-solid reaction, catalytic BESR must embroil gas composition variation and catalyst surface evolution. Therefore, in order to attain a complete view of the reaction, systematical investigation should be performed on both gas and solid phases. Gas chromatography (GC) and mass spectrometer (MS) are the two popular instruments used to monitor the gas phase composition and fourier transform infared spectroscopy (FTIR) can detect the surface species and their evolutions during BESR. In addition, using other characterization techniques including nuclear magnetic resonance (NMR) and laser Raman spectroscopy (LRS) can provide an alternative way to get better insight into the reaction mechanisms.

Based on the results reported in the literature, the dehydrogenation and dehydration reactions are the two pathways ethanol can go through first, the choice of which depends on the catalysts charged. If the catalyst has high acidity (e.g., Al_2O_3 and SiO_2 [105, 106] supported samples), dehydration reaction is favored, resulting in the formation of C_2H_4, a precursor of coking through polymerization. If the catalyst presents basic features (e.g., MgO and ZnO [107, 108] supported sample) instead, dehydrogenation reaction is preferred, leading to the production of acetaldehyde, an important reaction intermediate related to higher H_2 yield. Acetaldehyde can then be decomposed into CH_4 and CO [109] or undergoes steam reforming to generate CO and H_2 relying on the catalyst employed. These single carbon containing products (CH_4 and CO) can be further reformed to CO_2 and H_2 through methane steam reforming and water-gas shift reaction if sufficient water is supplied. Besides, two acetaldehyde molecules can react with each other to form acetone through aldol condensation reaction [35] or be oxidized to acetic acid [110]. Carbon can be formed at various stages from carbon-containing species via either cracking or Boudouard reaction [111].

Ethanol adsorption and subsequent surface reaction have been extensively studied over many different catalyst systems employing FTIR technique. Although the exact locations of the ethanol adsorption bands vary with catalysts tested, the identifications of surface species and its evolutions are well established. Ethanol can be adsorbed on the sample surface dissociatively and molecularly [112-114]. The ethoxide species is the result of ethanol dissociative adsorption. Then the surface acetate species is obtained from the oxidation of ethoxide by the lattice oxygen coming from the sample surface [115, 116]. The acetate species can then experience C-C breakage leading to the formation of single carbon fragments. Whether these fragments will be released directly from the surface or undergo further oxidation to carbonate species is closely linked to the sufficiency of oxygen stored in the sample. The adequate oxygen supplies benefit the formation of carbonate species. Finally CO_2 originates from the decomposition of carbonate species. However, compared with ethanol, water adsorption and its role in the subsequent surface reaction remain unclear for BESR. Therefore, the surface features need to be investigated during water adsorption and co-adsorption of water and ethanol.

^{13}C NMR technique has been applied into the study of ethanol adsorption behavior to track the evolution of carbon containing species over Cu/ZnO [117]. Different oxygenate species have been identified after integrating with the results obtained from MS. Unfortunately, just 1-C was labeled in the ethanol molecule, in order to get a comprehensive picture of the surface species and its evolution after ethanol adsorption, 2-C, even H and O labeled ethanol is also worth being considered. A similar approach is also applicable for water adsorption and co-adsorption of ethanol and water by choosing suitable isotopic labeled elements.

Compared with the kinetic studies focused on the steam reforming over single carbon containing reactants such as methanol [118-122] and methane [123-127] (MSR) which have been investigated intensively for tens of years, the kinetic investigations performed over ethanol steam reforming (ESR) reaction are still in their burgeoning stage, which might be due to the relatively complicated reaction networks involved originating from the increase of carbon atom. However, the knowledge accumulated during the systematic explorations of the kinetic mechanisms occurring during MSR provides a valuable starting point for ESR researchers to expand upon. In recent years, based on the observations obtained from both gas phase and sample surface, several kinetic models have been proposed to simulate the mechanistic behaviors of various catalyst systems [128-132], which will facilitate better understanding of the reaction mechanisms. If the estimated values are in good consistency with the reported experimental results, the assumed reaction pathways and rate-determining step (RDS) will uncover the actual reaction mechanisms to a certain level. Furthermore, the activation energy measured from this study provides the reference for molecular simulation. In addition, the outcomes from this kinetic analysis will benefit the reactor design which can promote mass and heat transfer during reaction.

Based on the TPD and DRIFTS results reported in [133], a possible reaction pathway for ethanol steam reforming over Co-based catalysts is proposed by our laboratories in Fig.6. In Scheme 1, the reactant molecules (EtOH and water) diffuse from gas phase to the surface of the catalyst. The ethanol molecules adsorb dissociatively on the Co sites, forming ethoxide species. Water, on the other hand, adsorbs on the support, forming hydroxyl groups. The first H abstracted from ethanol can either form OH groups with the surface oxygen species or combine with hydrogen from H_2O and form H_2 (Scheme 3). Ethoxide species move to the interface of metal and oxide support and be oxidized by an additional hydrogen abstraction forming aceteldehyde (Scheme 4). Acetaldehyde molecules may lead to the formation of acetone through an aldol-condensation type reaction and acetone molecules are observed only in the gas phase. Acetaldehyde species have a short surface residence time, converting readily to acetate species through further oxidation with surface oxygen or OH groups (Scheme 5). There are multiple routes for the acetate species once they are formed. In one of the routes, the metal may be involved in C-C bond cleavage leading to the formation of single carbon species (Scheme 7), leading to the formation of CH_4. The carbon-oxygen surface species may desorb or further oxidize to give carbonate species, especially on supports with high oxygen storage capacity (Scheme 8), which can desorb as CO_2 (Scheme 9). In a second route, especially, if oxygen accessibility is high, the CH_3 fragment will undergo oxidation through H subtraction and O addition (Scheme 10) to form formate, possibly through a formaldehyde intermediate (Scheme 11), and carbonate (Scheme 12). The catalyst surface is then regenerated through CO_2 desorption (Scheme 13) and ready for the next catalysis cycle regardless of the route followed.

If the surface is highly acidic, ethanol dehydration may dominate the reaction pathway and result in the formation of H_2O and C_2H_4 which is the major precursor to coke due to polymerization, as described in Scheme 2 and 6. If the oxygen mobility in the catalyst is not high enough, the acetate species may remain on the surface and lead to coke formation, as reported earlier [34, 134].

Briefly speaking, dissociative adsorption of ethanol and water leads to ethoxide species and hydroxyl groups, respectively. The active metal catalyzes the C-C bond cleavage and formation of single carbon species. BESR reaction could happen at the interface of the active metal and the oxide support, which could participate by providing oxygen from the lattice to facilitate the oxidation of carbon species. The resulting oxygen vacancies can be filled by the oxygen in the hydroxyl species formed from water adsorption. Therefore, it is necessary to have rapid oxygen delivery mechanism throughout the oxide support to prevent carbon deposition on the surface due to deficient oxidation of carbon species. High metal dispersion

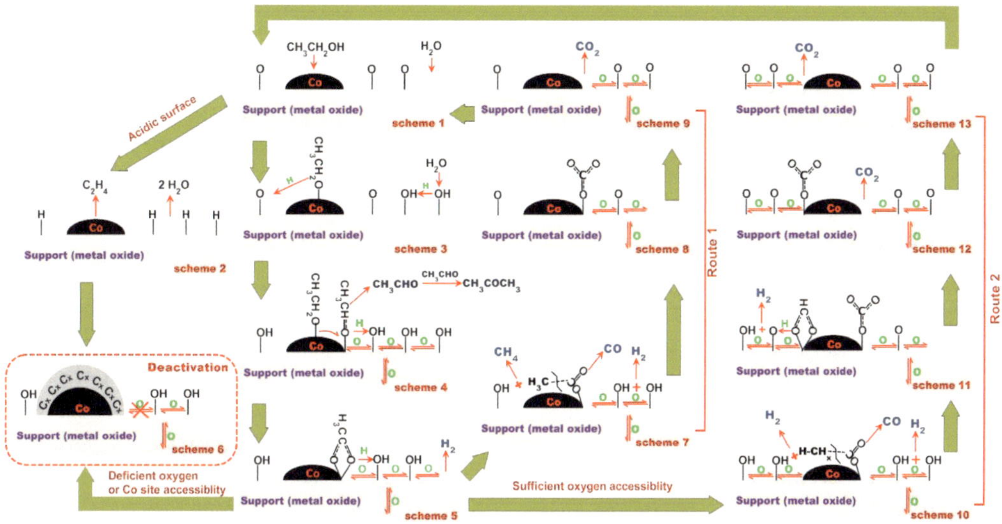

Fig. 6. Proposed Reaction Mechanism for Ethanol Steam Reforming over supported Co catalysts

will favor the ethanol adsorption and formation of more accessible metal/oxide interfaces as well as C-C cleavage. High oxygen storage capability and mobility will facilitate the oxygen delivery through the support and suppress coke deposition. The Co-based systems that incorporate oxides with high oxygen storage and oxygen mobility could deliver the required characteristics needed for active and stable BESR catalysts.

6. Computational approaches

Compared to significant amount of experimental efforts spent on catalytic BESR for surface reaction mechansim investigation, computational approach at molecular level still remains barely untouched in the past several decades probably due to its extreme complicacy and limited computation resources. However, recent years have witnessed the rapid development of computational technology, making the reaction simulation at catalyst surface technically feasible. For simplifying simulation work, many publications have purely focused on the ethanol or water alone adsorption and associated decomposition on single metal clusters [135-139].

Various methodologies have been developed to reasonably represent catalyst surface for obtaining more accurate simulation results. The slab geometry in contrast to cluster model is widely adopted to model the catalyst surface with certain thickness. In addition to the top atomic layer, several successive layers below are also included to simulate the bulk effect on the surface layer. The surface layer is thereafter allowed to be reconstructed in response to the constraint from bulk layers. Usually, a vacuum region with certain length is created right above the top layer of the slab model to prevent the interaction of adsorbed molecules with its periodic images [140]. The choice of supercell size comes from the compromise between computation accuracy and computation time span. "Nudged Elastic Band (NEB)" method [141, 142] is proven by many papers to be effective in transition state and associated energy barrier estimation and very useful in minimum energy pathway determination especially for complex chemical reactions. Most of recently published computational results are based on the self-consistent periodic density functional theory (DFT) calculation, which is more accurate than other commonly used computational methods such as *ab initio*, semi-empirical, and empirical methods.

According to the published papers, although there are some disagreements on the ethanol decomposition on model catalyst surface, the proposed pathways can still be generally classified into two routes. One is $CH_3CH_2OH \rightarrow CH_3CH_2O_{(a)} \rightarrow CH_2CH_2O_{(a)} \rightarrow CH_2CHO_{(a)} \rightarrow CH_2CO_{(a)} \rightarrow CH_{2(a)}+CO_{(a)}+4H_{(a)} \rightarrow CH_{4(g)}+CO_{(g)}+H_{2(g)}$. In this route, ethanol molecule first prefers to adsorb at atop sites and binds to the surface through the oxygen atom after O-H bond cleavage, followed by a six-membered ring of an oxametallacyclic compound formation through the elimination of the hydrogen atom attached to the β-carbon. This six-membered ring is usually located at the interface of active metal and support, creating a bridge between them. The ethanol decomposition process then continues with two consecutive eliminations of hydrogen atom attached to α-carbon. Scission of C-C bond then occurs under the facilitation of active metal, resulting in the formations of a series of adsorbates which subsequently desorb from substrate at elevated temperature to yield final gas products such as CH_4, CO, and H_2 [142-144]. The other suggested route follows the track of $CH_3CH_2OH \rightarrow CH_3CHOH_{(a)} \rightarrow CH_3CHO_{(a)} \rightarrow CH_3CO_{(a)} \rightarrow CH_2CO_{(a)} \rightarrow CHCO_{(a)} \rightarrow CH_{(a)}+CO_{(a)} \rightarrow CH_{4(g)}+CO_{(g)}+H_{2(g)}+C_{(s)}$ [145].

Unlike ethanol decomposition, water dissociation completes only in two steps (i.e., $H_2O \rightarrow H_{(a)}+OH_{(a)} \rightarrow 2H_{(a)}+O_{(a)}$), which is obviously due to its rather simple formulation. Compared to the second O-H bond breakage, the first one can take place with much lower activation energy [146]. Therefore, it can be easily predicted that hydroxyl group will have much higher chance to participate in BESR for ethanol oxidation than O* after water complete dissociation.

After a careful literature review, it is worth noting that the role of catalyst support and co-adsorption of ethanol and water are barely considered, which is probably attributed to its awful computational complicacy. In order to give a clear picture of what is really happening on catalyst surface during BESR and provide a theoretical support to our experimental observations and proposed reaction mechanism, we launched a computational task in collaboration with the Chemistry Department at Ohio State University. We employed plane-wave periodic DFT method implemented in the Vienna *ab initio* simulation program (VASP) to investigate the ethanol steam reforming reactions [147-149]. The projector augmented wave (PAW) method [150, 151], combined with a plane-wave basis set, was utilized to describe the core and valence electrons. The generalized gradient approximation (GGA) [152] of Perdew and Wang (PW91) [153] was applied for the exchange-correlation functional. The convergence of the plane-wave expansion was obtained with moderate truncation energy of 500 eV, while the electronic relaxation was converged to a tolerance of 1×10^{-4} eV. The Monkhorst–Pack grid [154] served in the generation of the k-points, and a ($4 \times 4 \times 1$) k-point grid was used for Brillouin zone sampling for surface calculations. Spin polarization was applied in all calculations.

The relaxed bulk structure of CeO_2 with a lattice parameter of 5.46 Å was used to construct the slab model. The CeO_2 (111) and Co/CeO_2 (111) surfaces were modeled as 2×1 super cells. A three molecular CeO_2 thick slab model was constructed, thus nine atomic layers in total. The super cell has dimensions: a = 7.72 Å, b = 6.69 Å, and c = 23.88 Å, and a 16 Å thick vacuum region is included to ensure that there is no interaction between the surface adsorbates of one layer and the next slab. To optimize the surface structure, the top three atomic layers of the slab with the adsorbates were allowed to relax. The bottom six atomic layers were fixed at the bulk positions of ceria. The NEB method [155-157] was employed to locate the transition states of various reactions over the catalyst surface. After numerical differentiation, each transition state was confirmed to have a single imaginary vibrational frequency.

Ethanol decomposition via steam reforming reaction was computationally studied on the $CeO_2(111)$ and $Co/CeO_2(111)$ surfaces. From our results, the most likely reaction pathway is described below. The decomposition of ethanol starts with the breaking of the O–H bond on the catalyst surface. The produced ethoxide unit prefers to be adsorbed on the catalyst surface by the $O_e \cdots Co$ interaction. With the assistance of a surface-bound hydroxyl moiety, derived from water dissociation, the C_α–H bond breaking of the ethoxide unit could proceed to yield the thermodynamically stable product (adsorbed acetaldehyde and hydrogen atom). The surface-bound hydroxyl group could act as a better hydrogen acceptor to assist the C_α–H bond-breaking reaction as compared to the surface oxygen atom of ceria. In the subsequent step, the surface-bound hydroxyl addition to acetaldehyde produces the hydroxyl adduct, $CH_3CH(O)(OH)$, as an intermediate. This $CH_3CH(O)(OH)$ intermediate further undergoes the loss of H from the C_α position to generate acetic acid. Acetic acid can then lose the acidic hydrogen from the hydroxyl unit, yielding an adsorbed acetate and hydrogen. The acetate could be further converted to the $CH_2(OH)COO$ intermediate via H-atom abstraction and subsequent surface-bound hydroxyl addition reactions. As suggested by the calculations, the C_α–C_β bond rupture from the chemisorbed $CH_2(O)COO$ intermediate generates formaldehyde and CO_2. Similar to acetaldehyde, the generated formaldehyde could react with a surface-bound hydroxyl group to produce the HCH(O)(OH) adduct that subsequently undergoes a H-atom abstraction reaction to yield formic acid. Then, formic acid loses the acidic hydrogen of the hydroxyl unit to generate surface-bound formate. Finally, formate could be converted to CO_2. Throughout the favorable reaction pathway from ethanol to CO_2, one of the most energetically costly steps on the potential energy surface is the C_β–H bond-breaking step of acetate for ethanol decomposition with the participation of surface-bound hydroxyl groups on the $Co/CeO_2(111)$ surface.

Our modeling indicates that surface-bound hydroxyl groups, which is formed from water dissociation, plays two critical roles in the ethanol steam reforming reaction. The first is to assist the hydrogen-abstraction reactions from carbon atoms. The second is their involvement in addition reactions to form the C=O or C=C double bond intermediates. Thus, a catalyst on which water could more effectively dissociate to form surface-bound hydroxyl and hydrogen might be a potentially better catalyst for steam reforming reactions.

On the $Co/CeO_2(111)$ surface, our computational work elucidates the formation of acetaldehyde and acetate intermediates and is consistent with extant experimental observations [133]. The present computational studies do not account for the generation of acetone, carbon monoxide, and methane, which are byproducts observed in experimental studies. A model that includes larger Co particles with some surface-bound hydroxyl groups would be more realistic and may account for the formation of other byproducts.

7. Economic considerations

Although recent years have witnessed an increasing number of studies in the literature on BESR reaction, the commercialization of a BESR process for hydrogen production still faces many obstacles before it can become a reality. The major obstacle is the cost associated with the process. While the cost of the catalyst, which is usually precious-metal based, can be an inhibitive factor, a detailed analysis of the economics involved in the process and an understanding of the contribution of many cost factors are still lacking.

An economic analysis model based on the cost structures in the United States was thereafter developed by our laboratories based on a process for hydrogen production from bio-ethanol steam reforming. The process includes upstream feedstock considerations as well as downstream hydrogen purification strategies and is analyzed for two different capacity levels, namely a central production scheme (150,000 kg H_2/day) and a distributed (forecourt) production scheme (1,500 kg H_2/day). The analysis was based on several assumptions and input parameters provided by the US Department of Energy and involved sensitivity analyses of several input parameters and their effects on the hydrogen selling price.

The detailed methodologies for performing economic analysis and associated results and discussions can be found in our recent publication [158]. Here we just give a brief summary of what we have obtained from this study. The hydrogen selling price is determined to be \$2.69/kg H_2 at central hydrogen production scale. According to cost breakdown analysis, ethanol feedstock contributes almost 70% of the total cost. Nevertheless, this technique is still economically competitive with other commonly used hydrogen generation technologies at same production scale such as methane steam reforming (\$1.5/kg H_2), and biomass gasification (\$1.77/kg H_2). When the production scale is downsized to forecourt level, the hydrogen selling price is significantly increased up to \$4.27/kg H_2, which is mainly attributed to the significant increase of capital cost contribution. A series of sensitivity analyses have been performed in order to determine the most significant factor influencing the final hydrogen selling price. From the analyses, hydrogen yield has a major effect on the estimated selling price through variation on ethanol feedstock cost contribution, which is reasonable since higher yield would require less feedstock to produce the same amount of hydrogen. Feed dilution is another important impact on hydrogen selling price, particularly at higher dilution percentage. The exponential escalation of hydrogen selling price is clearly observed when the dilution percentage is higher than 50%. Higher dilution percentage means that larger amount of gas should be processed to get the same amount of hydrogen. The effect of molar ratio of ethanol to water variation on hydrogen selling price has also been evaluated. As expected, hydrogen selling price is increased along with increasing molar ratio of water to ethanol, because larger amount of water is required to be evaporated to get the same amount of hydrogen, resulting in the capital and operation cost increase. However, another factor that is not reflected in this analysis is the fact that excess water (i.e., larger water-to-ethanol ratios) would inhibit coking on the surface and extend the active catalyst life time. So, choosing a higher water input may have additional advantages not captured by this analysis. Finally, the effect of catalyst cost and associated performance on hydrogen selling price has also been intensively explored. The estimations indicate the significance of using transition metal based catalyst for hydrogen production from BESR. If noble metal based catalyst is used instead, the hydrogen selling price will jump up to \$22.34/kg H_2 from \$4.27/kg H_2 where transition metal (e.g., Co) based catalyst is employed assuming that their catalytic performance is comparable. In order to get the same hydrogen selling price, the noble metal based catalyst has to either be operated under gas hourly space velocity 100 times higher or has lifetime 100 times longer than those of transition metal based catalyst, which is almost impossible from a realistic viewpoint.

8. Future development directions

The technical advantage of ethanol steam reforming over direct ethanol combustion for power generation is the improvement of thermal efficiency through hydrogen production exclusively used for fuel cell. In addition to stationary electricity generation, fuel cell is also designed for powering portable devices such as automobile. It is unsafe to travel around with compressed hydrogen tank on board. Therefore, there is a necessity for on-board steam reformer development where liquid ethanol rather than compressed hydrogen gas is fed into the storage tank. In order to get better mileage per gallon ethanol fed, the very important requirement of on-board steam reformer development is its light weight, which generates great demands on size reduction of on-board reformer. To fulfill the miniaturization and compactness requirements, various types of micro-structured reactors have been developed in recent years, which is typically composed of stacks of channeled blocks. Each micro-channel coated with active catalyst acts as the steam reformer for hydrogen production. Partial ethanol is combusted in the other side of the channel to supply heat required for reforming. Such design provides many technical advantages including rapid mass and heat transport due to large surface area to volume ratios, lower pressure drop, good structural and thermal stability, and precise control of reaction conditions leading to higher hydrogen yield [159, 160]. The main challenges faced by this technique before it becomes final commercialization are system integration, reactor fabrication process, and catalyst regeneration or replacement.

Combinatorial method originally developed for drug discovery has been introduced into the catalyst discovery field in the last decade to accelerate the catalyst screening process. By using this high-throughput approach, large and diverse libraries of inorganic materials can be prepared, processed, and tested simultaneously under the same reaction conditions for quickly obtaining potential candidates with desirable catalytic performance, which is beneficial for significant reduction of time and money spent on catalyst development [161, 162]. However, the relatively complicated algorithms for testing matrix determination, expensive testing instrument, and representability of the screening results should be better handled before it can be widely accepted as a standard catalyst development strategy.

The influence of external field (e.g., electric and magnetic field) on catalytic performance during BESR could be another interesting area to study. Because any chemical reaction involves electron transfer and rearrangement facilitated by the addition of catalyst, the application of external field which can exert impact on electron movement is expected to have influence on catalytic reactivity. Such effect has been recently evidenced by L. Yuan, et al. that hydrogen yield and selectivity were significantly enhanced when an AC current passed through Ni/Al_2O_3 catalyst [163].

According to LeChatelier's Principle, referring to Reaction (1), continuous removal CO_2 from product stream can shift the reaction equilibrium toward products side, leading to the improvement of hydrogen production. Based on literature review, there are mainly two methods for CO_2 in-situ removal: addition of CO_2 sorbent and CO_2 selective membrane. The CO_2 sorbent used for this purpose has to be regenerated at temperature higher than reaction temperature for reuse. For doing so, the high temperature CO_2 sorbent has to be circulated between reactor and regenerator [164]. The CO_2 sorbent is usually regenerated under the hot air environment and has good resistance to high temperature and attrition. According to literature reporting, CaO and lithium silicate are among the most commonly used CO_2 sorbents for hydrogen production. For CO_2 selective membrane, CO_2 is either rejected by the membrane and stays in the retentate side, or diffuses through the membrane and swept out as permeate. In order to in-situ remove CO_2 or perform hydrogen purification within the reformer, various types of membrane reactors have been developed in recent years to obtain hydrogen rich gas stream. Moreover, catalytic membrane reactor has also been invented to perform water-gas shift (WGS) and separation simultaneously through applying certain catalyst onto the membrane surface, among which Pd-impregnated membrane is the most reported one for getting purfied hydrogen product [165, 166]. Nevertheless, many technical problems including cost reduction, selectivity and permeation efficiency improvement, and rigidity enhancement have to be solved before it becomes economically attractive.

The high cost of ethanol feedstock for steam reforming mainly comes from the downstream distillation and purification steps of the crude ethanol obtained from fermentation. If the crude ethanol can be directly used as the feedstock for hydrogen production from BESR, the large amount of energy wasted during distillation for water and other impurities removal can be eliminated, leading to the significant cost reduction of ethanol feedstock and in turn hydrogen produced from BESR. In addition, other oxygenated hydrocarbons contained in the fermentation broth can also be steam reformed to generate extra 7% hydrogen if crude ethanol is employed compared to steam reforming of pure ethanol. Although this approach sounds promising for final commercialization of BESR technique, the challenge still remains at the catalyst's tolerance to the impurities present in the crude ethanol solution. According to related publications, several researchers have conducted such study to evaluate the impact of impurities on catalytic performance toward hydrogen production. A. Akande and his coworkers investigated the influence of crude ethanol simulated through adding small amount of lactic acid, glycerol, and maltose to ethanol aqueous solution on the catalytic performance of Ni/Al_2O_3 [128, 167]. Initial catalyst deactivation was observed followed by stable run within 12 hours test. Similar study has also been performed by our group over Co/CeO_2. ~90% hydrogen yield is achieved and well maintained within 100 hours run. A more systematic research has been recently implemented by A. Valant, et al. over $Rh/MgAl_2O_4$ [168]. More oxygenated hydrocarbons including esters, aldehydes, amine, acetic acid, methanol, and linear or branched alchols have been tested for its influences on catalytic performance of BESR. Catalyst deactivation is observed for certain impurity additions. Through catalyst modification, much better stability has been achieved using Rh-$Ni/Y\text{-}Al_2O_3$.

Although high pressure operation will result in inhibition of hydrogen production, as predicted thermodynamically referring to Section 2.7, it is still worth investigating, because high pressure operation will significantly lower down the hydrogen compression cost for storage and transporation. In order to compensate the hydrogen production loss, hydrogen selective membrane reactor has been recently proposed in combination with high pressure operation by Argonne National Laboratory [169]. By doing so, the formed hydrogen can be continuously removed leading to the thermodynamic equilibrium shift toward hydrogen production.

9. Acknowledgment

We gratefully acknowledge funding from the U.S. Department of Energy through grant DE-FG36-05GO15033. The Ohio Supercomputer Center (OSC) is also acknowledged for generous computational support of this research.

10. References

[1] A National Vision of America's Transition to A Hydrogen Economy - To 2030 and Beyond; U.S. Department of Energy (DOE): Washington, D.C, 2002

[2] Idriss, H. (2004). Ethanol Reactions over the Surfaces of Noble Metal/Cerium Oxide Catalysts. *Platinum Metals Review*, Vol.48, No.3, (July 2004), pp. 105-115, ISSN 0032-1400

[3] Das, D., Veziroğlu, T.N. (2001). Hydrogen Production by Biological Processes: A Survey of Literature. *International Journal of Hydrogen Energy*, Vol.26, No.1, (January 2001), pp. 13-28, ISSN 0360-3199

[4] Nath, K., Das, D. (2004). Improvement of Fermentative Hydrogen Production: Various Approaches. *Applied Microbiology and Biotechnology*, Vol.65, No.5, (July 2004), pp. 520-529, ISSN 0175-7598

[5] Hallenbeck, P.C., Benemann, J.R. (2002). Biological Hydrogen Production; Fundamentals and limiting processes. *International Journal of Hydrogen Energy*, Vol.27, No.11-12, (November-December 2002), pp. 1185-1193, ISSN 0360-3199

[6] Hwang, M.H., Jang, N.J., Hyun, S.H., Kim, I.S. (2004). Anaerobic Bio-Hydrogen Production from Ethanol Fermentation: the Role of pH. *Journal of Biotechnology*, Vol.111, No.3, (August 2004), pp. 297-309, ISSN 0168-1656

[7] Maness, P.C., Weaver, P.F. (1999). Biological H_2 from Fuel Gases and from H_2O. *Proceedings of the 1999 US DOE Hydrogen Program Review*

[8] Nada, A.A., Barakat, M.H., Hamed, H.A., Mohamed, N.R., Veziroglu, T.N. (2005). Studies on the Photocatalytic Hydrogen Production Using Suspended Modified TiO_2 Photocatalysts. *International Journal of Hydrogen Energy*, Vol.30, No.7, (July 2005), pp. 687-691, ISSN 0360-3199

[9] Vorontsov, A.V., Dubovitskaya, V.P. (2004). Selectivity of Photocatalytic Oxidation of Gaseous Ethanol over Pure and Modified TiO_2. *Journal of Catalysis*, Vol.221, No.1, (January 2004), pp.102-109, ISSN 0021-9517

[10] Yu, Z., Chuang, S. (2007). In-situ IR Study of Adsorbed Species and Photogenerated Electrons During Photocatalytic Oxidation of Ethanol on TiO_2. *Journal of Catalysis*, Vol.246, No.1, (February 2007), pp.118-126, ISSN 0021-9517

[11] Kasata, T., Kawai, T. (1981). Heterogeneous Photocatalytic Production of Hydrogen and Methane from Ethanol and Water. *Chemical Physics Letters*, Vol.80, No.2, (June 1981), pp. 341-344, ISSN 0009-2614

[12] Bamwenda, G.R., Tsubota, S., Nakamura, T., Haruta, M. (1995). Photoassisted Hydrogen Production from A Water-Ethanol Solution: A Comparison of Activities of Au-TiO_2 and Pt-TiO_2. *Journal of Photochemistry and Photobiology A: Chemistry*, Vol.89, No.2, (July 1995), pp. 177-189, ISSN 1010-6030

[13] Klosek, S., Raftery, D. (2001). Visible Light Driven V-Doped TiO_2 Photocatalyst and Its Photooxidation of Ethanol. *Journal of Physical Chemistry B*, Vol.105, No.14, (March 2001), pp. 2815-2819, ISSN 1520-6106

[14] Yang, Y.Z., Chang, C.H., Idriss, H. (2006). Photo-Catalytic Production of Hydrogen from Ethanol over M/TiO_2 Catalysts (M=Pd, Pt or Rh). *Applied Catalysis B: Environmental*, Vol.67, No.3-4, (October 2006), pp. 217-222, ISSN 0926-3373

[15] Strataki, N., Antoniadou, M., Dracopoulos, V., Lianos, P. (2010). Visible-Light Photocatalytic Hydrogen Production from Ethanol-Water Mixtures Using A Pt-CdS-TiO_2 Photocatalyst. *Catalysis Today*, Vol.151, No.1-2, (April 2010), pp. 53-57, ISSN 0920-5861

[16] Wang, Y., Zhang, Z., Zhu, Y., Li, Z., Vajtai, R., Ci, L., Ajayan, P.M. (2008). Nanostructured VO_2 Photocatalysts for Hydrogen Production. *ACS Nano*, Vol.2, No.7, (July 2008), pp. 1492-1496, ISSN 1936-0851

[17] Baeck, S., Choi, K., Jaramillo, T., Stucky, G., McFarland, E. (2003). Enhancement of Photocatalytic and Electrochromic Properties of Electrochemically Fabricated Mesoporous WO_3 Thin Films. *Advanced Materials*, Vol. 15, No.15, (August 2003), pp. 1269-1273, ISSN 1521-4095

[18] Fu, X., Leung, D., Wang, X., Xue, W., Fu, X. (2011). Photocatalytic Reforming of Ethanol to H_2 and CH_4 over $ZnSn(OH)_6$ nanotubes. *International Journal of Hydrogen Energy*, Vol.36, No.2, (January 2011), pp. 1524-1530, ISSN 0360-3199

[19] Mizukoshi, Y., Makise, Y., Shuto, T., Hu, J., Tominaga, A., Shironita, S., Tanabe, S. (2007). Immobilization of Noble Metal Nanoparticles on the Surface of TiO_2 by the Sonochemical Method: Photocatalytic Production of Hydrogen from An Aqueous Solution of Ethanol. *Ultrasonics Sonochemistry*, Vol.14, No.3, (March 2007), pp. 387-392, ISSN 1350-4177

[20] Davda, R.R., Shabaker, J.W., Huber, G.W., Cortright, R.D., Dumesic, J.A. (2005). A Review of Catalytic Issues and Process Conditions for Renewable Hydrogen and Alkanes by Aqueous-Phase Reforming of oxygenated Hydrocarbons over Supported Metal Catalysts. *Applied Catalysis B: Environmental*, Vol.56, No.1-2, (March 2005), pp. 171-186, ISSN 0926-3373

[21] Tokarev, A.V., Kirilin, A.V., Murzina, E.V., Eränen, K., Kustov, L.M., Murzin, D.Y., Mikkola, J.P. (2010). The Role of Bio-Ethanol in Aqueous Phase Reforming to Sustainable Hydrogen. *International Journal of Hydrogen Energy*, Vol.35, No.22, (November 2010), pp. 12642-12649, ISSN 0360-3199

[22] Wang, W., Wang, Y. (2009). Dry Reforming of Ethanol for Hydrogen Production: Thermodynamic Investigation. *International Journal of Hydrogen Energy*, Vol.34, No.13, (July 2009), pp. 5382-5389, ISSN 0360-3199

[23] Hu, X., Lu, G. (2009). Syngas Production by CO_2 Reforming of Ethanol over Ni/Al_2O_3 Catalyst. *Catalysis Communications*, Vol.10, No.13, (July 2009), pp. 1633-1637, ISSN 1566-7367

[24] Silva, A., Souza, K., Jacobs, G., Graham, U., Davis, B., Mattos, L., Noronha, F. (2011). Steam and CO_2 Reforming of Ethanol over Rh/CeO_2 Catalyst. *Applied Catalysis B: Environmental*, Vol.102, No.1-2, (February 2011), pp. 94-109, ISSN 0926-3373

[25] Jankhah, S., Abatzoglou, N., Gitzhofer, F. (2008). Thermal and Catalytic Dry Reforming and Cracking of Ethanol for Hydrogen and Carbon Nanofilaments' Production. *International Journal of Hydrogen Energy*, Vol.33, No.18, (September 2008), pp. 4769-4779, ISSN 0360-3199

[26] Oliveira-Vigier, K.D., Abatzoglou, N., Gitzhofer, F. (2005). Dry-Reforming of Ethanol in the Presence of A 316 Stainless Steel Catalyst. *The Canadian Journal of Chemical Engineering*, Vol.83, No.6, (December 2005), pp. 978-984, ISSN 1939-019X

[27] Petitpas, G., Rollier, J.D., Darmon, A., Gonzalez-Aguilar, J., Metkemeijer, R., Fulcheri, L. (2007). A Comparative Study of Non-Thermal Plasma Assisted Reforming Technologies. *International Journal of Hydrogen Energy*, Vol.32, No.14, (September 2007), pp. 2848-2867, ISSN 0360-3199

[28] Aubry, O., Met, C., Khacef, A., Cormier, J.M. (2005). On the Use of A Non-Thermal Plasma Reactor for Ethanol Steam Reforming. *Chemical Engineering Journal*, Vol.106, No.3., (February 2005), pp. 241-247, ISSN 1385-8947

[29] Wang, W., Wang, Z., Ding, Y., Xi, J., Lu, G. (2002). Partial Oxidation of Ethanol to Hydrogen over Ni-Fe Catalysts. *Catalysis Letters*, Vol.81, No.1-2, (January 2002), pp. 63-68, ISSN 1011-372X

[30] Mattos, L.V., Noronha, F.B. (2005). Hydrogen production for fuel cell applications by ethanol partial oxidation on Pt/CeO_2 catalysts: the effect of the reaction conditions and reaction mechanism. *Journal of Catalysis*, Vol.233, No.2, (July 2005), pp.453-463, ISSN 0021-9517

[31] Hsu, S., Bi, J., Wang, W., Yeh, C., Wang, C. (2008). Low Temperature Partial Oxidation of Ethanol over Supported Platinum Catalysts for Hydrogen Production. *International Journal of Hydrogen Energy*, Vol.33, No.2, (January 2008), pp. 693-699, ISSN 0360-3199

[32] Vaidya, P.D., Rodrigues, A.E. (2006). Insight into Steam Reforming of Ethanol to Produce Hydrogen for Fuel Cells. *Chemical Engineering Journal*, Vol.117, No.1., (March 2006), pp. 39-49, ISSN 1385-8947

[33] Ni, M., Leung, D., Leung, M. (2007). A Review on Reforming Bio-Ethanol for Hydrogen Production. *International Journal of Hydrogen Energy*, Vol.32, No.15, (October 2007), pp. 3238-3247, ISSN 0360-3199

[34] Erdőhelyi, A., Raskó, J., Kecskés, T., Tóth, M., Dőmők, M., Baán, K. (2006). Hydrogen Formation in Ethanol Steam Reforming on Supported Noble Metal Catalysts. *Catalysis Today*, Vol.116, No.3, (August 2006), pp. 367-376, ISSN 0920-5861

[35] Liguras, D.K., Kondarides, D.I., Verykios, X.E. (2003). Production of Hydrogen for Fuel Cells by Steam Reforming of Ethanol over Supported Noble Metal Catalysts. *Applied Catalysis B: Environmental*, Vol.43, No.4, (July 2003), pp. 345-354, ISSN 0926-3373

[36] Koh, A., Leong, W.K., Chen, L., Ang, T.P., Lin, J. (2008). Highly Efficient Ruthenium and Ruthenium-Platinum Cluster-Derived Nanocatalysts for Hydrogen Production via Ethanol Steam Reforming. *Catalysis Communications*, Vol.9, No.1, (January 2008), pp. 170-175, ISSN 1566-7367

[37] Bi, J., Hong, Y., Lee, C., Yeh, C., Wa, C. (2007). Novel Zirconia-Supported Catalysts for Low-Temperature Oxidative Steam Reforming of Ethanol. *Catalysis Today*, Vol.129, No.3-4, (December 2007), pp. 322-329, ISSN 0920-5861

[38] Breen, J.P., Burch, R., Coleman, H.M. (2002). Metal-Catalyzed Steam Reforming of Ethanol in the Production of Hydrogen for Fuel Cell Applications. *Applied Catalysis B: Environmental*, Vol.39, No.1, (November 2002), pp. 65-74, ISSN 0926-3373

[39] Sheng, P.Y., Idriss, H. (2004). Ethanol Reactions over Au-Rh/CeO_2 Catalysts. Total Decomposition and H_2 Formation. *Journal of Vacuum Science and Technology A*, Vol.22, No.4, (July 2004), pp. 1652-1658, ISSN 0734-2101

[40] Kuga, J., Subramani, V., Song, C., Engelhard, M.H., Chin, Y. (2006). Effects of Nanocrystalline CeO_2 Supports on the Properties and Performance of Ni-Rh Bimetallic Catalyst for Oxidative Steam Reforming of Ethanol. *Journal of Catalysis*, Vol.238, No.2, (March 2006), pp.430-440, ISSN 0021-9517

[41] Cavallaro, S., Chiodo, V., Freni, S., Mondello, N., Frusteri, F. (2003). Performance of Rh/Al_2O_3 Catalyst in the Steam Reforming of Ethanol: H_2 Production for MCFC. *Applied Catalysis A: General*, Vol.249, No.1, (August 2003), pp. 119-128, ISSN 0926-860X

[42] Rogatis, L.D., Montini, T., Casula, M.F., Fornasiero, P. (2008). Design of $Rh@Ce_{0.2}Zr_{0.8}O_2$-Al_2O_3 Nanocomposite for Ethanol Steam Reforming. *Journal of Alloys and Compounds*, Vol.451, No.1-2, (February 2008), pp. 516-520, ISSN 0925-8388

[43] Frusteri, F., Freni, S., Spadaro, L., Chiodo, V., Bonura, G., Donato, S. (2004). H_2 Production for MC Fuel Cell by Steam Reforming of Ethanol over MgO Supported Pd, Rh, Ni, and Co Catalysts. *Catalysis Communications*, Vol.5, No.10, (October 2004), pp. 611-615, ISSN 1566-7367

[44] Montini, T., Rogatis, L.D., Gombac, V., Fornasiero, P. (2007). Rh(1%)@$Ce_xZr_{1-x}O_2$-Al_2O_3 Nanocomposites : Active and Stable Catalysts for Ethanol Steam Reforming. *Applied Catalysis B: Environmental*, Vol.71, No.3-4, (February 2007), pp. 125-134, ISSN 0926-3373

[45] Diagne, C., Idriss, H., Kiennemann, A. (2002). Hydrogen Production by Ethanol Reforming over Rh/CeO_2-ZrO_2 Catalysts. *Catalysis Communications*, Vol.3, No.12, (December 2002), pp. 565-571, ISSN 1566-7367

[46] Auprêtre, F., Descorme, C., Duprez, D. (2002). Bio-Ethanol Catalytic Steam Reforming over Supported Metal Catalysts. *Catalysis Communications*, Vol.3, No.6, (June 2002), pp. 263-267, ISSN 1566-7367

[47] Yang, Y., Ma, J., Wu, F. (2006). Production of Hydrogen by Steam Reforming of Ethanol over a Ni/ZnO Catalyst. *International Journal of Hydrogen Energy*, Vol.31, No.7, (June 2006), pp. 877-882, ISSN 0360-3199

[48] Sun, J., Qiu, X., Wu, F., Zhu, W. (2005). H2 from Steam Reforming of Ethanol at Low Temperature over Ni/Y2O3, Ni/La2O3, and Ni/Al2O3 Catalysts for Fuel Cell Application. *International Journal of Hydrogen Energy*, Vol.30, No.4, (March 2005), pp. 437-445, ISSN 0360-3199

[49] Fatsikostas, A.N., Kondarides, D.I., Verykios, X.E. (2002). Production of Hydrogen for Fuel Cells by Reformation of Biomass-Derived Ethanol. *Catalysis Today*, Vol.75, No.1-4, (July 2002), pp. 145-155, ISSN 0920-5861

[50] Fatsikostas, A.N., Verykios, X.E. (2004). Reaction Network of Steam Reforming of Ethanol over Ni-Based Catalysts. *Journal of Catalysis*, Vol.225, No.2, (July 2004), pp.439-452, ISSN 0021-9517

[51] Frusteri, F., Freni, S., Chiodo, V., Spadaro, L., Bonura, G., Cavallaro, S. (2004). Patassium Improved Stability of Ni/MgO in the Steam Reforming of Ethanol for the Production of Hydrogen for MCFC. *Journal of Power Sources*, Vol.132, No.1-2, (May 2004), pp. 139-144, ISSN 0378-7753

[52] Biswas, P., Kunzru, D. (2008). Oxidative Steam Reforming of Ethanol over Ni/CeO_2-ZrO_2 Catalyst. *Chemical Engineering Journal*, Vol.136, No.1., (February 2008), pp. 41-49, ISSN 1385-8947

[53] Biswas, P., Kunzru, D. (2007). Steam Reforming of Ethanol for Production of Hydrogen over Ni/CeO_2-ZrO_2 Catalyst: Effect of Support and Metal Loading. *International Journal of Hydrogen Energy*, Vol.32, No.8, (June 2007), pp. 969-980, ISSN 0360-3199

[54] Srinivas, D., Satyanarayana, C.V.V., Potdar, H.S., Ratnasamy, P. (2003). Structural Studies on Ni-CeO_2-ZrO_2 Catalysts for Steam Reforming of Ethanol. *Applied Catalysis A: General*, Vol.246, No.2, (June 2003), pp. 323-334, ISSN 0926-860X

[55] Benito, M., Padilla, R., Rodríguez, L., Sanz, J.L., Daza, L. (2007). Zirconia Supported Catalysts for Bioethanol Steam Reforming : Effect of Active Phase and Zirconia Structure. *Journal of Power Sources*, Vol.169, No.1, (June 2007), pp. 167-176, ISSN 0378-7753

[56] Fierro, V., Akdim, O., Mirodatos, C. (2003). On-Board Hydrogen Production in A Hybrid Electric Vehicle by Bio-Ethanol Oxidative Steam Reforming over Ni and Noble Metal Based Catalysts. *Green Chemistry*, Vol.5, No.1., (January 2003), pp. 20-24, ISSN 1463-9270

[57] Marino, J.F., Cerrella, E.G., Duhalde, S., Jobbagy, M., Laborde, M.A. (1998). Hydrogen from Steam Reforming of Ethanol. Characterization and Performance of Copper-Nickel Supported Catalysts. *International Journal of Hydrogen Energy*, Vol.23, No.12, (December 1998), pp. 1095-1101, ISSN 0360-3199

[58] Velu, S., Satoh, N., Gopinath, C.S., Suzuki, K. (2002). Oxidative Reforming of Bio-Ethanol over CuNiZnAl Mixed Oxide Catalysts for Hydrogen Production. *Catalysis Letters*, Vol.82, No.1-2, (September 2002), pp. 145-152, ISSN 1011-372X

[59] Haga, F., Nakajima, T., Miya, H., Mishima, S. (1997). Catalytic Properties of Supported Cobalt Catalysts for Steam Reforming of Ethanol. *Catalysis Letters*, Vol.48, No.3-4, (October 1997), pp. 223-227, ISSN 1011-372X

[60] Haga, F., Nakajima, T., Miya, H., Mishima, S. (1998). Effect of Crystallite Size on the Catalysis of Alumina-Supported Cobalt Catalyst for Steam Reforming of Ethanol. *Reaction Kinetics and Catalysis Letters*, Vol.63, No.2, (March 1998), pp. 253-259, ISSN 0133-1736

[61] Llorca, J., Piscina, P.R., Sales, J., Homs, N. (2001). Direct Production of Hydrogen from Ethanolic Aqueous Solutions over Oxide Catalysts. *Chemical Communications*, Vol.7, No.1, (March 2001), pp. 641-642, ISSN 1359-7345

[62] Mielenz, J.R. (2001). Ethanol Production from Biomass : Technology and Commercialization Status. *Current Opinion in Microbiology*, Vol.4, No.3, (June 2001), pp. 324-329, ISSN 1369-5274

[63] Homs, N., Llorca, J., Piscina, P.R. (2006). Low-Temperature Steam-Reforming of Ethanol over ZnO-Supported Ni and Cu Catalysts : The Effect of Nickel and Copper Addition to ZnO-Supported Cobalt-Based Catalysts. *Catalysis Today*, Vol.116, No.3, (August 2006), pp. 361-366, ISSN 0920-5861

[64] Galetti, A.E., Gomez, M.F., Arrua, L.A., Marchi, A.J., Abello, M.C. (2008). Study of CuCoZnAl Oxide as Catalyst for the Hydrogen Production from Ethanol

Reforming. *Catalysis Communications*, Vol.9, No.6, (March 2008), pp. 1201-1208, ISSN 1566-7367

[65] Llorca, J., Homs, N., Sales, J., Piscina, P.R. (2002). Efficient Production of Hydrogen over Supported Cobalt Catalysts from Ethanol Steam Reforming. *Journal of Catalysis*, Vol.209, No.2, (July 2002), pp.306-317, ISSN 0021-9517

[66] Wang, H., Ye, J.L., Liu, Y., Li, Y.D., Qin, Y.N. (2007). Steam Reforming of Ethanol over Co3O4/CeO2 Catalysts Prepared by Different Methods. *Catalysis Today*, Vol.129, No.3-4, (December 2007), pp. 305-312, ISSN 0920-5861

[67] Zhang, B., Tang, X., Li, Y., Xu, Y., Shen, W. (2007). Hydrogen Production from Steam Reforming of Ethanol and Glycerol over Ceria-Supported Metal Catalysts. *International Journal of Hydrogen Energy*, Vol.32, No.13, (September 2007), pp. 2367-2373, ISSN 0360-3199

[68] Zhang, B., Tang, X., Li, Y., Xu, Y., Shen, W. (2006). Steam Reforming of Bio-Ethanol for the Production of Hydrogen over Ceria-Supported Co, Ir, and Ni Catalysts. *Catalysis Communications*, Vol.7, No.6, (June 2006), pp. 367-372, ISSN 1566-7367

[69] Song, H., Zhang, L., Watson, R.B., Braden, D., Ozkan, U.S. (2007). Investigation of Bio-Ethanol Steam Reforming over Cobalt-Based Catalysts. *Catalysis Today*, Vol.129, No.3-4, (December 2007), pp. 346-354, ISSN 0920-5861

[70] Bellido, J.D.A., Assaf, E.M. (2008). Nickel Catalysts Supported on ZrO_2, Y_2O_3-Stablized ZrO_2 and CaO-Stablized ZrO_2 for the Steam Reforming of Ethanol: Effect of the Support and Nickel Load. *Journal of Power Sources*, Vol.177, No.1, (February 2008), pp. 24-32, ISSN 0378-7753

[71] Rodriguez, J.A., Wang, X., Hanson, J.C., Liu, G. (2003). The Behavior of Mixed-Metal Oxides : Structural and Electronic Properties of $Ce_{1-x}Ca_xO_2$ and $Ce_{1-x}Ca_xO_{2-x}$. *The Journals of Chemical Physics*, Vol.119, No.11, (September 2003), pp. 5659-5669, ISSN 0021-9606

[72] Urasaki, K., Tokunaga, K., Sekine, Y., Matsukata, M., Kikuchi, E. (2008). Production of Hydrogen by Steam Reforming of Ethanol over Cobalt and Nickel Catalysts Supported on Perovskite-Type Oxides. *Catalysis Communications*, Vol.9, No.5, (March 2008), pp. 600-604, ISSN 1566-7367

[73] Natile, M. M., Poletto, F., Galenda, A., Glisenti, A., Montini, T., Rogatis, L. De and Fornasiero, P. (2008). $La_{0.6}Sr_{0.4}Co_{1-y}Fe_yO_{3-\delta}$ Perovskites : Influence of the Co/Fe Atomic Ratios on Properties and Catalytic Activity toward Alchol Steam-Reforming. *Chemistry of Materials*, Vol.20, No.6, (February 2008), pp. 2314-2327, ISSN 0897-4756

[74] Llorca, J., Piscina, P.R., Dalmon, J., Sales, J., Homs, N. (2003). CO-Free Hydrogen from Steam-Reforming of Bioethanol over ZnO-Supported Cobalt Catalysts: Effect of the Metallic Precursor. *Applied Catalysis B: Environmental*, Vol.43, No. 4, (July 2003), pp. 355-369, ISSN 0926-3373

[75] Panpranot, J., Kaewkun, S., Praserthdam, P., Goodwin, J.G. (2003). Effect of Cobalt Precursors on the Dispersion of Cobalt on MCM-41. *Catalysis Letters*, Vol.91, No.1-2, (November 2003), pp. 95-102, ISSN 1011-372X

[76] Batista, M.S., Santos, R.K.S., Assaf, E.M., Assaf, J.M., Ticianelli, E.A. Characterization of the Activity and Stability of Supported Cobalt Catalysts for the Steam Reforming of

Ethanol. *Journal of Power Sources*, Vol.124, No.1, (October 2003), pp. 99-103, ISSN 0378-7753

[77] Song, S., Akande, A.J., Idem, R.O., Mahinpey, N. (2007). Inter-Relationshiop Between Preparation Methods, Nickel Loading, Characteristics and Performance in the Reforming of Crude Ethanol over Ni/Al_2O_3 Catalysts: A Neural Network Approach. *Engineering Applications of Artificial Intelligence*, Vol.20, No.2, (March 2007), pp. 261-271, ISSN 0952-1976

[78] Ho, S., Su, Y. (1997). Effect of Ethanol Impregnation on the Properties of Silica-Supported Cobalt Catalysts. *Journal of Catalysis*, Vol.168, No.1, (May 1997), pp. 51-59, ISSN 0021-9517

[79] Enache, D.I., Rebours, B., Auberger, M.R., Revel, R. (2002). In-Situ XRD Study of the Influence of Thermal Treatment on the Characteristics and the Catalytic Properties of Cobalt-Based Fischer-Tropsch Catalysts. *Journal of Catalysis*, Vol.205, No.2, (January 2002), pp. 346-353, ISSN 0021-9517

[80] Ruckenstein, E., Wang, H.Y. (2000). Effect of Calcinations on the Species Formed and the Reduction Behavior of the Cobalt-Magnesia Catalysts. *Catalysis Letters*, Vol.70, No.1-2, (December 2000), pp. 15-21, ISSN 1011-372X

[81] Llorca, J., Homs, N., Sales, J., Fierro, J.G., Piscina, P.R. (2004). Effect of Sodium Addition on the Performance of Co-ZnO-Based Catalysts for Hydrogen Production from Bioethanol. *Journal of Catalysis*, Vol.222, No.2, (March 2004), pp. 470-480, ISSN 0021-9517

[82] Jacobs, G., Das, T.K., Zhang, Y., Li, J. (2002). Fischer-Tropsch Synthesis : Support, Loading, and Promoter Effects on the Reduciblity of Cobalt Catalysts. *Applied Catalysis A: General*, Vol.233, No.1-2, (July 2002), pp. 263-281, ISSN 0926-860X

[83] Huang, L., Xu, Y. (2002). Studies on the Interaction Between Ruthenium and Cobalt in Supported Catalysts in Favor of Hydroformylation. *Catalysis Letters*, Vol.69, No.3-4, (November 2002), pp. 145-151, ISSN 1011-372X

[84] Profeti, L.P.R., Ticianelli, E.A., Assaf, E.M. (2008). Production of Hydrogen by Ethanol Steam Reforming on Co/Al_2O_3 Catalysts: Effect of Addition of Small Quantities of Noble Metals. *Journal of Power Sources*, Vol.175, No.1, (January 2008), pp. 482-489, ISSN 0378-7753

[85] Soled, S.L., Iglesia, E., Fiato, R.A., Baumgartner, J.E., Vroman, H. (2003). Control of Metal Dispersion and Structure by Changes in the Solid-State Chemistry of Supported Cobalt Fischer-Tropsch Catalysts. Topics in Catalysis, Vol.26, No.1-4, (December 2003), pp. 101-109, ISSN 1022-5528

[86] Hu, X., Lu, G. (2007). Investigation of Steam Reforming of Acetic Acid to Hydrogen over Ni-Co Metal Catalyst. *Journal of Molecular Catalysis A : Chemical*, Vol.261, No.1, (January 2007), pp. 43-48, ISSN 1381-1169

[87] Torres, J.A., Llorca, J., Casanovas, A., Domínguez, M., Salvadó, J., Montané, D. (2007). Steam Reforming of Ethanol at Moderate Temperature: Multifactorial Design Analysis of Ni/La_2O_3-Al_2O_3, and Fe- and Mn-Promoted Co/ZnO Catalysts. *Journal of Power Sources*, Vol.169, No.1, (July 2007), pp. 158-166, ISSN 0378-7753

[88] Mattos, L.V., Noronha, F.B. (2005). The Influence of the Nature of the Metal on the Performance of Cerium Oxide Supported Catalysts in the Partial Oxidation of

Ethanol. *Journal of Power Sources*, Vol.152, No.1, (December 2005), pp. 50-59, ISSN 0378-7753

[89] Llorca, J., Dalmon, J., Piscina, P.R., Homs, N. (2003). In Situ Magnetic Characterization of Supported Cobalt Catalysts under Steam Reforming of Ethanol. *Applied Catalysis A: General*, Vol.243, No.2, (April 2003), pp. 261-269, ISSN 0926-860X

[90] Guil, J.M., Homs, N., Llorca, J., Piscina, P.R. (2005). Microcalorimetric and Infrared Studies of Ethanol and Acetaldehyde Adsorption to Investigate the Ethanol Steam Reforming on Supported Cobalt Catalysts. *Journal of Physical Chemistry B*, Vol.109, No.21, (May 2005), pp. 10813-10819, ISSN 1520-6106

[91] Batista, M.S., Santos, R.K.S., Assaf, E.M., Assaf, J.M., Ticianelli, E.A. (2004). High Efficiency Steam Reforming of Ethanol by Cobalt-Based Catalysts. *Journal of Power Sources*, Vol.134, No.1, (July 2004), pp. 27-32, ISSN 0378-7753

[92] Idriss, H., Diagne, C., Hindermann, J.P., Kiennemann, A., Barteau, M.A. (1995). Reactions of Acetaldehyde on CeO_2 and CeO_2-Supported Catalysts. *Journal of Catalysis*, Vol.155, No.2, (September 1995), pp. 219-237, ISSN 0021-9517

[93] Tuti, S., Pepe, F. (2008). On the Catalytic Activity of Cobalt Oxide for the Steam Reforming of Ethanol. *Catalysis Letters*, Vol.122, No.1-2, (April 2008), pp. 196-203, ISSN 1011-372X

[94] Kaddouri, A., Mazzocchia, C. (2004). A Study of the Influence of the Synthesis Conditions upon the Catalytic Properties of Co/SiO_2 or Co/Al_2O_3 Catalysts Used for Ethanol Steam Reforming. *Catalysis Communications*, Vol.5, No.6, (June 2004), pp. 339-345, ISSN 1566-7367

[95] Vargas, J.C., Libs, S., Roger, A., Kiennemann, A. (2005). Study of Ce-Zr-Co Fluorite-Type Oxide as Catalysts for Hydrogen Production by Steam Reforming of Bioethanol. *Catalysis Today*, Vol.107-108, No.1, (October 2005), pp. 417-425, ISSN 0920-5861

[96] Hsiao, W., Lin, Y., Chen, Y., Lee, C. (2007). The Effect of the Morphology of Nanocrystalline CeO_2 on Ethanol Reforming. *Chemical Physics Letters*, Vol.441, No.4-6, (June 2007), pp. 294-299, ISSN 0009-2614

[97] Sun, C., Sun, J., Xiao, G., Zhang, H., Qiu, X., Li, H., Chen, L. (2006). Mesoscale Organization of Nearly Monodisperse Flowerlike Ceria Microspheres. *Journal of Physical Chemistry B*, Vol.110, No.27, (June 2006), pp. 13445-13452, ISSN 1520-6106

[98] Gu, F., Wang, Z., Han, D., Shi, C., Guo, G. (2007). Reverse Micelles Directed Synthesis of Mesoporous Ceria Nanostructures. *Materials Science and Engineering: B*, Vol.139, No.1, (April 2007), pp. 62-68, ISSN 0921-5107

[99] Song, H., Zhang, L., Ozkan, U.S. (2007). Effect of Synthesis Parameters on the Catalytic Activity of Co-ZrO_2 for Bio-Ethanol Steam Reforming. *Green Chemistry*, Vol.9, No.6., (June 2007), pp. 686-694, ISSN 1463-9270

[100] Song, H., Ozkan, U.S. (2009). Ethanol Steam Reforming over Co-Based Catalysts: Role of Oxygen Mobility. *Journal of Catalysis*, Vol.261, No.1, (January 2009), pp. 66-74, ISSN 0021-9517

[101] Song, H., Ozkan, U.S. (2010). The Role of Impregnation Medium on the Activity of Ceria-Supported Cobalt Catalysts for Ethanol Steam Reforming. *Journal of Molecular Catalysis A : Chemical*, Vol.318, No.1-2, (March 2010), pp. 21-29, ISSN 1381-1169

[102] Song, H., Ozkan, U.S. (2010). Changing the Oxygen Mobility in Co/Ceria Catalysts by Ca Incorporation: Implications for Ethanol Steam Reforming. *Journal of Physical Chemistry A*, Vol.114, No.11, (May 2010), pp. 3796-3801, ISSN 1089-5639

[103] Song, H., Mirkelamoglu, B., Ozkan, U.S. (2010). Effect of Cobalt Precursor on the Performance of Ceria-Supported Cobalt Catalysts for Ethanol Steam Reforming. *Applied Catalysis A: General*, Vol.382, No.1, (June 2010), pp. 58-64, ISSN 0926-860X

[104] Song, H., Tan, B., Ozkan, U.S. (2009). Novel Synthesis Techniques for Preparation of Col/CeO_2 as Ethanol Steam Reforming Catalysts. *Catalysis Letters*, Vol.132, No.3-4, (October 2009), pp. 422-429, ISSN 1011-372X

[105] Sánchez, M.C., Navarro, R.M., Fierro, J.L.G. (2007). Ethanol Steam Reforming over Ni/M_xO_y-Al_2O_3 (M=Ce, La, Zr and Mg) Catalysts: Influence of Support on the Hydrogen Production. *International Journal of Hydrogen Energy*, Vol.32, No.10-11, (July-August 2007), pp. 1462-1471, ISSN 0360-3199

[106] Sánchez, M.C., Navarro, R.M., Fierro, J.L.G. (2007). Ethanol Steam Reforming over Ni/La-Al_2O_3 Catalysts: Influence of Lanthanum Loading. *Catalysis Today*, Vol.129, No.3-4, (December 2007), pp. 336-345, ISSN 0920-5861

[107] Freni, S., Cavallaro, S., Mondello, N., Spadaro, L., Frusteri, F. (2003). Production of Hydrogen for MC Fuel Cell by Steam Reforming of Ethanol over MgO Supported Ni and Co Catalysts. *Catalysis Communications*, Vol.4, No.6, (June 2003), pp. 259-268, ISSN 1566-7367

[108] Casanovas, A, Gerons, M.S., Griffon, F., Llorca, J. (2008). Autothermal Generation of Hydrogen from Ethanol in a Microreactor. *International Journal of Hydrogen Energy*, Vol.33, No.7, (April 2008), pp. 1827-1833, ISSN 0360-3199

[109] Deluga, G.A., Salge, J.R., Schmidt, L.D., Verykios, X.E. (2004). Renewable Hydrogen from Ethanol by Autothermal Reforming. *Science*, Vol.303, No.13, (February 2004), pp. 993-997, ISSN 0036-8075

[110] Mariño, F, Baronetti, M.G., Laborde, M. (2004). Hydrogen Production via Catalytic Gasification of Ethanol. A Mechanism Proposal over Copper-Nickel Catalysts. *International Journal of Hydrogen Energy*, Vol.29, No.1, (January 2004), pp. 67-71, ISSN 0360-3199

[111] Benito, M., Sanz, J.L., Isabel, R., Padilla, R., Arjona, R., Dazaa, L. (2005). Bio-Ethanol Steam Reforming: Insights on the Mechanism for Hydrogen Production. *Journal of Power Sources*, Vol.151, No.1, (October 2005), pp. 11-17, ISSN 0378-7753

[112] Dömök, M., Tóth, M., Raskó, J., Erdőhelyi, A. (2007). Adsorption and Reactions of Ethanol and Ethanol-Water Mixture on Alumina-Supported Pt Catalysts. *Applied Catalysis B: Environmental*, Vol.69, No.3-4, (January 2007), pp. 262-272, ISSN 0926-3373

[113] Erdőhelyi, A., Raskó, J., Kecskés, T., Tóth, M. (2006). Hydrogen Formation in Ethanol Reforming on Supported Noble Metal Catalysts. *Catalysis Today*, Vol.116, No.3, (August 2006), pp. 367-376, ISSN 0920-5861

[114] Jacobs, G., Keogh, R.A., Davis, B.H. (2007). Steam Reforming of Ethanol over Pt/Ceria with Co-Fed Hydrogen. *Journal of Catalysis*, Vol.245, No.2, (January 2007), pp. 326-337, ISSN 0021-9517

[115] Raskó, J., Dömök, M., Baán, K., Erdőhelyi, A. (2006). FTIR and Mass Spectrometric Study of the Interaction of Ethanol and Ethanol-Water with Oxide-Supported Platinum Catalysts. *Applied Catalysis A: General*, Vol.299, No.1, (January 2006), pp. 202-211, ISSN 0926-860X

[116] Resinia, C., Cavallaro, S., Frusteri, F., Freni, S. (2007). Initial Steps in the Production of H_2 from Ethanol: A FT-IR Study of Adsorbed Species on Ni/MgO Catalyst Surface. *Reaction Kinetics and Catalysis Letters*, Vol.90, No.1, (February 2007), pp. 117-126, ISSN 0133-1736

[117] Chung, M., Moon, D., Kim, H., Park, K., Ihm, S. (1996). Higher Oxygenate Formation from Ethanol on Cu/ZnO Catalysts: Synergism and Reaction Mechanism. *Journal of Molecular Catalysis A : Chemical*, Vol.113, No.3, (December 1996), pp. 507-515, ISSN 1381-1169

[118] Pfeifer, P., Kölbl, A., Schubert, K. (2005). Kinetic Investigations on Methanol Steam Reforming on PdZn Catalysts in Microchannel Reactors and Model Transfer into the Pressure Gap Region. *Catalysis Today*, Vol.110, No.1-2, (December 2005), pp. 76-85, ISSN 0920-5861

[119] Frank, B., Jentoft, F.C., Soerijanto, H., Kröhnert, J., Schlögl, R., Schomäcker, R. (2007). Steam Reforming of Methanol over Copper-Containing Catalysts: Influence of Support Material on Microkinetics. *Journal of Catalysis*, Vol.246, No.1, (February 2007), pp. 177-192, ISSN 0021-9517

[120] Patel, S., Pant, K.K. (2007). Experimental Study and Mechanistic Kinetic Modeling for Selective Production of Hydrogen via Catalytic Steam Reforming of Methanol. *Chemical Engineering Science*, Vol.62, No.18-20, (September-October 2007), pp. 5425-5435, ISSN 0009-2509

[121] Mastalir, A., Frank, B., Szizybalski, A., Soerijanto, H., Deshpande, A., Niederberger, M., Schomäcker, R., Schlögl, R., Ressler, T. (2005). Steam Reforming of Methanol over $Cu/ZrO_2/CeO_2$ Catalysts: A Kinetic Study. *Journal of Catalysis*, Vol.230, No.2, (March 2005), pp. 464-475, ISSN 0021-9517

[122] Peppley, B.A., Amphlett, J.C., Kearns, L.M., Mann, R.F. (1999). Methanol-Steam Reforming on $Cu/ZnO/Al_2O_3$ Catalysts. Part 2. A Comprehensive Kinetic Model. *Applied Catalysis A: General*, Vol.179, No.1-2, (April 1999), pp. 31-49, ISSN 0926-860X

[123] Hou, K., Hughes, R. (2001). The Kinetics of Methane Steam Reforming over A Ni/α-Al_2O_3 Catalyst. *Chemical Engineering Science*, Vol.82, No.1-3, (March 2001), pp. 311-328, ISSN 0009-2509

[124] Craciun, R., Shereck, B., Gorte, R.J. (1998). Kinetic Studies of Methane Steam Reforming on Ceria-Supported Pd. *Catalysis Letters*, Vol.51, No.3-4, (May 1998), pp. 149-153, ISSN 1011-372X

[125] Hoang, D.L., Chan, S.H., O.L. Ding, O.L. (2005). Kinetics and Modeling Study of Methane Steam Reforming over Sulfide Nickel Catalyst on A Gamma Alumina Support. *Chemical Engineering Science*, Vol.112, No.1-3, (September 2005), pp. 1-11, ISSN 0009-2509

[126] Laosiripojana, N., Assabumrungrat, S. (2005). Methanol Steam Reforming over Ni/Ce-ZrO_2 Catalyst: Influences of Ce-ZrO_2 Support on Reactivity, Resistance toward

Carbon Formation, and Intrinsic Reaction Kinetics. *Applied Catalysis A: General*, Vol.290, No.1-2, (August 2005), pp. 200-211, ISSN 0926-860X

[127] Berman, A., Karn, R.K., Epstein, M. (2005). Kinetics of Steam Reforming of Methane on Ru/Al_2O_3 Catalyst Promoted with Mn Oxides. *Applied Catalysis A: General*, Vol.282, No.1-2, (March 2005), pp. 73-83, ISSN 0926-860X

[128] Akande, A., Aboudheir, A., Idema, R., Dalai, A. (2006). Kinetic Modeling of Hydrogen Production by the Catalytic Reforming of Crude Ethanol over A Co-Precipitated Ni-Al2O3 Catalyst in A Packed Bed Tubular Reactor. *International Journal of Hydrogen Energy*, Vol.31, No.12, (September 2006), pp. 1707-1715, ISSN 0360-3199

[129] Therdthianwong, A., Sakulkoakiet, T., Therdthianwong, S. (2001). Hydrogen Production by Catalytic Ethanol Steam Reforming. *ScienceAsia*, Vol.27, No.3, (September 2001), pp. 193-198, ISSN 1513-1874

[130] Vaidya, P.D., Rodrigues, A.E. (2006). Kinetics of Steam Reforming of Ethanol over A Ru/Al_2O_3 Catalyst. *Industrial & Engineering Chemistry Research*, Vol.45, No.19, (September 2006), pp. 6614-6618, ISSN 0888-5885

[131] Sahoo, D.R., Vajpai, S., Patel, S., Pant, K.K. (2007). Kinetic Modeling of Steam Reforming of Ethanol for the Production of Hydrogen over Co/Al_2O_3 Catalyst. *Chemical Engineering Science*, Vol.125, No.3, (January 2007), pp. 139-147, ISSN 0009-2509

[132] Mathure, P.V., Ganguly, S., Patwardhan, A.V., Saha, R.K. (2007). Steam Reforming of Ethanol Using a Commercial Nickel-Based Catalyst. *Industrial & Engineering Chemistry Research*, Vol.46, No.25, (December 2007), pp. 8471-8479, ISSN 0888-5885

[133] Song, H., Bao, X., Hadad, C., Ozkan, U.S. (2011). Adsorption/Desorption Behavior of Ethanol Steam Reforming Reactants and Intermediates over Supported Cobalt Catalysts. *Catalysis Letters*, Vol.141, No.1, (January 2011), pp. 43-54, ISSN 1011-372X

[134] Lima, S.M., Silva, A.M., Graham, U.M., Jacobs, G., Davis, B.H., Mattos, L.V., Noronha, F.B. (2009). Ethanol Decomposition and Steam Reforming of Ethanol over $CeZrO_2$ and $Pt/CeZrO_2$ Catalyst: Reaction Mechanism and Deactivation. *Applied Catalysis A: General*, Vol.352, No.1-2, (January 2009), pp. 95-113, ISSN 0926-860X

[135] Vesselli, E., Coslovich, G., Comelli, G., Rosei, R. (2005). Modelling of Ethanol Decomposition on Pt(111): A Comparison with Experiment and Density Functional Theory. *Journal of Physics: Condensed Matter*, Vol.17, No.39, (October 2005), pp. 6139-6148, ISSN 0953-8984

[136] Yang, M., Bao, X., Li, W. (2007). First Principle Study of Ethanol Adsorption and Formation of Hydrogen Bond on Rh(111) Surface. *Journal of Physical Chemistry C*, Vol.111, No.20, (May 2007), pp. 7403-7410, ISSN 1932-7447

[137] Fartaria, R., Freitas, F., Fernandes, F. (2007). A Force Field for Simulating Ethanol Adsorption on Au(111) Surfaces. A DFT Study. *International Journal of Quantum Chemistry*, Vol.107, No.11, (May 2007), pp. 2169-2177, ISSN 1097-461X

[138] Pozzo, M., Carlini, G., Rosei, R., Alfè, D. (2007). Comparative Study of Water Dissociation on Rh(111) and Ni(111) Studied with First Principles Calculations. *The Journal of Chemical Physics*, Vol.126, No.16, (April 2007), pp. 164706-164712, ISSN 0021-9606

[139] Andersson, K., Nikitn, A., Pettersson, L.G.M., Nilsson, A., Ogasawara, H. (2004). Water Dissociation on Ru(001): An Activated Process. *Physical Review Letters*, Vol.93, No.19, (November 2004), pp. 196101-196104, ISSN 0031-9007

[140] Chen, H., Liu, S., Ho, J. (2006). Theoretical Calculation of the Dehydrogenation of Ethanol on A Rh/CeO_2 Surface. *Journal of Physical Chemistry B*, Vol.110, No.30, (August 2006), pp. 14816-14823, ISSN 1520-6106

[141] Wang, J., Lee, C.S., Lin, M.C. (2009). Mechanism of Ethanol Reforming : Theoretical Foundations. *Journal of Physical Chemistry C*, Vol.113, No.16, (April 2009), pp. 6681-6688, ISSN 1932-7447

[142] Chiang, H., Wang, C., Cheng, Y., Jiang, J., Hsieh, H. (2010). Density Functional Theory Study of Ethanol Decomposition on 3Ni/α-Al_2O_3(0001) Surface. *Langmuir*, Vol.26, No.20, (October 2010), pp. 15845-15851, ISSN 0742-7463

[143] Li, H., Chen, H., Peng, S., Ho, J. (2009). Dehydrogenation of Ethanol on An O_2-4Rh/CeO_{2-x}(111) Surface: A Computational Study. *Chemical Physics*, Vol.359, No.1-3, (May 2009), pp. 141-150, ISSN 0301-0104

[144] Wu, S., Lia, Y., Ho, J., Hsieh, H. (2009). Density Functional Studies of Ethanol Dehydrogenation on A 2Rh/γ-Al_2O_3(110) Surface. *Journal of Physical Chemistry C*, Vol.113, No.36, (September 2009), pp. 16181-16187, ISSN 1932-7447

[145] Li, M., Guo, W., Jiang, R., Zhao, L., Shan, H. (2010). Decomposition of Ethanol on Pd(111): A Density Functional Theory Study. *Langmuir*, Vol.26, No.3, (February 2010), pp. 1879-1888, ISSN 0742-7463

[146] Phatak, A., Delgass, W., Ribeiro, F., Schneider, W. (2009). Density Functional Theory Comparison of Water Dissociation Steps on Cu, Au, Ni, Pd, and Pt. *Journal of Physical Chemistry C*, Vol.113, No.17, (April 2009), pp. 7269-7276, ISSN 1932-7447

[147] Kresse, G, Hafner, J. (1993). *Ab Initio* Molecular Dynamics for Liquid Metals. *Physical Review B*, Vol.47, No.1, (January 1993), pp. 558-561, ISSN 1098-0121

[148] Kresse, G., Furthmüller, J. (1996). Efficiency of *Ab-Initio* Total Energy Calculations for Metals and Semiconductors Using A Plane-Wave Basis Set. *Computational Materials Science*, Vol.6, No.1, (July 1996), pp. 15-50, ISSN 0927-0256

[149] Kresse, G., Furthmüller, J. (1996). Efficient Iterative Schemes for Ab-Initio Total-Energy Calculations Using A Plane-Wave Basis Set. *Physical Review B*, Vol.54, No.16, (October 1996), pp. 11169-11186, ISSN 1098-0121

[150] Blöchl, P.E. (1994). Projector Augmented-Wave Method. *Physical Review B*, Vol.50, No.24, (December 1994), pp. 17953-17979, ISSN 1098-0121

[151] Kresse, G., Joubert, D. (1999). From Ultrasoft Pseudopotentials to the Projector Augumented-Wave Method. *Physical Review B*, Vol.59, No.3, (January 1999), pp. 1758-1775, ISSN 1098-0121

[152] White, J.A., Bird, D.M. (1994). Implementation of Gradient-Corrected Exchange-Correlation Potentials in Car-Parrinello Total-Energy Calculations. *Physical Review B*, Vol.50, No.7, (August 1994), pp. 4954-4957, ISSN 1098-0121

[153] Perdew, J.P., Chevary, J.A., Vosko, S.H., Jackson, K.A., Pederson, M.R., Singh, D.J., Fiolhais, C. (1992). Atoms, Molecules, Solids, and Surfaces: Applications of the Generalized Gradient Approximation for Exchange and Correlation. *Physical Review B*, Vol.46, No.11, (September 1992), pp. 6671-6687, ISSN 1098-0121

[154] Clotet, A., Pacchioni, G. (1996). Acetylene on Cu and Pd(111) Surfaces: A Comparative Theoretical Study of Bonding Mechanism, Adsorption Sites, and Vibrational Spectra. *Surface Science*, Vol.346, No.1-3, (February 1996), pp. 91-107, ISSN 0039-6028

[155] Ulitsky, A., Elber, R. (1990). A New Technique to Calculate Steepest Descent Paths in Flexible Polyatomic Systems. *Journal of Chemical Physics*, Vol.92, No.2, (January 1990), pp. 1510-1511, ISSN 0021-9606

[156] Mills, G., Jónsson, H., Schenter, G.K. (1995). Reversible Work Transition State Theory: Application to Dissociative Adsorption of Hydrogen. *Surface Science*, Vol.324, No.2-3, (February 1995), pp. 305-337, ISSN 0039-6028

[157] Henkelman, G., Uberuaga, B.P., Jónsson, H. (2000). A Climbing Image Nudged Elastic Band Method for Finding Saddle Points and Minimum Energy Paths. *Journal of Chemical Physics*, Vol.113, No.22, (December 2000), pp. 9901-9904, ISSN 0021-9606

[158] Song, H, Ozkan, U.S. (2010). Economic Analysis of Hydrogen Production Through A Bio-Ethanol Steam Reforming Process: Sensitivity Analyses and Cost Estimations. *International Journal of Hydrogen Energy*, Vol.35, No.1, (January 2010), pp. 127-134, ISSN 0360-3199

[159] Cai, W., Wang, F., Veen, A., Descorme, C., Schuurman, Y., Shen, W., Mirodatos, C. (2010). Hydrogen Production from Ethanol Steam Reforming in A Micro-Channel Reactor. *International Journal of Hydrogen Energy*, Vol.35, No.3, (February 2010), pp. 1152-1159, ISSN 0360-3199

[160] Casanovas, A., Domínguez, M., Ledesma, C., López, E., Llorca, J. (2009). Catalytic Walls and Micro-Devices for Generating Hydrogen by Low Temperature Steam Reforming of Ethanol. *Catalysis Today*, Vol.143, No.1-2, (May 2009), pp. 32-37, ISSN 0920-5861

[161] Szijjártó, G., Tompos, A., Margitfavi, J. (2011). High-Throughput and Combinatorial Development of Multicomponent Catalysts for Ethanol Steam Reforming. *Applied Catalysis A: General*, Vol.391, No.1-2, (January 2011), pp. 417-426, ISSN 0926-860X

[162] Duan, S., Senkan, S. (2005). Catalytic Conversion of Ethanol to Hydrogen Using Combinatorial Methods. *Industrial & Engineering Chemistry Research*, Vol.44, No.16, (August 2005), pp. 6381-6386, ISSN 0888-5885

[163] Yuan, L., Ye, T., Gong, F., Guo, Q., Torimoto, Y., Yamamoto, M., Li, Q. (2009). Hydrogen Production from the Current-Enhanced Reforming and Decomposition of Ethanol. *Energy & Fuels*, Vol.23, No.6, (June 2009), pp. 3103-3112, ISSN 0887-0624

[164] Kinoshita, C.M., Turn, S.Q. (2003). Production of Hydrogen from Bio-Oil Using CaO as A CO_2 Sorbent. *International Journal of Hydrogen Energy*, Vol.28, No.10, (October 2003), pp. 1065-1071, ISSN 0360-3199

[165] Yu, C., Lee, D., Park, S., Lee, K., Lee, K. (2009). Ethanol Steam Reforming in A Membrane Reactor with Pt-impregnated Knudsen Membranes. *Applied Catalysis B: Environmental*, Vol.86, No.3-4, (February 2009), pp. 121-126, ISSN 0926-3373

[166] Tosti, S., Basile, A., Borgognoni, F., Capaldo, V., Cordiner, S., Cave, S., Gallucci, F., Rizzello, C., Santucci, A., Traversa, E. (2008). Low-Temperature Ethanol Steam Reforming in A Pd-Ag Membrane Reactor: Part 2. Pt-Based and Ni-Based Catalysts